Analytical and Numerical Aspects of Partial Differential Equations

Analytical and Numerical Aspects of Partial Differential Equations

Notes of a Lecture Series

Editors
Etienne Emmrich
Petra Wittbold

Walter de Gruyter · Berlin · New York

Editors

Etienne Emmrich
Institut für Mathematik
Technische Universität Berlin
Straße des 17. Juni 136
10623 Berlin
Germany
E-mail: emmrich@math.tu-berlin.de

Petra Wittbold
Institut für Mathematik
Technische Universität Berlin
Straße des 17. Juni 136
10623 Berlin
Germany
E-mail: wittbold@math.tu-berlin.de

Mathematics Subject Classification 2000: 35-02, 35-06, 35A05, 35A15, 35B35, 35B65, 35K55, 35K65, 35K90, 35L65, 35Q30, 35Q55, 35Q51, 35R05, 47D06, 65K10, 65M12, 65M50, 65N15, 65Z05, 70-08, 70C20, 70G75, 74G70, 74Qxx, 82D10

Keywords: Partial differential equations, parabolic equation, hyperbolic equation, nonlinear Schrödinger equation, Navier–Stokes equation, Vlasov equation, non-autonomous Cauchy problem, scalar hyperbolic conservation laws, entropy, maximal regularity, complexity of numerical methods

⊗ Printed on acid-free paper which falls within the guidelines of the ANSI to ensure permanence and durability.

ISBN 978-3-11-020447-6

Bibliographic information published by the Deutsche Nationalbibliothek

The Deutsche Nationalbibliothek lists this publication in the Deutsche Nationalbibliografie; detailed bibliographic data are available in the Internet at http://dnb.d-nb.de.

Printed in Germany.
Cover design: Thomas Bonnie, Hamburg.
Printing and binding: Hubert & Co. GmbH & Co. KG, Göttingen.

Preface

This book grew out of a series of lectures at the Technische Universität Berlin held by young mathematicians from France who followed an invitation for a short-term stay during the academic years 2007–2009.

The lecture series was made possible by the financial support of the foundation *Stiftung Luftbrückendank*, which was founded in 1959 by Willy Brandt, former Mayor of Berlin and Chancellor of Germany, commemorating the Berlin Airlift 1948/49.

The book addresses students of mathematics in their last year, PhD students as well as younger researchers working in the field of partial differential equations and their numerical treatment. Most of the contributions only require some basic knowledge in functional analysis, partial differential equations, and numerical analysis.

The topics of the contributions range from an investigation of the qualitative behaviour of solutions of particular partial differential equations to the study of the complexity of numerical methods for the reliable and efficient solution of partial differential equations. A main focus is on nonlinear problems and theoretical studies reflecting recent developments of topical interest.

Although the contributions deal with rather different topics, it indeed turns out that the questions to be answered, the methods and the mathematical techniques are always of the same or at least of similar type. It is, therefore, the editors' hope that this book not only provides an introduction into several distinct topics but also serves as a guide to advanced techniques and main ideas in the analysis of (nonlinear) partial differential equations and their discretisation methods.

The editors are pleased to include the contribution of Gregory A. Chechkin and Andrey Yu. Goritsky (Moscow) on scalar hyperbolic conservation laws that arose from Stanislav N. Kruzhkov's lectures given at the Moscow University. This contribution was translated and taught in Berlin by our colleague Boris Andreianov (Besançon) who is one of Kruzhkov's last students.

The editors wish to express their gratitude to Heinz-Gerd Reese (Berlin), Executive Director of the Stiftung Luftbrückendank, to the authors for making the effort to write up the lectures, and, last but not least, to Dr. Robert Plato (Berlin), Editorial Director at the Walter de Gruyter's publishing house, for his thorough and careful assistance, his patience, and a very pleasant collaboration.

Berlin, May 2009

Etienne Emmrich
Petra Wittbold

Table of contents

S. N. Kruzhkov's lectures on first-order quasilinear PDEs

Gregory A. Chechkin and Andrey Yu. Goritsky

Abstract. The present contribution originates from short notes intended to accompany the lectures of Professor Stanislav Nikolaïevich Kruzhkov given for the students of the Moscow State Lomonosov University during the years 1994–1997. Since then, they were enriched by many exercises which should allow the reader to assimilate more easily the contents of the lectures and to appropriate the fundamental techniques. This text is prepared for graduate students studying PDEs, but the exposition is elementary, and no previous knowledge of PDEs is required. Yet a command of basic analysis and ODE tools is needed. The text can also be used as an exercise book.

The lectures provide an exposition of the nonlocal theory of quasilinear partial differential equations of first order, also called conservation laws. According to S. N. Kruzhkov's "ideology", much attention is paid to the motivation (from both the mathematical viewpoint and the context of applications) of each step in the development of the theory. Also the historical development of the subject is reflected in these notes.

We consider questions of local existence of smooth solutions to Cauchy problems for linear and quasilinear equations. We expose a detailed theory of discontinuous weak solutions to quasilinear equations with one spatial variable. We derive the Rankine–Hugoniot condition, motivate in various ways admissibility conditions for generalized (weak) solutions and relate the admissibility issue to the notions of entropy and of energy. We pay special attention to the resolution of the so-called Riemann problem. The lectures contain many original problems and exercises; many aspects of the theory are explained by means of examples. The text is completed by an afterword showing that the theory of conservation laws is yet full of challenging questions and awaiting for new ideas.[0]

Keywords. first-order quasilinear PDE, characteristics, generalized solution, shock wave, rarefaction wave, admissibility condition, entropy, Riemann problem.

AMS classification. 35F20, 35F25, 35L65.

[0] *Note added by the translator (NT)* — The authors, the translator and the editors made an effort to produce a readable English text while preserving the flavour of S. N. Kruzhkov's expression and his original way of teaching. The reading of the lectures will surely require some effort (for instance, many comments and precisions are given in parentheses). In some cases, we kept the original "russian" terminology (usually accompanied by footnote remarks), either because it does not have an exact "western" counterpart, or because it was much used in the founding works of the Soviet researchers, including S. N. Kruzhkov himself.

We hope that the reader will be recompensed for her or his effort by the vivacity of the exposition and by the originality of the approach. Indeed, while at the mid-1990th, only few treaties on the subject of conservation laws were available (see [20, 48, 49]), the situation changed completely in the last ten years. The textbooks and monographs [11, 14, 22, 32, 33, 35, 47] are mainly concerned with conservation laws and systems. With respect to the material covered, the present notes can be compared with the introductory chapters of [11, 22, 33] and with the relevant chapters of the already classical PDE textbook [16]. Yet in the present lecture notes the exposition is quite different, with a strong emphasis on examples and motivation of the theory.

This is a beginner's course on conservation laws; in a sense, it stops just where the modern theory begins, before advanced analysis techniques enter the stage. For further reading, we refer to any of the above textbooks.

Introduction

The study of first-order partial differential equations is almost as ancient as the notion of the partial derivative. PDEs of first order appear in many mechanical and geometrical problems, due to the physical meaning of the notion of derivative (the velocity of motion) and to its geometrical meaning (the tangent of the angle). Local theory of such equations was born in the 18th century.

In many problems of this type one of the variables is the time variable, and processes can last for a sufficiently long time. During this period, some singularities of classical solutions can appear. Among these singularities, we consider only weak discontinuities (which are jumps of derivatives of the solution) and strong discontinuities (which are jumps of the solutions themselves). We do not deal with the "blow up"-type singularities.

It is clear that after the singularities have appeared, in order to give a meaning to the equation under consideration one has to define weak derivatives and weak solutions. These notions were introduced into mathematical language only in the 20th century. The first mathematical realization of this "ideology" was the classical paper of E. Hopf [23] (1950). In this paper, a nonlocal theory for the Cauchy problem was constructed for the equation

$$u_t + \left(u^2/2\right)_x = 0 \tag{0.1}$$

with initial datum

$$u\big|_{t=0} = u_0(x), \tag{0.2}$$

where $u_0(x)$ is an arbitrary bounded measurable function. The equation

$$u_t + (f(u))_x = 0 \tag{0.3}$$

is a natural generalization of equation (0.1). Important results for the nonlocal theory of this equation were obtained (in the chronological order of the papers) by O. A. Oleĭnik [36, 37], A. N. Tikhonov, A. A. Samarskiĭ [50], P. D. Lax [31], O. A. Ladyzhenskaya [29], I. M. Gel'fand [18].[1] The most complete theory of the Cauchy problem (0.3), (0.2) in the space of bounded measurable functions was achieved in the papers by S. N. Kruzhkov [25, 26] (see also [27]).[2]

1 Derivation of the equations

The Hopf equation. Consider a one-dimensional medium consisting of particles moving without interaction in the absence of external forces. Denote by $u(t,x)$ the velocity of the particle located at the point x at the time instant t. If $x = \varphi(t)$ is the

[1] *NT* — Throughout the lectures, no attempt is made to give a complete account on the works on the subject of first-order quasilinear equations; the above references were those that most influenced S. N. Kruzhkov's work.

[2] *NT* — Also should be mentioned the contribution by A. I. Vol'pert [52], who constructed a complete well-posedness theory in the smaller class BV of all functions of bounded variation. As shown in [52], this class is a convenient generalization of the class of piecewise smooth functions widely used in the present lectures.

trajectory of a fixed particle, then the velocity of this particle is $\dot{\varphi}(t) = u(t, \varphi(t))$, and the acceleration $\ddot{\varphi}(t)$ is equal to zero for all t. Hence,

$$0 = \frac{d^2\varphi}{dt^2} = \frac{d}{dt}u(t, \varphi(t)) = \frac{\partial u}{\partial t} + \frac{\partial u}{\partial x}\dot{\varphi} = \frac{\partial u}{\partial t} + \frac{\partial u}{\partial x}u.$$

The obtained equation

$$u_t + uu_x = 0, \tag{1.1}$$

which describes the velocity field u of non-interacting particles, is called the Hopf equation.

The continuity (or mass conservation) equation. This equation, usually presented in a course on the mechanics of solids, describes the movement of a fluid (a liquid or a gas) in $\mathbb{R}^n$ if there are no sinks nor sources. Denote the velocity vector of the fluid by $v(x,t) = (v_1, \ldots, v_n)$ and its density by $\rho(x,t)$. Let us fix a domain $V \subset \mathbb{R}^n$. At the moment t, the mass of the fluid contained in this domain is equal to

$$M_V(t) = \int_V \rho(x,t)\,dx;$$

this mass is changing with the rate dM_V/dt. On the other hand, in the absence of sources and sinks inside V, the change of mass M_V is only due to movements of the fluid through the boundary ∂V of the domain, i.e., the rate of change of the mass $M_V(t)$ is equal to the flux of the fluid through ∂V:

$$\frac{dM_V}{dt} = -\int_{\partial V} (v(x,t), \nu) \cdot \rho(x,t)\,dS_x.$$

Here (v, ν) is the scalar product of the velocity vector v and the outward unit normal vector ν to the boundary ∂V at the point $x \in \partial V$; dS_x is an element of area on ∂V.

Hence, we have

$$\frac{d}{dt}\int_V \rho(x,t)\,dx = -\int_{\partial V} (v(x,t), \nu) \cdot \rho(x,t)\,dS_x. \tag{1.2}$$

Under the assumption that ρ and v are sufficiently smooth, we rewrite the right-hand side of the formula (1.2) with the help of the divergence theorem (the Gauss–Green formula), i.e., using the fact that the integral of the divergence over a domain is equal to the flux through the boundary of this domain:

$$\int_V \frac{\partial \rho}{\partial t}\,dx = -\int_V \operatorname{div}(\rho v)\,dx. \tag{1.3}$$

Here div is the divergence operator with respect to the spatial variables. Let us remind that the divergence of the vector field $a(x) = (a_1, \ldots, a_n) \in \mathbb{R}^n$ is the scalar

$$\operatorname{div} a = (a_1)_{x_1} + \cdots + (a_n)_{x_n}.$$

Since the domain $V \subset \mathbb{R}^n$ is arbitrary, using (1.3) we get the so-called continuity equation, well known in hydrodynamics:

$$\frac{\partial \rho}{\partial t} + \operatorname{div}(\rho v) = 0. \tag{1.4}$$

Equation of fluid infiltration through sand. For the sake of simplicity, we introduce several natural assumptions. Suppose that the fluid moves under the sole action of the gravity, i.e., the direction of the movement is vertical and there is no dependence on horizontal coordinates. Neither sources nor sinks are present. The speed of infiltration v is a function of the density $\rho \equiv u(t,x)$, i.e., $v = v(u)$.

It is experimentally verified that the dependence $v(u)$ has a form as in Figure 1. On the segment $[0, u_0]$ one can assume that the dependence is almost parabolic, i.e., $v(u) = Cu^2$.

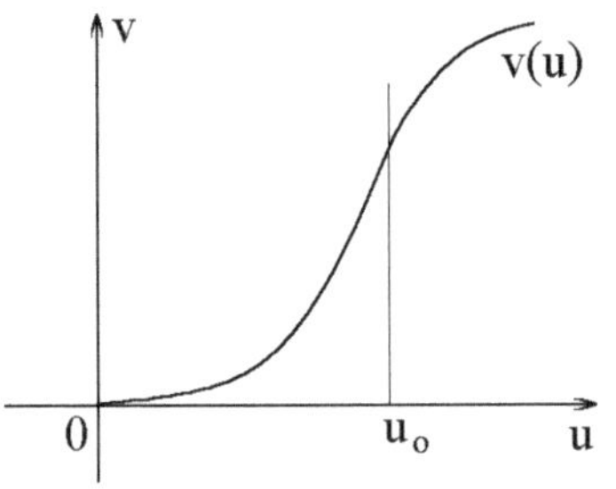

Figure 1. Experimental dependence $v = v(u)$.

In the one-dimensional case under consideration, the equation (1.4) will be rewritten as follows :

$$u_t(t,x) + \left[u(t,x) \cdot v\left(u(t,x)\right)\right]_x = 0, \tag{1.5}$$

or

$$u_t + p(u)u_x = 0, \quad \text{where} \quad p(u) = v(u) + v'(u)u.$$

Keeping in mind the experimental dependence of the speed of infiltration on the density, we assume that $v(u) = u^2/3$, and finally we get

$$u_t + u^2 u_x = 0.$$

The traffic equation. This equation can also be derived from the one-dimensional (in x) continuity equation (1.4). In traffic problems, $u(t,x)$ represents the density of cars on the road (at point x at time t); and the dependence of the velocity v of cars on the density u is linear:

$$v(u) = C - ku, \qquad C, k = \text{const} > 0.$$

In this case, equation (1.5) reads as follows:

$$u_t + (Cu - ku^2)_x = 0.$$

2 The local classical theory

First order PDEs can be solved locally by means of methods of the theory of ordinary differential equations, using the so-called *characteristic system*. From the physical point of view this fact can be considered as an expression of the duality of the wave theory and the particle theory of media. The field satisfies a PDE of first order; and the behaviour of the particles constituting the field is described by a system of ODEs. The connection between the first-order PDE and the corresponding system of ODEs allows to study the behaviour of particles instead of studying the evolution of waves.

It should be noted that the majority of questions in this chapter are considered in the textbooks on ODEs (for instance, [3, Chapter 2]). Different exercises on linear and quasilinear equations of first order can be found in [17, §20].

Below we remind basic notions of the aforementioned local theory for linear and quasilinear equations.

2.1 Linear equations

Let $v = v(x)$ be a smooth vector field in a domain $\Omega \subset \mathbb{R}^n$.

Definition 2.1. The equation

$$L_v[u] \equiv v_1(x)\frac{\partial u}{\partial x_1} + \cdots + v_n(x)\frac{\partial u}{\partial x_n} = 0. \tag{2.1}$$

is said to be a *linear homogeneous* PDE of first-order.

A continuously differentiable function $u = u(x)$ is called *classical* solution of this equation if u satisfies the equation at any point of its domain.

Recall that in the ODE theory, the operator $L_v \equiv v_1\frac{\partial}{\partial x_1} + \cdots + v_n\frac{\partial}{\partial x_n}$ is called the derivation operator along the vector field v. Geometrically, equation (2.1) means that the gradient $\nabla u \equiv (\frac{\partial u}{\partial x_1}, \ldots, \frac{\partial u}{\partial x_n})$ of the unknown function $u = u(x)$ is orthogonal to the vector field v in all points of the domain Ω.

A smooth function $u = u(x)$ is a solution of the equation (2.1) if and only if u is constant along the phase curves of the field v, i.e., it is the first integral of the system of equations

$$\begin{cases} \dot{x}_1 &= v_1(x_1, \ldots, x_n), \\ \dot{x}_2 &= v_2(x_1, \ldots, x_n), \\ \ldots & \\ \dot{x}_n &= v_n(x_1, \ldots, x_n). \end{cases} \tag{2.2}$$

The system (2.2), which can be written in vector form $\dot{x} = v(x)$, is called *the characteristic system of the linear equation* (2.1). A solution of the characteristic system is called *a characteristic*, the vector field $v = v(x)$ over the n-dimensional space of x is called *the characteristic vector field of the linear equation.*

Definition 2.2. A *linear inhomogeneous* first-order PDE is the equation

$$L_v[u] = f(x), \tag{2.3}$$

where $f = f(x)$ is a given function.

Equation (2.3) expresses the fact that if we move along the characteristic $x = x(t)$ (i.e., along the solution $x = x(t)$ of the system (2.2)), then $u(x(t))$ is changing with the given speed $f(x(t))$. Thus, in the case of an inhomogeneous linear equation, the characteristic system (2.2) should be supplemented with the additional equation on u:

$$\dot{u} = f(x_1, \dots, x_n). \tag{2.4}$$

2.2 The Cauchy problem

Definition 2.3. The *Cauchy problem* for a first-order partial differential equation is the problem of finding the solution $u = u(x)$ of this equation satisfying the initial condition

$$u\big|_\gamma = u_0(x), \tag{2.5}$$

where $\gamma \subset \mathbb{R}^n$, $\dim \gamma = n - 1$, is a fixed smooth hypersurface in the x-space, and $u_0 = u_0(x)$ is a given smooth function defined on γ.

In order to solve the Cauchy problem (2.1), (2.5) for a linear homogeneous equation, it is sufficient to continue the function $u(x)$ from the surface γ along the characteristics $x(t)$ by a constant. In the case of the problem (2.3), (2.5) for the inhomogeneous equation, the initial data should be extrapolated according to the law (2.4).

Note two important features of the Cauchy problem, specified above.

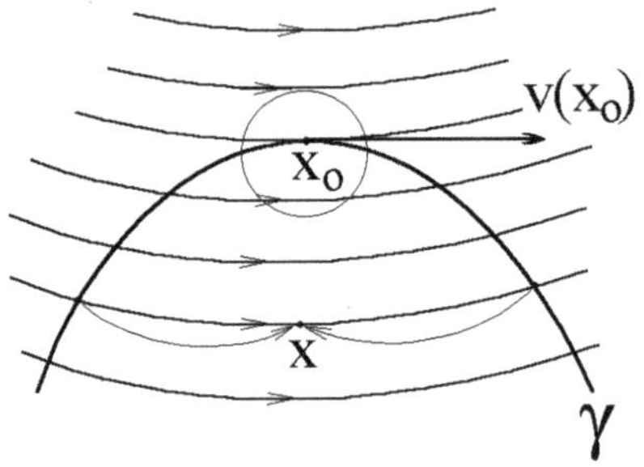

Figure 2. Example of a characteristic point.

Remark 2.4. The Cauchy problem is set locally (i.e., in a neighbourhood of a point x_0 on γ). Otherwise, as it can be seen in Figure 2, characteristics passing through a given point x may cross γ twice (or even several times), carrying different values of u to this point. Thus the solution to the problem (2.1), (2.5) exists only for specially selected initial data u_0.

Moreover, it can happen that the set of all the characteristics which have common points with the initial surface γ do not cover the whole domain where we want to solve the Cauchy problem. In this case, we have no uniqueness of a solution to the Cauchy problem.

Remark 2.5. If in the point $x_0 \in \gamma$ the vector $v(x_0)$ is parallel to the surface γ (such points x_0 are called *characteristic points*, see Figure 2), then, even choosing a very small neighbourhood of this point, we cannot guarantee that we shall not have the same difficulties as we mentioned in Remark 2.4. Hence, the existence and the uniqueness of a solution to a Cauchy problem can be guaranteed only in a neighbourhood of a non-characteristic point on γ.

Linear first-order PDEs can be impossible to solve in a neighbourhood of a characteristic point even in the case when each characteristic has exactly one point of intersection with the initial surface γ.

Example 2.6. Consider the following Cauchy problem:

$$\frac{\partial u}{\partial x} = 0, \qquad u\big|_{y=x^3} = x^2. \tag{2.6}$$

The characteristic vector field is the constant field $(1,0)$, the characteristics are the straight lines $y = C$; each of them has only one intersection point with the curve $\gamma = \{(x,y) \mid y = x^3\}$. If we extend the initial function $u_0(x) = x^2$ (which is equal to $y^{2/3}$ on γ) so that it is constant along the characteristics, we get the x-independent "solution" $u(x,y) = y^{2/3}$ which is not a classical solution because it is not a continuously differentiable function on the line $y = 0$.

The possible objection that, nevertheless, the function constructed above has a partial derivative with respect to x (and hence satisfies the equation in the classical sense) is easy to remove. It is sufficient to change the variables in problem (2.6) according to the formula $x = x' + y'$, $y = x' - y'$. After this rotation and rescaling on the axes, we obtain the following Cauchy problem:

$$\frac{\partial u}{\partial x'} + \frac{\partial u}{\partial y'} = 0, \qquad u\big|_{\gamma} = (x' + y')^2,$$

the curve γ being defined by the equation $x' - y' = (x' + y')^3$. The transformed "solution" $u(x', y') = (x' - y')^{2/3}$ has no partial derivatives in x' nor in y' on the line $x' - y' = 0$.

2.3 Quasilinear equations

Definition 2.7. The equation

$$L_{v(x,u)}[u] \equiv v_1(x,u)\frac{\partial u}{\partial x_1} + \cdots + v_n(x,u)\frac{\partial u}{\partial x_n} = f(x,u) \tag{2.7}$$

is called a *quasilinear* first-order PDE. If in the equation (2.7) all the coefficients v_i are independent of u, i.e., $v_i = v_i(x)$, then the PDE is called *semilinear*.

As for the linear equation, we write down the system (2.2), (2.4):

$$\left\{ \begin{array}{rcl} \dot{x}_1 &=& v_1(x_1, \ldots, x_n, u), \\ & \cdots & \\ \dot{x}_n &=& v_n(x_1, \ldots, x_n, u), \\ \dot{u} &=& f(x_1, \ldots, x_n, u). \end{array} \right. \tag{2.8}$$

This system is called the *characteristic system of the quasilinear equation* (2.7); solutions $(x, u) = (x(t), u(t)) \in \mathbb{R}^{n+1}$ to the system (2.8) are called *characteristics* of this equation; a *characteristic vector field of a quasilinear equation* (2.7) is a smooth vector field with components $(v_1(x,u), \dots, v_n(x,u), f(x,u))$ in the $(n+1)$-dimensional space with coordinates $(x_1, \dots, x_n, u)$.

Remark 2.8. If a linear equation is considered as being quasilinear, and also in the case of a semilinear equation, the projection $(v_1, \dots, v_n)$ on the x-space of the vector $(v_1, \dots, v_n, f)$ in the point (x_0, u_0) does not depend on u_0, since the coefficients v_i do not depend on u. Hence in these cases the projections on the x-space of the characteristics that lie at "different heights" coincide (here we mean that the vertical axis represents the variable u).

If the smooth hypersurface $M \subset \mathbb{R}^{n+1}$ is the graph of a function $u = u(x)$, then the normal vector to this surface in the coordinates (x, u) has the form $(\nabla_x u, -1) = (\partial u/\partial x_1, \dots, \partial u/\partial x_n, -1)$. Therefore, geometrically, the equation (2.7) expresses the orthogonality of the characteristic vector $(v(x,u), f(x,u))$ and the normal vector to M. Thus, we have the following theorem.

Theorem 2.9. *A smooth function $u = u(x)$ is a solution to the equation (2.7) if and only if the graph $M = \{(x, u(x))\}$, which is a hypersurface in the space $\mathbb{R}^{n+1}$, is tangent, in all its points, to the characteristic vector field $(v_1, \dots, v_n, f)$.*

Corollary 2.10. *The graph of any solution $u = u(x)$ to the equation (2.7) is spanned by characteristics.*

Indeed, by definition, the characteristics $(x(t), u(t))$ are tangent to the characteristic vector field (see (2.8)); therefore any characteristics having a point in common with the graph of u lies entirely on this graph. (Here and in the sequel, we always assume that the characteristic system complies with the assumptions of the standard existence and uniqueness theorems of the theory of ODEs.)

For the case of a quasilinear equation, the Cauchy problem (2.7), (2.5) can be solved geometrically as follows. Let

$$\Gamma = \{(x, u_0(x)) \mid x \in \gamma\} \subset \mathbb{R}^{n+1}, \qquad \dim \Gamma = n - 1,$$

be the graph of the initial function $u_0 = u_0(x)$. Issuing a characteristic from each point of Γ, we obtain some surface M of codimension one. Below we show that, whenever the point $(x_0, u_0(x_0))$ is non-characteristic, at least locally (in some neighbourhood of the point $(x_0, u_0(x_0)) \in \Gamma$) the hypersurface M represents the graph of the unknown solution $u = u(x)$.

Definition 2.11. A point $(x_0, u_0) \in \Gamma$ is called *a characteristic point*, if the vector $v(x_0, u_0)$ is tangent to γ at this point.

Remark 2.12. In the case of a quasilinear equation, one does not ask whether a point $x_0 \in \gamma \subset \mathbb{R}^n$ is a characteristic point. Indeed, the characteristic vector field also depends on u. In this case, one should ask whether a point $(x_0, u_0(x_0)) \in \Gamma \subset \mathbb{R}^{n+1}$ is a characteristic point.

If $(x_0, u_0(x_0)) \in \Gamma$ is a non-characteristic point, then the hyperplane T tangent to M at this point projects isomorphically onto the x-space. Indeed, the hyperplane T is spanned by the directions tangent to Γ (their projections span the hyperplane in $\mathbb{R}^n$ tangent to γ) and by the characteristic vector $(v(x_0, u_0(x_0)), f(x_0, u_0(x_0)))$ (its projection is the vector $v(x_0, u_0(x_0))$ transversal to γ). Consequently, locally in a neighbourhood of the point $(x_0, u_0(x_0)) \in \Gamma$, the hypersurface M constructed above represents the graph of a smooth function $u = u(x)$, which is the desired solution.

3 Classical (smooth) solutions of the Cauchy problem and formation of singularities

3.1 Quasilinear equations with one space variable

In the sequel, we will always consider the following equation in the unknown function $u = u(t, x)$ depending on two variables (t has the meaning of time, and $x \in \mathbb{R}^1$ represents the one-dimensional space coordinate):

$$u_t + (f(u))_x \equiv u_t + f'(u)u_x = 0. \tag{3.1}$$

Here $f \in C^2$ is a given function, which will be called the *flux function*. The initial data is prescribed at time $t = 0$:

$$u\big|_{t=0} = u(0, x) = u_0(x). \tag{3.2}$$

In this section, we investigate the possibility to construct solutions of the problem (3.1)–(3.2) within the class of smooth functions defined in the strip

$$\Pi_T \equiv \{(t, x) \mid -\infty < x < +\infty,\ 0 < t < T\}.$$

Let us apply the results of the general theory, as exposed above, to this concrete case.

We see that the equation (3.1) is quasilinear; for this case, the characteristic system (2.8) takes the form

$$\begin{cases} \dot{t} &= 1, \\ \dot{x} &= f'(u), \\ \dot{u} &= 0. \end{cases} \tag{3.3}$$

The first equation in system (3.3) together with the initial condition $t(0) = 0$ (we take this condition because of (3.2)) means exactly the following: the independent variable in system (3.3) (the differentiation with respect to this variable is denoted by a dot ($\dot{}$)) coincides with the time variable t of the equation (3.1). Thus it is natural to exclude the first equation from the characteristic system (3.3) associated with the Cauchy problem (3.1)–(3.2).

In the case considered, the initial curve $\gamma \in \mathbb{R}^2_{t,x}$ is the straight line $t = 0$, i.e., $\gamma = \{(t, x) \mid t = 0\}$, and the curve $\Gamma \in \mathbb{R}^3_{t,x,u}$ is the set of points

$$\Gamma = \{(t, x, u) \mid t = 0,\ x = y,\ u = u_0(y)\},$$

parameterized by the space variable y. Let us stress that in this case, all the points of Γ are non-characteristic, since the vector $(\dot{t}, \dot{x}) = (1, f'(u))$ is transversal to $\gamma = \{t = 0\}$.

Thus in our case, we can rewrite the characteristic system (3.3) (with the initial data corresponding to (3.2)) in the form

$$\begin{cases} \dot{x} = f'(u), & x(0) = y, \\ \dot{u} = 0, & u(0) = u_0(y). \end{cases} \tag{3.4}$$

Solutions of this system (i.e., the characteristics of equation (3.1)) are the straight lines

$$u \equiv u_0(y), \qquad x = y + f'(u_0(y))t \tag{3.5}$$

in the three-dimensional space of points (t, x, u).

As was pointed out in Section 2.3, the graph of the solution $u = u(t, x)$ of problem (3.1)–(3.2) is the union of the characteristics issued from the points of the initial curve Γ; thus, the graph of u consists of the straight lines (3.5). Therefore, the solution of problem (3.1)–(3.2) at different time instants $t > 0$ (i.e., the sections of the graph of the solution $u = u(t, x)$ of this problem by different hyperplanes $t = \text{const}$) can be constructed as follows. The graph of the initial function $u = u_0(x)$ should be transformed by displacing each point (x, u) of this graph horizontally (i.e., in the direction of the x-axis) with the speed $f'(u)$. If $f'(u) = 0$ then the point (x, u) does not move. If $f'(u) > 0$, then the point moves to the right; and, the greater $f'(u)$ is, the quicker it moves. Similarly, in the case $f'(u) < 0$, the point (x, u) moves to the left (see Fig. 3).

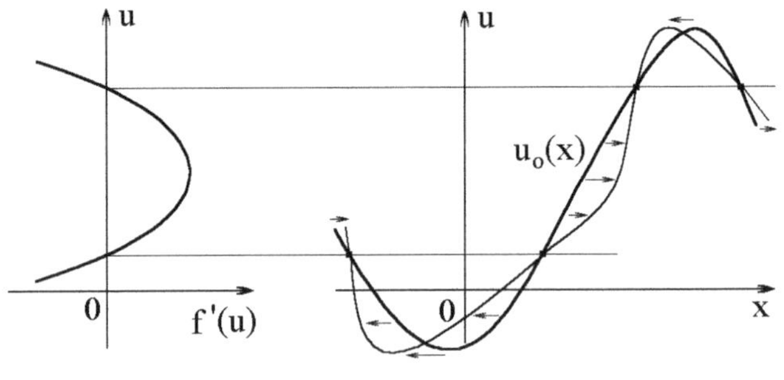

Figure 3. Evolution from initial graph.

Remark 3.1. Assume that the graph of the initial function $u_0 = u_0(x)$ delimits a finite area (this is the case, for instance, when u_0 has finite support). Then the aforementioned transformation of the graph leaves the area invariant. Indeed, all the points of the graph of u_0 lying on the same horizonal line move with the same speed; consequently, the lengths of the horizontal segments joining the points of the graph remain unchanged.

The fact that the area under the graph remains constant can also be obtained by a direct calculation. Let $S(t) = \int_{-\infty}^{+\infty} u(t, x)\, dx$ be the area in question, i.e., the area delimited by the graph of $u = u(t, x)$ of problem (3.1)–(3.2) (here $t \geqslant 0$ is fixed). Then

$$\frac{d}{dt}S(t) = \int_{-\infty}^{+\infty} u_t(t,x)\,dx = -\int_{-\infty}^{+\infty} (f(u(t,x)))_x\,dx = -f\,(u(t,x))\Big|_{x=-\infty}^{x=+\infty}$$
$$= f(0) - f(0) = 0,$$

which means that $S(t) \equiv \text{const}$.

While the graph of the solution evolves as described above, at a certain moment $T > 0$ it may happen that the transformed curve ceases to represent the graph of a smooth function $u(T, x)$ of variable x.

Consider, for instance, the Hopf equation, i.e., the equation (3.1) with $f(u) = u^2/2$. This equation describes the evolution of the velocity field of a medium consisting of non-interacting particles (see Section 1). Each particle moves in absence of forces and thus conserves its initial speed.

Consider two particles located, at the initial instant $t = 0$, at points x_1 and x_2 with $x_1 < x_2$. If the initial velocity distribution $u_0 = u_0(x)$ is a monotone non-decreasing function, then the initial velocity $u_0(x_1)$ of the first particle (which is its velocity for all subsequent instants of time) is less than or equal to the velocity $u_0(x_2)$ of the second particle: $u_0(x_1) \leqslant u_0(x_2)$. Since also the initial locations of the two particles obey the inequality $x_1 < x_2$, at any time instant $t > 0$ the two particles will never occupy the same space location; i.e., no particle collision happens in this case.

On the contrary, if the initial velocity distribution $u_0 = u_0(x)$ is not a monotone non-decreasing function, then the quicker particles will overtake the slower ones (or, possibly, particles can move towards each other), and at some instant $T > 0$ collisions should occur. Starting from this time instant T, our model does not reflect the physical reality any more, because the particles "passing through each other" should interact (collide) in one way or another. Mathematically, such interaction is usually accounted for by adding a term of the form εu_{xx} onto the right-hand side of equation (3.1), where $\varepsilon > 0$ has the meaning of a viscosity coefficient. We will encounter this model in Section 5.2.

Exercise 3.1. *For the Hopf equation, represent approximatively the velocity distribution $u = u(t, x)$ at different time instants $t > 0$, if the initial velocity distribution is given by the function*

(i) $u_0(x) = \arctan x$,

(ii) $u_0(x) = -\arctan x$,

(iii) $u_0(x) = \sin x$,

(iv) $u_0(x) = -\sin x$,

(v) $u_0(x) = x^3$,

(vi) $u_0(x) = -x^3$.

For the initial data prescribed above, find the maximal time instant $T > 0$ such that a smooth solution of the Cauchy problem (for the Hopf equation)

$$u_t + uu_x = 0, \qquad u\big|_{t=0} = u_0(x),$$

exists in the strip $\Pi_T = \{(t, x) \mid 0 < t < T,\ x \in \mathbb{R}\}$.

Exercise 3.2. *Represent approximatively the sections of the graph of the solution of the Cauchy problem*

$$u_t + \left(f(u)\right)_x = 0, \qquad u\big|_{t=0} = u_0(x),$$

at different time instants $t > 0$ for

(i) $f(u) = \cos u, \quad u_0(x) = x,$

(ii) $f(u) = \cos u, \quad u_0(x) = \sin x,$

(iii) $f(u) = u^3/3, \quad u_0(x) = \sin x.$

3.2 Reduction of the Cauchy problem to an implicit functional equation

One can solve the Cauchy problem for the quasilinear equation (3.1) directly, making no reference to the local theory of first-order quasilinear PDEs exposed above. This is the goal of the present section.

Assume that we already have a smooth solution $u = u(t, x)$ of the problem (3.1)–(3.2) under consideration.

Proposition 3.2. *The function $u = u(t, x)$ is constant along the integral curves of the ordinary differential equation*

$$\frac{dx}{dt} = f'\left(u(t, x)\right). \tag{3.6}$$

Proof. Differentiate the function $u = u(t, x)$ in the direction of the integral curves $(t, x(t))$ of equation (3.6):

$$\frac{du}{dt} = \frac{\partial u}{\partial t} + \frac{\partial u}{\partial x} \cdot \frac{dx}{dt} = u_t + u_x \cdot f'(u) = u_t + \left(f(u)\right)_x = 0. \qquad \square$$

As u remains constant along these integral curves, it follows that the solutions of (3.6) are the linear functions $x = f'(u)t + C_1$. (The straight lines $x - f'(u)t = C_1$, lying in the hyperplanes $u = C_2$, are exactly the characteristics of the quasilinear equation (3.1).)

Consequently, the value $u(t_0, x_0)$ of the solution $u = u(t, x)$ at the point (t_0, x_0) is conserved along the whole line

$$x - f'\left(u(t_0, x_0)\right) \cdot t = C = x_0 - f'\left(u(t_0, x_0)\right) \cdot t_0. \tag{3.7}$$

Extending this line until it intersects the x-axis at some point $(0, y_0)$, we take the value $u_0(y_0)$ at this point. Since the point $(0, y_0)$ lies on the straight line (3.7), we have $y_0 = x_0 - f'\left(u(t_0, x_0)\right) \cdot t_0$. Thus,

$$u(t_0, x_0) = u_0(y_0) = u_0\left(x_0 - f'\left(u(t_0, x_0)\right) \cdot t_0\right).$$

As the point (t_0, x_0) is arbitrary, we obtain the following identity for the solution u of the Cauchy problem (3.1)–(3.2):

$$u = u_0\left(x - f'(u)t\right). \tag{3.8}$$

Thus, the problem of finding the domain into which the solution $u = u(t,x)$ of (3.1)–(3.2) can be extended amounts to finding the domain where equation (3.8) with the unknown u has one and only one solution.

Remark 3.3. Formula (3.8) can also be obtained while solving practically the Cauchy problem for the quasilinear equation, according to [17, §20]. The characteristic system

$$\frac{dt}{1} = \frac{dx}{f'(u)} = \frac{du}{0}$$

associated with the equation (3.1) possesses two first integrals:

$$I_1(t,x,u) \equiv u, \qquad I_2(t,x,u) \equiv x - f'(u)t. \tag{3.9}$$

On the initial curve $\Gamma = \{(0, y, u_0(y))\} \in \mathbb{R}^3_{t,x,u}$, these two first integrals take the values

$$I_1\big|_\Gamma = u_0(y), \qquad I_2\big|_\Gamma = y.$$

Consequently, I_1 and I_2 are linked on Γ by the relation

$$I_1 = u_0(I_2). \tag{3.10}$$

The first integrals remain constant on the characteristics (i.e., on the integral curves of the characteristic system). Thus, relation (3.10) remains valid on all characteristics issued from the surface Γ. It remains to notice that, upon substituting (3.9) into (3.10), we get exactly the equation (3.8).

On the other hand, the Cauchy problem (3.1)–(3.2) can be solved by extending the solution $u = u(t,x)$ from the initial point $(0, y)$ by the constant value (the value $u_0(y)$ of the solution at this initial point) along the line

$$x - f'\left(u_0(y)\right) \cdot t = C = y - f'\left(u_0(y)\right) \cdot 0 = y, \tag{3.11}$$

that is, by setting $u(t,x) = u_0(y)$ for all x and t which satisfy (3.11). Expressing the variable y in equation (3.11) through x and t, we get a function $y = y(t,x)$; consequently,

$$u(t,x) = u_0\left(y(t,x)\right). \tag{3.12}$$

In this case, extending the solution is reduced to the problem of finding the domain in which equation (3.11), with y for the unknown, can be solved in a unique way.

3.3 Condition for existence of a smooth solution in a strip

Let us find the maximal value among all time instants $T > 0$ for which equation (3.8) determines a smooth solution $u = u(t, x)$ in the strip Π_T. In fact, we have to determine the greatest possible value of T such that the equation

$$\Phi(t, x, u) \equiv u - u_0(x - f'(u)t) = 0, \tag{3.13}$$

with unknown u, has a unique solution for all fixed t in the interval $[0, T)$ and all $x \in \mathbb{R}$. For $t = 0$, the function $\Phi = \Phi(0, x, u)$ is monotone increasing in u. Thus, by the implicit function theorem the time instant T in question is restricted by the relation

$$\Phi_u(u, x, t) = 1 + u_0'(x - f'(u)t) \cdot f''(u) \cdot t > 0 \tag{3.14}$$

for all points (t, x, u) such that $\Phi(t, x, u) = 0$ and $t \in [0, T)$.

If $|f''(u)| \leqslant L$ on the range of the function $u_0 = u_0(x)$, and if, in addition, $|u_0'| \leqslant K$, then (3.14) is certainly satisfied whenever $1 - KL \cdot t > 0$. Therefore, there exists a smooth solution of problem (3.1)–(3.2) in the strip

$$0 < t < \frac{1}{KL}\,.$$

Problem 3.1. *Show that if the functions u_0' and f'' keep constant signs (i.e., the function u_0 is monotone, and the function f is either convex or concave) and if the two signs coincide, then a smooth solution $u = u(t, x)$ exists in the whole half-space $t > 0$.*

Starting from inequality (3.14), we can also obtain the exact value of the maximal time instant T which delimits the time interval of existence of a smooth solution. To do this, denote $y = x - f'(u)t$ and notice that $u = u_0(y)$ because of (3.13). Then (3.14) is rewritten as

$$1 + u_0'(y) \cdot f''(u_0(y)) \cdot t > 0.$$

Hence,

$$T = \frac{1}{-\inf\limits_{y\in\mathbb{R}} [u_0'(y) f''(u_0(y))]} = \frac{1}{-\inf\limits_{y\in\mathbb{R}} \left[\frac{d}{dy} f'(u_0(y))\right]} \tag{3.15}$$

if only the above infimum is negative. Otherwise, if $\inf_{y\in\mathbb{R}} [u_0'(y) f''(u_0(y))] \geqslant 0$, then $T = +\infty$ (see Problem 3.1).

Problem 3.2. *Check that a function $u = u(t, x)$, which is smooth in a strip Π_T and which satisfies* (3.8), *is a solution of the Cauchy problem* (3.1)–(3.2).

Problem 3.3. *Show that the function $u = u(t, x)$ given by* (3.12), *where $y = y(t, x)$ is a smooth function in Π_T such that* (3.11) *holds, is a solution to the Cauchy problem* (3.1)–(3.2).

Problem 3.4. *Show that the formulas* (3.8) *and* (3.12) *define the same solution of the Cauchy problem* (3.1)–(3.2).

Problem 3.5. *Show that, whenever* $\inf_{y\in\mathbb{R}}[u_0'(y)f''(u_0(y))] = -\infty$, *there is no strip* $\Pi_T = \{(t,x) \mid 0 < t < T, x \in \mathbb{R}\}$, $T > 0$, *such that a smooth solution to problem* (3.1)–(3.2) *exists.*

Exercise 3.3. *Find the maximal value* $T > 0$ *for which there exists a smooth solution to the Cauchy problem*

$$u_t + f'(u)u_x = 0, \qquad u\big|_{t=0} = u_0(x), \tag{3.16}$$

in the strip $\Pi_T = \{(t,x) \mid 0 < t < T, x \in \mathbb{R}\}$, *for*

(i) $f(u) = u^2/2, \quad u_0(x) = \arctan x,$

(ii) $f(u) = u^2/2, \quad u_0(x) = -\arctan x,$

(iii) $f(u) = \cos u, \quad u_0(x) = x,$

(iv) $f(u) = \cos u, \quad u_0(x) = \sin x,$

(v) $f(u) = u^3/3, \quad u_0(x) = \sin x.$

Exercise 3.4. *Which of the Cauchy problems of the form* (3.16), *with the data prescribed below, admit a smooth solution* $u = u(t,x)$ *in the whole half-space* $t > 0$, *and, in contrast, which of them do not possess a smooth solution in any strip* Π_T, $T > 0$:

(i) $f(u) = u^2/2, \quad u_0(x) = x^3,$

(ii) $f(u) = u^2/2, \quad u_0(x) = -x^3,$

(iii) $f(u) = u^4, \quad u_0(x) = x,$

(iv) $f(u) = u^4, \quad u_0(x) = -x?$

3.4 Formation of singularities

To fix the ideas, consider the following Cauchy problem for the Hopf equation (1.1), i.e., for the equation of the form (3.1) with $f(u) = u^2/2$:

$$u_t + uu_x = 0, \qquad u\big|_{t=0} = u_0(x), \tag{3.17}$$

the initial datum u_0 being the smooth function given by

$$u_0(x) = \begin{cases} 2 & \text{for } x \leqslant -3, \\ \psi_1(x) & \text{for } -3 < x < -1, \\ -x & \text{for } -1 \leqslant x \leqslant 1, \\ \psi_2(x) & \text{for } 1 < x < 3, \\ -2 & \text{for } x \geqslant 3 \end{cases}$$

(see Fig. 4a). Here the functions ψ_1 and ψ_2 connect, in a smooth way, the two constant values taken by u_0 as $|x| \geqslant 3$ with the linear function representing u_0 as $|x| \leqslant 1$. While

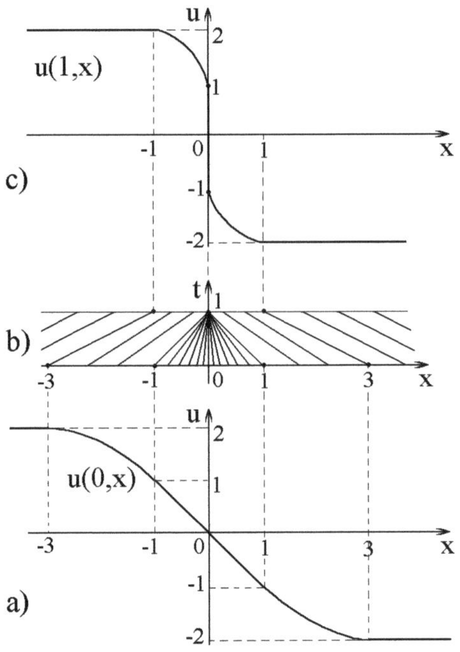

Figure 4. Formation of a strong discontinuity.

doing this, we can choose ψ_1 and ψ_2 in such a way that $-1 < \psi_i'(x) \leqslant 0$, $i = 1, 2$, as $1 < |x| < 3$.

As we have $|u_0'| \leqslant 1$ and $f'' = 1$, the results of the previous section imply the existence of a unique smooth solution $u = u(t,x)$ to problem (3.17) in the strip $0 < t < 1$. As was shown in Section 3.2, in order to construct this solution one has to issue the straight line (see (3.11))

$$x - u_0(y) \cdot t = y, \tag{3.18}$$

starting at every point $(t,x) = (0,y)$ of the line $t = 0$, and one has to assign $u(t,x) = u_0(y)$ at all the points (t,x) of this line.

For $y \leqslant -3$ (for $y \geqslant 3$, respectively) the equation (3.18) determines (see Fig. 4b) the family of parallel straight lines $x = 2t + y$ (or $x = -2t + y$, respectively). Consequently,

$$u(t,x) = 2 \quad \text{for} \quad 0 \leqslant t \leqslant 1, \quad x \leqslant 2t - 3,$$
$$u(t,x) = -2 \quad \text{for} \quad 0 \leqslant t \leqslant 1, \quad x \leqslant 3 - 2t.$$

Further, for $|y| \leqslant 1$ the corresponding straight lines are given by $x + yt = y$, i.e., by $x = y(1-t)$; on these lines, $u = -y = -x/(1-t)$. This means that

$$u(t,x) = -x/(1-t) \quad \text{for} \quad 0 \leqslant t < 1, \quad |x| \leqslant 1 - t.$$

On the set $0 \leqslant t \leqslant 1$, $1-t < |x| < 3-2t$, we cannot write down an explicit formula for $u = u(t,x)$ without defining explicitly the functions ψ_i. Nevertheless, we can guarantee that the straight lines of the form (3.18), corresponding to different values of y from the set $(-3,-1) \cup (1,3)$, do not intersect inside the strip $0 \leqslant t \leqslant 1$ because $|\psi_i'| < 1$ on this set.

For $t = 1$, through each point $(t,x) = (1,x)$ with $x \neq 0$ there passes one and only one straight line (3.18), corresponding to some value y with $|y| > 1$ (see Fig. 4b). Such a line carries the value $u = u_0(y)$ for the solution at the point $(1,x)$. Moreover, if $x \to -0$, then the corresponding value of y tends to -1; and if $x \to +0$, then $y \to 1$. Consequently, at the time instant $t = 1$, we obtain a function $x \mapsto u(1,x)$ which is smooth for $x < 0$ and for $x > 0$, according to the implicit function theorem. As has been pointed out,

$$\lim_{x\to\pm 0} u(1,x) = \lim_{y\to\pm 1} u_0(y) = \mp 1.$$

As to the point $(1,0)$, different characteristics bring different values of u to this point. More precisely, all the lines of the form (3.18) with $|y| \leqslant 1$ (i.e., the lines $x = y(1-t)$) pass through this point; each line carries the corresponding value $u = -y$, so that all the values contained within the segment $[-1,1]$ are brought to the point $(1,0)$.

The graph of the function $u = u(1,x)$ is depicted in Fig. 4c.

To summarize, starting from a smooth function $u(0,x) = u_0(x)$ at the initial instant of time $t = 0$, at time $t = 1$ we obtain the function $x \mapsto u(1,x)$ which turns out to be discontinuous at the point $x = 0$. This kind of discontinuity, where $u(t_0, x_0+0) \neq u(t_0, x_0-0)$, is called a strong one. Consequently, we can say that the solution of problem (3.17) forms a *strong discontinuity* at the time $t_0 = 1$ at the point $x_0 = 0$.

For the general problem (3.1)–(3.2), whenever $\inf_{y\in\mathbb{R}} [u_0'(y) f''(u_0(y))]$ is negative and it is attained on a non-trivial segment $[y_-, y_+]$, strong discontinuity occurs at the time instant T given by (3.15). In this situation, like in the example just analyzed, all the straight lines (3.11) corresponding to $y \in [y_-, y_+]$ intersect at some point (T, x_0); they bring different values of u to this point.

Problem 3.6. *Show that if*

$$u_0'(y) f''(u_0(y)) = I \quad \forall y \in [y_-, y_+], \quad \textit{where} \quad I = \inf_{y\in\mathbb{R}} [u_0'(y) f''(u_0(y))], \quad I < 0,$$

then the family of straight lines (3.11) *corresponding to* $y \in [y_-, y_+]$ *crosses at one point.*

Instead of a strong discontinuity, a so-called *weak discontinuity* may occur in a solution $u = u(t,x)$ at the time instant T. This term simply means that the function $x \mapsto u(T,x)$ is continuous in x, but fails to be differentiable in x.

Problem 3.7. *Let the infimum* $I = \inf_{y\in\mathbb{R}} [u_0'(y) f''(u_0(y))]$ *be a negative minimum, attained at a single point* y_0. *Let* T *be given by* (3.15). *Show that in this situation, the solution* $u = u(t,x)$, *which is smooth for* $t < T$, *has a weak discontinuity at the point* $(T, y_0 + f'(u_0(y_0))T)$; *in addition, for each* $t > T$ *some of the lines given by* (3.11) *cross.*

4 Generalized solutions of quasilinear equations

As has been shown in the previous section, whatever the smoothness of the initial data is, classical solutions of first-order quasilinear PDEs can develop singularities as time grows. Furthermore, in applications one often encounters problems with discontinuous initial data. The nature of the equations we consider (here, the role of the characteristics is important, because they "carry" the information from the initial datum) is such that we cannot expect that the initial singularities smooth out automatically for $t > 0$. Therefore, it is necessary to extend the notion of a classical solution by considering so-called generalized solutions, i.e., solutions lying in classes of functions which contain functions with discontinuities.

4.1 The notion of generalized solution

There exists a general approach leading to a notion of generalized solution; it has its origin in the theory of distributions. In this approach, the pointwise differential equation is replaced by an appropriate family of integral identities. When restricted to classical (i.e., sufficiently smooth) solutions, these identities are equivalent to the original differential equation. However the integral identities make sense for a much wider class of functions. A function satisfying such integral identities is often called a generalized solution.[3]

The approach we will now develop exploits the Green–Gauss formula.

Theorem 4.1 (The Green–Gauss (Ostrogradskiĭ–Gauss) formula). *Let Ω be a bounded domain of $\mathbb{R}^n$ with smooth boundary $\partial\Omega$ and $w \in C^1(\overline{\Omega})$. Then*

$$\int_\Omega \frac{\partial w}{\partial x_i}\, dx = \int_{\partial\Omega} w \cos(\nu, x_i)\, dS_x.$$

Here $\cos(\nu, x_i)$ is the i-th component of the outward unit normal vector ν (this is the cosine of the angle formed by the direction of the outward normal vector to $\partial\Omega$ and the direction of the i-th coordinate axis Ox_i); and dS_x is the infinitesimal area element on $\partial\Omega$.

Let us apply Theorem 4.1 to the function $w = uv$, $u, v \in C^1(\overline{\Omega})$. Passing one of the terms from the left-hand to the right-hand side, we get the following corollary.

Corollary 4.2 (Integration-by-parts formula). *For any $u, v \in C^1(\overline{\Omega})$,*

$$\int_\Omega v \frac{\partial u}{\partial x_i}\, dx = \int_{\partial\Omega} uv \cos(\nu, x_i)\, dS_x - \int_\Omega u \frac{\partial v}{\partial x_i}\, dx. \tag{4.1}$$

The first term in the right-hand side of (4.1) is analogous to the non-integral term which appears in the well-known one-dimensional integration-by-parts formula.

[3] *NT* — In the literature, these solutions are most usually called "weak" solutions. In the present lectures, the authors have kept the terminology and the approach of S. N. Kruzhkov, designed in order to facilitate the assimilation of the idea of a weak (generalized) solution, and to stress, throughout all the lectures, the distinction and the connections between the classical solutions and the generalized ones.

Assume that a function $u = u(t,x) \in C^1(\Omega)$ is a classical solution of the equation

$$u_t + (f(u))_x = 0, \tag{4.2}$$

$f \in C^1(\mathbb{R})$, in some domain $\Omega \subset \mathbb{R}^2$, e.g., in the strip $\Omega = \Pi_T := \{-\infty < x < +\infty,\ 0 < t < T\}$. This means that substituting $u(t,x)$ into equation (4.2), we obtain a correct identity for all points $(t,x) \in \Omega$. Let us multiply this equation by a compactly supported infinitely differentiable function $\varphi = \varphi(t,x)$. Saying that φ is compactly supported means that $\varphi = 0$ outside of some bounded domain G such that, in addition, $\overline{G} \subset \Omega$. (The space of all compactly supported infinitely differentiable functions on Ω is denoted by $C_0^\infty(\Omega)$.) Since the functions $u = u(t,x)$, $f = f(u(t,x))$, $\varphi = \varphi(t,x)$ are smooth, we can use the integration-by-parts formula (4.1):

$$\begin{aligned}
0 &= \int_\Omega [u_t + (f(u))_x]\,\varphi\, dtdx = \int_G u_t\varphi\, dtdx + \int_G (f(u))_x\varphi\, dtdx \\
&= \int_{\partial G} (u\cos(\nu,t) + f(u)\cos(\nu,x))\,\varphi\, dS - \int_G (u\varphi_t + f(u)\varphi_x)\ dtdx \\
&= -\int_\Omega (u\varphi_t + f(u)\varphi_x)\ dtdx.
\end{aligned}$$

Here we took advantage of the fact that $\varphi(t,x) = 0$ for $(t,x) \in \Omega \setminus G$, which is the case, in particular, for $(t,x) \in \partial G$.

Consequently, we have obtained the following assertion: if $u = u(t,x)$ is a smooth solution of equation (4.2) in the domain Ω, then

$$\int_\Omega (u\varphi_t + f(u)\varphi_x)\ dtdx = 0 \quad \forall \varphi \in C_0^\infty(\Omega). \tag{4.3}$$

The relation (4.3) is taken for the definition of a generalized solution (sometimes called a solution in the sense of integral identity or distributional solution) of the equation (4.2). A generalized solution of the equation we consider need not to be smooth. But any classical solution $u = u(t,x)$ of equation (4.2) is also its generalized solution.

The converse fact is also easy to establish: if a function $u = u(t,x)$ is a generalized solution of equation (4.2) which turns out to be smooth (i.e., u belongs to $C^1(\Omega)$ and it satisfies (4.3)), then it is also a classical solution of this equation (i.e, substituting it into equation (4.2) yields a correct equality). Indeed, the calculations above remain true when carried out in the reversed order. Moreover, the fact that the continuous function $[u_t + (f(u))_x]$ satisfies

$$\int_\Omega [u_t + (f(u))_x]\,\varphi\, dtdx = 0 \quad \forall \varphi \in C_0^\infty(\Omega)$$

implies that $u_t(t,x) + [f(u(t,x))]_x = 0$ for all $(t,x) \in \Omega$.

Problem 4.1. *Justify the latter assertion rigorously.*

4.2 The Rankine–Hugoniot condition

Consider a smooth function $u = u(t, x)$ in a domain $\Omega \subset \mathbb{R}^2_{t,x}$, and associate to this function the vector field $\vec{v} = (u, f(u))$ defined on the same domain. The function u is a classical solution of the equation (4.2) if and only if $\operatorname{div} \vec{v} = 0$; in turn, the latter condition means that the flux of the vector field $\vec{v}$ through the boundary of any domain $G \subset \Omega$ equals zero:

$$\int_{\partial G} (\vec{v}, \nu)\, dS = 0 \qquad \forall G \subset \Omega. \tag{4.4}$$

Here ν is the outward unit normal vector to ∂G, and $(\vec{v}, \nu)$ denotes the scalar product of the vectors $\vec{v}$ and ν. The identity (4.4) is called *a conservation law*.

Now assume we have a piecewise smooth function $u = u(t, x)$ that satisfies equation (4.2) in a neighbourhood of each of its smoothness points. In this case, the conservation law (4.4) need not hold in general (the flux of $\vec{v}$ may be non-zero, if the domain G contains a curve across which $u = u(t, x)$ is discontinuous). We now show that, nevertheless, for any piecewise smooth generalized solution of equation (4.2) (solution in the sense of the integral identity (4.3)), this important physical law does hold. In a sense, the essential feature of the differential equation (4.2) is to express the law (4.4); and this feature is "inherited" by the generalized formulation (4.3).

The proof amounts to the fact that, on every discontinuity curve, a generalized solution satisfies the so-called Rankine–Hugoniot condition. For a piecewise smooth function $u = u(t, x)$ that satisfies equation (4.2) in a neighbourhood of each point of smoothness, this condition is necessary and sufficient for u to be a generalized solution in the sense of the integral identity (4.3). The present section is devoted to the deduction of the aforementioned Rankine–Hugoniot condition.

Let $u = u(t, x)$ be a piecewise smooth generalized solution of equation (4.2) in the domain $\Omega \subset \mathbb{R}^2$, i.e., a solution in the sense of the integral identity (4.3). To be specific, let us assume that Ω is divided into two parts Ω_- and Ω_+, separated by some curve Γ (see Fig. 5); we further assume that in each of these two parts, the function $u = u(t, x)$ is smooth, i.e., $u \in C^1(\Omega_-) \cap C^1(\Omega_+)$, and that there exist one-sided limits u_- and u_+ of the function u as one approaches Γ from the side of Ω_- and from the side of Ω_+, respectively.

Consequently, at each point $(t_0, x_0) \in \Gamma$ of the discontinuity curve Γ, one can define

$$u_-(t_0, x_0) = \lim_{\substack{(t,x)\to(t_0,x_0)\\ (t,x)\in\Omega_-}} u(t, x) \quad \text{and} \quad u_+(t_0, x_0) = \lim_{\substack{(t,x)\to(t_0,x_0)\\ (t,x)\in\Omega_+}} u(t, x).$$

Such discontinuities are called discontinuities of the first kind, or strong discontinuities, or jumps.

Notice that $u = u(t, x)$ is a generalized solution of (4.2) in each of the two subdomains Ω_- and Ω_+, in view of the fact that $C_0^\infty(\Omega_\pm) \subset C_0^\infty(\Omega)$. Moreover, this function is smooth in Ω_- and Ω_+. Therefore, according to what has already been proved, in each of the two subdomains, the function $u = u(t, x)$ is a classical solution of equation (4.2). Let us derive the conditions satisfied by $u = u(t, x)$ along the discontinuity curve Γ.

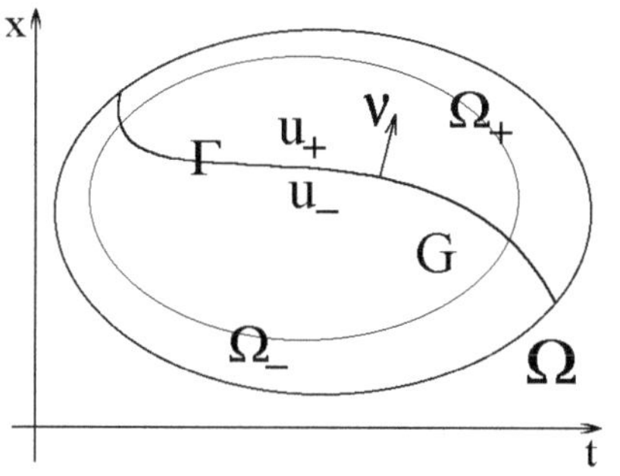

Figure 5. Strong discontinuity (jump).

Proposition 4.3. *Assume that the curve Γ contained within the domain Ω is represented by the graph of a smooth function $x = x(t)$. Then the piecewise smooth generalized solution $u = u(t, x)$ of equation (4.2) satisfies the following condition on Γ, called* the Rankine–Hugoniot condition:

$$\frac{dx}{dt} = \frac{[f(u)]}{[u]} = \frac{f(u_+) - f(u_-)}{u_+ - u_-}, \tag{4.5}$$

where $[u] = u_+ - u_-$ is the jump of the function u on the discontinuity curve Γ, and $[f(u)] = f(u_+) - f(u_-)$ is the jump of $f = f(u)$.

Taking into account the relation $dx/dt = -\cos(\nu, t)/\cos(\nu, x)$, where $\cos(\nu, t)$ and $\cos(\nu, x)$ are the components of the unit normal vector ν to the curve $\Gamma = \{(t, x(t))\}$ (the vector is oriented to point from Ω_- to Ω_+; notice that $\cos(\nu, x) \neq 0$), the equality (4.5) can be rewritten in the equivalent form

$$[u]\cos(\nu, t) + [f(u)]\cos(\nu, x) = 0. \tag{4.6}$$

Definition 4.4. *A shock wave* is a discontinuous generalized solution of equation (4.2).

Thus we can say that the Rankine–Hugoniot condition (4.5) relates the speed $\dot{x}$ of propagation of a shock wave with the flux function $f = f(u)$ and the limit states u_+ and u_- of the shock-wave solution $u = u(t, x)$.

Proof of Proposition 4.3. Let us prove the formula (4.6). By the definition of a generalized solution, for any "test" function $\varphi \in C_0^\infty(\Omega)$ such that $\varphi(t, x) = 0$ for $(t, x) \notin G$, $\overline{G} \subset \Omega$, we have

$$\begin{aligned} 0 &= \int_\Omega (u\varphi_t + f(u)\varphi_x)\ dtdx \\ &= \int_{\Omega_- \cap G} (u\varphi_t + f(u)\varphi_x)\ dtdx + \int_{\Omega_+ \cap G} (u\varphi_t + f(u)\varphi_x)\ dtdx. \end{aligned}$$

The functions $u = u(t, x)$, $f = f(u(t, x))$, and $\varphi = \varphi(t, x)$ are smooth in the domains $\Omega_- \cap G$ and $\Omega_+ \cap G$. Since these domains are bounded, while integrating on these domains we can transfer derivatives according to the multi-dimensional integration-by-parts formula (4.1). Notice that the boundaries of these domains consist of Γ and of parts of ∂G. The integrals over ∂G are equal to zero due to the fact that $\varphi(t, x) = 0$ for $(t, x) \in \partial G$. Thus, we have

$$
\begin{aligned}
0 &= -\int_{\Omega_- \cap G} (u_t\varphi + (f(u))_x\varphi)\, dtdx + \int_{\Gamma\cap G} (u_- \cos(\nu, t) + f(u_-)\cos(\nu, x))\,\varphi\, dS \\
&\quad - \int_{\Omega_+ \cap G} (u_t\varphi + (f(u))_x\varphi)\, dtdx + \int_{\Gamma\cap G} (u_+ \cos(-\nu, t) + f(u_+)\cos(-\nu, x))\,\varphi\, dS \\
&= -\int_{\Omega_-} (u_t + (f(u))_x)\,\varphi\, dtdx - \int_{\Omega_+} (u_t + (f(u))_x)\,\varphi\, dtdx \\
&\qquad - \int_\Gamma \Big((u_+ - u_-)\cos(\nu, t) + (f(u_+) - f(u_-))\cos(\nu, x) \Big)\varphi\, dS.
\end{aligned}
$$

Here we used the fact that ν is the outward unit normal vector to the part Γ of the boundary of the domain $\Omega_- \cap G$; thus $-\nu$ is the outward unit normal vector to the part Γ of the boundary of $\Omega_+ \cap G$. As was already mentioned, $u = u(t, x)$ is a classical solution in both domains Ω_- and Ω_+, i.e., equation (4.2) holds for $(t, x) \in \Omega_- \cup \Omega_+$. Therefore, we have

$$
\int_\Gamma ([u]\cos(\nu, t) + [f(u)]\cos(\nu, x))\,\varphi\, dS = 0 \qquad \forall \varphi \in C_0^\infty(\Omega). \tag{4.7}
$$

Consequently, the equality (4.6) is satisfied at all points $(t, x) \in \Gamma$ where the discontinuity curve Γ is smooth (i.e., at the points $(t, x) \in \Gamma$ where the normal vector $\nu = (\cos(\nu, t), \cos(\nu, x))$ depends continuously on the point of Γ). □

The converse of the statement of the above theorem also holds true. Precisely, let a function $u = u(t, x)$ be a classical solution of equation (4.2) in each of the domains Ω_- and Ω_+. Assume that the function u has a discontinuity of the first kind on the curve Γ separating Ω_- from Ω_+ and that the Rankine–Hugoniot condition holds on the discontinuity curve Γ. Then u is a generalized solution of equation (4.2) in the domain $\Omega = \Omega_- \cup \Gamma \cup \Omega_+$. Indeed, starting from (4.7) and using the fact that

$$
u_t + (f(u))_x = 0 \qquad \text{for} \quad (t, x) \in \Omega_- \cup \Omega_+,
$$

we can reverse all the calculations of the above proof. This eventually leads to the integral identity (4.3), which is the definition of a generalized solution.

Problem 4.2. *Justify rigorously the above statement.*

Theorem 4.5. *Assume that $u = u(t, x)$ is a piecewise smooth function*[4] *defined in a domain Ω with a finite number of components $\Omega_1, \Omega_2, \ldots, \Omega_m$ where u is smooth, and, ac-*

[4] *NT* — Throughout the lectures, the term "piecewise smooth" refers exactly to the situation described in the assumption formulated in the present paragraph. This framework is sufficient to illustrate the key ideas of generalized solutions. In general, there may exist discontinuous generalized solutions with a much more complicated structure, but they are far beyond our scope.

cordingly, with a finite number of curves of discontinuity of the first kind $\Gamma_1, \Gamma_2, \ldots, \Gamma_k$, *so that we have*

$$\Omega = \Big(\bigcup_{i=1}^{m} \Omega_i\Big) \bigcup \Big(\bigcup_{i=1}^{k} \Gamma_i\Big)$$

(*see Fig.* 6 *which corresponds to the case of a strip domain* $\Omega = \Pi_T$).

The function $u = u(t,x)$ *is a generalized solution of equation* (4.2) *in the domain* Ω *in the sense of the integral identity* (4.3) *if and only if* u *is a classical solution of this equation in a neighbourhood of each smoothness point of* u (*i.e., on each of the sets* Ω_i, $i = 1, \ldots, m$) *and, moreover, the Rankine–Hugoniot condition* (4.6) *is satisfied on each discontinuity curve* Γ_i, $i = 1, \ldots, k$ *except for the finite number of points where some of the curves* Γ_i *intersect one another.*

For the proof, it is sufficient to consider the restriction of the function u to each discontinuity curve Γ_i and the two smoothness components Ω_{i_1}, Ω_{i_2} adjacent to Γ_i; then we can exploit the assertions already shown in Proposition 4.3 and in Problem 4.2.

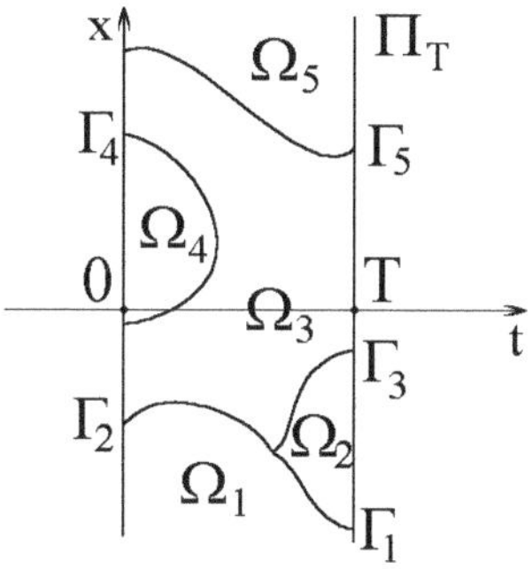

Figure 6. Piecewise smooth solution.

Proposition 4.6. *Let* $u = u(t,x)$ *be a piecewise smooth generalized solution of equation* (4.2) *in the domain* Ω *in the sense of the integral identity* (4.3). *Then the vector field* $\vec{v} = (u, f(u))$ *satisfies the conservation law* (4.4).

Proof. Assume that Ω_i are the components of smoothness of u. Let G be an arbitrary subdomain of the domain Ω. For all i, the flux of the vector field $\vec{v} = (u, f(u))$ through $\partial(\Omega_i \cap G)$ is equal to zero, because u is a classical solution of equation (4.2) in the subdomain Ω_i and thus also in the subdomain $\Omega_i \cap G$. Therefore, we can represent zero as the sum of these fluxes over all boundaries $\partial(\Omega_i \cap G)$. Thanks to the Rankine–Hugoniot condition (4.6), on each discontinuity curve Γ_j the total flux (i.e., the sum of the fluxes from the two sides of Γ_j) of the vector field $\vec{v}$ across the curve $\Gamma_j \cap G$ is equal to zero. Consequently, the sum of the fluxes across all the boundaries $\partial(\Omega_i \cap G)$ is equal to the flux of the vector field $\vec{v}$ through ∂G. This proves (4.4). □

As has been mentioned in Remark 3.1, the area delimited by the graph of a classical solution $u = u(t,x)$ of the problem (3.1)–(3.2) remains constant as a function of time

$t \geqslant 0$, whenever this area is finite. It turns out that also the generalized solutions obey this property. Thus the process of formation of a shock wave (a process that can be visualized as an "overturning" of the graph) occurs in such a way that the part which is "cut off" has area equal to the area of the "extra" part (see Fig. 7); this equality of the two areas is a direct consequence of the Rankine–Hugoniot condition.

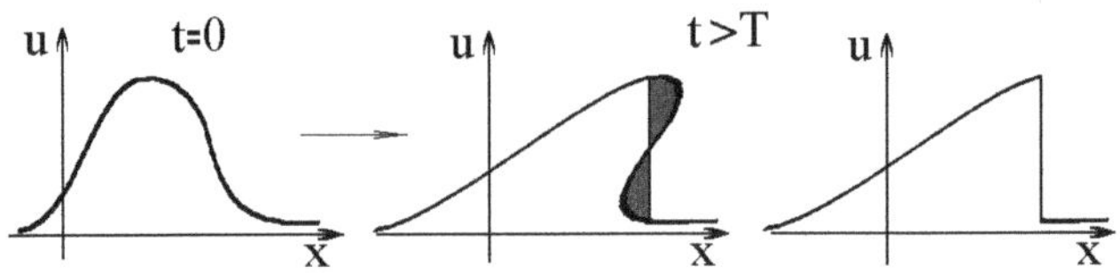

Figure 7. Area-preserving "overturning" of the graph.

Proposition 4.7. *Assume that $u = u(t,x)$ is a piecewise smooth function with compact support in x, such that $x = x(t)$ is the unique discontinuity curve of u and such that u is a generalized solution of equation* (4.2). *Denote*

$$S(t) = \int_{-\infty}^{+\infty} u(t,x)\,dx.$$

Then the function $S = S(t)$ is independent of t, i.e., $S(t) \equiv \mathrm{const}$.

Proof. Indeed, we can write

$$S(t) = \int_{-\infty}^{x(t)} u(t,x)\,dx + \int_{x(t)}^{+\infty} u(t,x)\,dx,$$

where $x = x(t)$ is the curve of discontinuity of the generalized solution $u = u(t,x)$. As previously, we denote by $u_\pm = \lim_{x\to x(t)\pm 0} u(t,x)$ the one-sided limits (limits along the x-axis) of the solution u on the discontinuity curve. Then

$$\begin{aligned} \frac{dS}{dt} &= u(t,x(t)-0)\cdot \dot{x}(t) + \int_{-\infty}^{x(t)} u_t(t,x)\,dx \\ &\quad - u(t,x(t)+0)\cdot \dot{x}(t) + \int_{x(t)}^{+\infty} u_t(t,x)\,dx \\ &= (u_- - u_+)\cdot \dot{x}(t) - \int_{-\infty}^{x(t)} \Big(f(u(t,x))\Big)_x\,dx - \int_{x(t)}^{+\infty} \Big(f(u(t,x))\Big)_x\,dx \\ &= (u_- - u_+)\cdot \dot{x}(t) \\ &\quad - f(u(t,x(t)-0)) + f(u(t,-\infty)) - f(u(t,+\infty)) + f(u(t,x(t)+0)) \\ &= (f(u_+) - f(u_-)) - (u_+ - u_-)\cdot \dot{x}(t). \end{aligned} \tag{4.8}$$

In these calculations, in addition to the equation (4.2) itself, we took advantage of the fact that u has compact support in x, so that $f(u(t,-\infty)) = f(u(t,+\infty)) = f(0)$.

Now if $u_+ = u_-$, then from (4.8) we clearly have

$$\frac{dS}{dt} = 0.$$

In the case $u_+ \neq u_-$, we have the same conclusion thanks to the Rankine–Hugoniot condition (4.5). □

Problem 4.3. *Prove the analogous result for the case where a piecewise smooth generalized (in the sense of the integral identity (4.3)) solution $u = u(t,x)$ of equation (4.2) has a finite number of discontinuity curves $x = x_j(t)$, $j = 1,\dots,N$.*

Remark 4.8. If a function $u = u(t,x)$ has a weak discontinuity on the curve Γ, i.e, u is continuous across Γ and only its derivatives u_t, u_x are discontinuous on Γ, then the Rankine–Hugoniot condition (4.6) is trivially satisfied (indeed, $[u] = 0$ and, consequently, also $[f(u)] = 0$). Therefore, a continuous function $u = u(t,x)$, which is piecewise smooth in a domain Ω and is a classical solution of equation (4.2) in a neighbourhood of each smoothness point, is also a generalized solution of (4.2) in the whole domain Ω (it is clear that the function $u = u(t,x)$ is not a classical solution in Ω, since it is not differentiable at the points $(t,x) \in \Gamma \subset \Omega$).

Remark 4.9. Formally, passing to the limit in (4.5) as $u_\pm \to u$, we infer that

$$\frac{dx}{dt} = f'(u(t,x)), \tag{4.9}$$

on a weak discontinuity curve $\Gamma = \{(t,x) \mid x = x(t)\}$ of $u = u(t,x)$; this means that a weak discontinuity propagates along a characteristic.

Let us provide a rigorous justification of this fact.

Let $\Gamma = \{(t,x) \mid x = x(t)\}$ be a weak discontinuity curve separating two classical solutions $u = u(t,x)$ and $v = v(t,x)$ of equation (4.2). Then

$$u(t,x(t)) \equiv v(t,x(t)). \tag{4.10}$$

Differentiating (4.10) with respect to t, we obtain

$$u_t(t,x(t)) + u_x(t,x(t)) \cdot \frac{dx}{dt} = v_t(t,x(t)) + v_x(t,x(t)) \cdot \frac{dx}{dt}$$

Here and in the sequel, u_x, v_x, u_t, v_t denote the corresponding limits of the derivatives as the point (t,x) tends to the weak discontinuity curve Γ. (The existence of these limits follows from the definition of a weak discontinuity.) Expressing the t-derivatives from the equation (4.2), we have

$$u_x(t,x(t)) \cdot \frac{dx}{dt} - f'(u(t,x(t)))u_x = v_x(t,x(t)) \cdot \frac{dx}{dt} - f'(v(t,x(t)))v_x.$$

Hence, taking into account (4.10), we obtain

$$\Big(u_x(t,x(t)) - v_x(t,x(t))\Big)\left(\frac{dx}{dt} - f'(u(t,x(t))\right) = 0.$$

Since the curve $x = x(t)$ is a weak discontinuity curve, the relation $u_x(t,x) \neq v_x(t,x)$ holds on this curve; thus (4.9) follows.

Exercise 4.1. *Is it true that the following functions* $u = u(t,x)$ *are generalized solutions (in the sense of the integral identity* (4.3)) *of equation* (4.2) *in the strip* Π_T *(remind that* $\Pi_T = \{-\infty < x < +\infty,\ 0 < t < T\}$), *for*

(i) $$f(u) = u^2/2, \qquad u(t,x) = \begin{cases} 0 & \text{for } x < t, \\ 1 & \text{for } x > t; \end{cases}$$

(ii) $$f(u) = u^2/2, \qquad u(t,x) = \begin{cases} 0 & \text{for } x < t, \\ 2 & \text{for } x > t; \end{cases}$$

(iii) $$f(u) = u^2/2, \qquad u(t,x) = \begin{cases} 2 & \text{for } x < t, \\ 0 & \text{for } x > t; \end{cases}$$

(iv) $$f(u) = -u^2, \qquad u(t,x) = \begin{cases} 1 & \text{for } x < 0, \\ -1 & \text{for } x > 0; \end{cases}$$

(v) $$f(u) = -u^2, \qquad u(t,x) = \begin{cases} -1 & \text{for } x < 0, \\ 1 & \text{for } x > 0; \end{cases}$$

(vi) $$f(u) = u^3, \qquad u(t,x) = \begin{cases} 1 & \text{for } x < 0, \\ -1 & \text{for } x > 0; \end{cases}$$

(vii) $$f(u) = u^3, \qquad u(t,x) = \begin{cases} -1 & \text{for } x < t, \\ 1 & \text{for } x > t; \end{cases}$$

(viii) $$f(u) = u^3, \qquad u(t,x) = \begin{cases} 1 & \text{for } x < t, \\ -1 & \text{for } x > t? \end{cases}$$

Exercise 4.2. *Construct some non-trivial generalized solutions in the strip* Π_T *for the equations*

(i) $u_t - (u^3)_x = 0,$

(ii) $u_t - u^2 \cdot u_x = 0,$

(iii) $u_t + \sin u \cdot u_x = 0,$

(iv) $u_t - (e^u)_x = 0,$

(v) $u_t + (e^u)_x = 0,$

(vi) $u_t + u_x/u = 0$

(by non-trivial, we mean a generalized solution that cannot be identified with a classical solution upon modifying its values on a set of Lebesgue measure zero).

4.3 Example of non-uniqueness of a generalized solution

It turns out that extending the notion of solution of equation (4.2) by replacing this equation with the integral identity (4.3) (let us stress again that this identity expresses in a generalized way the conservation law (4.4) for the vector field $\vec{v} = (u, f(u))$) may result in non-uniqueness of a generalized solution to a Cauchy problem. In order to observe this loss of uniqueness of a solution, let us consider equation (4.2) with the flux function $f(u) = u^2$ and with the zero initial datum:

$$u_t + 2uu_x = 0, \qquad x \in \mathbb{R}, \quad 0 < t < T, \tag{4.11}$$

$$u\big|_{t=0} = 0. \tag{4.12}$$

The function $u(t, x) \equiv 0$ is a classical solution, and thus it is also a generalized solution of the above problem. Nonetheless, we can construct non-zero generalized solutions of the problem considered. Assign (see Fig. 8)

$$u_\delta(t,x) = \begin{cases} 0 & \text{for } x < -\delta t, \\ -\delta & \text{for } -\delta t < x < 0, \\ +\delta & \text{for } 0 < x < +\delta t, \\ 0 & \text{for } x > +\delta t, \end{cases} \qquad \text{where } \delta > 0. \tag{4.13}$$

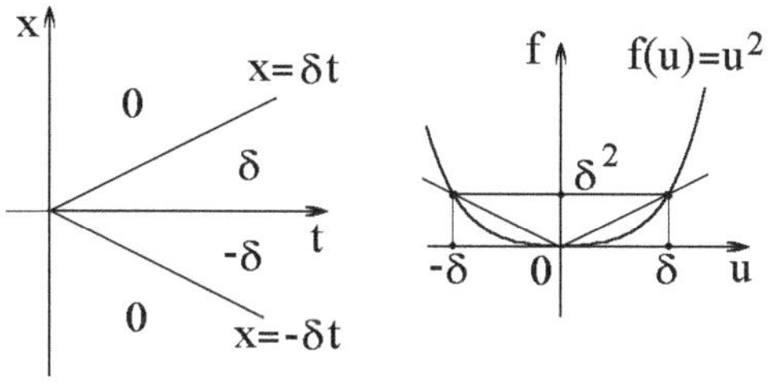

Figure 8. One-parameter family of "wrong" solutions.

Formula (4.13) defines the function $u_\delta = u_\delta(t, x)$ with four components of smoothness; on each of these, u_δ is a classical solution of equation (4.11) (it is clear that, in general, any constant satisfies equation (4.2) whatever be the flux function $f = f(u)$). Let us check the Rankine–Hugoniot condition on each of the three lines of discontinuity of the first kind (which are $x = 0$ and $x = \pm\delta t$):

as $x = 0$, we have $u_- = -\delta$, $u_+ = \delta$, and

$$\frac{dx}{dt} = 0 = \frac{\delta^2 - (-\delta)^2}{\delta - (-\delta)} = \frac{f(u_+) - f(u_-)}{u_+ - u_-};$$

as $x = -\delta t$, we have $u_- = 0$, $u_+ = -\delta$, and

$$\frac{dx}{dt} = -\delta = \frac{(-\delta)^2 - 0^2}{(-\delta) - 0} = \frac{f(u_+) - f(u_-)}{u_+ - u_-};$$

as $x = \delta t$, we have $u_- = \delta$, $u_+ = 0$, and

$$\frac{dx}{dt} = \delta = \frac{0^2 - \delta^2}{0 - \delta} = \frac{f(u_+) - f(u_-)}{u_+ - u_-}.$$

Notice that, in the case of piecewise constant solutions, the Rankine–Hugoniot condition has a simple geometrical interpretation. Let us draw the graph of the flux function $f = f(u)$ respective to the axes (u, f), oriented parallel to the axes (t, x). Next, mark the points $(u_-, f(u_-))$ and $(u_+, f(u_+))$ on the graph (see Fig. 9). Then the segment connecting the two points must be parallel to the discontinuity line $x = x(t) = kt$. Indeed, the slope of this segment is equal to $\frac{f(u_+)-f(u_-)}{u_+-u_-}$, while the slope of the discontinuity line equals $\frac{dx}{dt} = k$; the equality between the two slopes is exactly what the Rankine–Hugoniot condition (4.5) expresses.

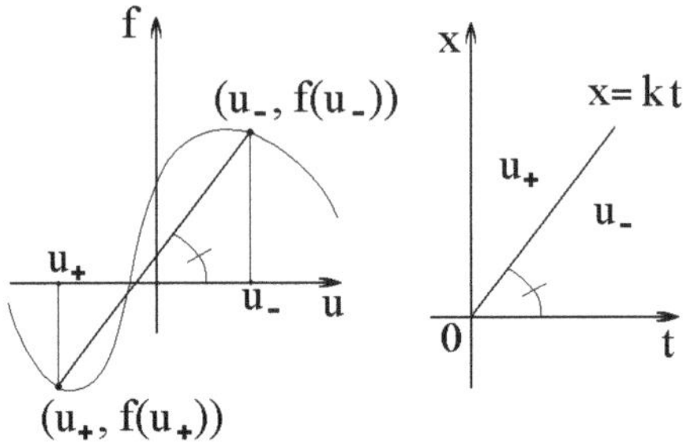

Figure 9. Geometrical interpretation of the Rankine–Hugoniot condition.

This geometrical point of view facilitates the graphical representation of the generalized solutions $u_\delta(t, x)$ of equation (4.11), as constructed above. Marking the points $(0, 0)$, $(\pm\delta, \delta^2)$ and joining them by segments in the way Fig. 8 shows, we obtain the slopes of the discontinuity lines in u_δ.

Exercise 4.3. *Construct a generalized solution of the problem* (4.11)–(4.12) *which is piecewise constant and has three discontinuity lines* (*as in the solution* $u_\delta = u_\delta(t, x)$), *different from any of the solutions* (4.13). *For the solution constructed, verify analytically the Rankine–Hugoniot relation on all the discontinuity lines.*

Let us point out that it is not possible to construct a piecewise constant generalized solution of problem (4.11)–(4.12) with exactly two discontinuity lines. Indeed, such a solution would have two distinct jumps: a jump from the state 0 (on the left from the discontinuity line) to some constant state δ (on the right), and the jump from δ (now on the left) to 0 (now on the right). According to the Rankine–Hugoniot condition, these jumps can only occur along straight lines of the form $x = \frac{f(\delta)-f(0)}{\delta-0}t + C$, $C \in \mathbb{R}$. Since the solution also obeys the zero initial datum, the constant C should be the same for the two jumps. Thus both jumps cancel each other, because they occur along one and the same line; thus our piecewise constant solution is in fact equal to zero.

Exercise 4.4. *Construct piecewise constant generalized solutions of* (4.11)–(4.12) *with more than three discontinuity lines.*

Exercise 4.5. *Is it possible to construct a solution as in the previous exercise but with an even number of discontinuity lines, each of these lines being a ray originating from the point* $(0, 0)$ *of the* (t, x)*-plane?*

In order to construct a non-zero generalized solution of the Cauchy problem

$$u_t + (f(u))_x = 0, \qquad u\big|_{t=0} = 0, \tag{4.14}$$

with an arbitrarily chosen flux function $f = f(u)$, it is sufficient to pick two numbers α and β, $\alpha < 0 < \beta$, in such a way that the points $(0, f(0))$, $(\alpha, f(\alpha))$ and $(\beta, f(\beta))$ are not aligned. Then we join these points pairwise by straight line segments, as it was described above for the case $f(u) = u^2$ (see Fig. 8), and obtain the slopes of the discontinuity rays in the plane (t, x) for the solution to be constructed. Since $\alpha < 0 < \beta$, the slope of the segment joining $(\alpha, f(\alpha))$ with $(\beta, f(\beta))$ is always the intermediate one among the three slopes. Thus the construction produces a piecewise constant solution with the zero initial datum and the two intermediate states α, β.

Exercise 4.6. *Justify carefully that the above construction leads to a piecewise constant generalized solution of problem* (4.14). *Show that if, e.g.,* $0 < \alpha < \beta$, *then the analogous construction yields a non-trivial generalized solution with the initial datum* $u_0(x) \equiv \alpha$.

The above construction breaks down in the case where such non-aligned points on the graph of $f = f(u)$ cannot be found. This corresponds exactly to the case of an affine flux function, i.e., $f(u) = au + b$, $a, b \in \mathbb{R}$. In the latter case, our quasilinear problem is in fact linear:

$$u_t + au_x = 0, \qquad u|_{t=0} = u_0(x). \tag{4.15}$$

In the case where u_0 is smooth (this applies, in particular, to $u_0 \equiv 0$), the unique classical solution of this problem is easily constructed by the method of Section 2; the solution takes the form $u(t, x) = u_0(x - at)$.

Problem 4.4. *Show that for any piecewise smooth solution of equation* $u_t + au_x = 0$, $a =$ const, *the curves of discontinuity are the characteristics of the equation, i.e., the lines* $x = at + C$. *Then, prove the uniqueness of a piecewise smooth solution of the Cauchy problem* (4.15) *with a piecewise smooth initial datum* u_0. *Precisely, show that this solution is given by the equality* $u(t, x) = u_0(x - at)$.

It can be shown that this solution is unique not only within the class of classical solutions, but also within the class of generalized ones; but this is beyond the scope of these notes. In particular, the zero solution is the unique generalized solution of problem (4.14) in the case of a linear flux function $f = f(u)$.

Exercise 4.7. *Construct non-trivial generalized solutions of the problem* (4.14) *with* $f(u) = u^3$, *then with* $f(u) = \sin u$. *Is it possible to construct such solutions with more than three discontinuity lines?*

It should be understood that, from the physical point of view, all the non-trivial generalized solutions to the problem (4.11)–(4.12) or to the problem (4.14) are "wrong"; notwithstanding the fact that these functions satisfy the PDE in the sense of the integral identity (4.3) and comply with the conservation law (4.4), the only "physically correct" solution of the above problems should be, unquestionably, the solution $u(t, x) \equiv 0$. Consequently, we should also devise a mathematical condition which would select, among all the generalized solutions, the unique "correct" solution. This condition, called the entropy increase condition, will now be formulated.

5 The notion of generalized entropy solution

As exposed in the previous sections, in the study of the Cauchy problem for the equation

$$u_t + (f(u))_x = 0 \tag{5.1}$$

with the initial data

$$u\big|_{t=0} = u_0(x), \tag{5.2}$$

we encounter the following situation:

1) There exist some bounded smooth (infinitely differentiable) initial data u_0 such that the unique classical solution $u = u(t, x)$ remains a smooth function up to some critical instant of time T, but the limit

$$u(T, x) = \lim_{t \to T-0} u(t, x)$$

is only a piecewise smooth function with discontinuities of the first kind. The equation (5.1) is one of the so-called "hyperbolic" equations, and the smooth solutions of these equations are determined by the "information" propagated from the initial manifold along the characteristics. Thus it happens that this "information" itself leads to the appearance of discontinuities of the first kind. In this case, it is natural to expect that the solution remains discontinuous as well on some time interval $[T, T + \delta]$. This means that, in order to construct a nonlocal theory of the Cauchy problem (5.1)–(5.2), discontinuous solutions must be introduced into our consideration.

2) One natural approach for introducing such generalized solutions relies on the ideas of the theory of distributions (this approach was discussed in Section 4.1). Even

in a class as wide as the class of all locally bounded measurable functions in Π_T, one could consider generalized solutions $u = u(t,x)$ in the sense of the integral identity

$$\int_{\Pi_T} [u\varphi_t + f(u)\varphi_x]\, dx\, dt = 0, \tag{5.3}$$

which should hold for all "test" functions $\varphi \in C_0^\infty(\Pi_T)$; the initial datum (5.2) should be taken, say, "in the $L_{1,\text{loc}}$ sense" (see (5.31) in Section 5.5 for the exact definition).

Nonetheless, as we have demonstrated in the previous section, so defined generalized solutions of the Cauchy problem may fail to be unique (even for the case $u_0(x) \equiv 0$). It is clear that the non-uniqueness stems from the fact that the "wrong" solutions u_δ, $\delta \neq 0$, have discontinuities. One could guess that not all the discontinuities are admissible; but how can we find the appropriate restrictions on the discontinuities?

5.1 Admissibility condition on discontinuities: the case of a convex flux function

Let us make the additional assumption

$$f'' \geqslant 0, \qquad f \in C^3(\mathbb{R}), \qquad u_0 \in C^2(\mathbb{R}).$$

Problem 5.1. *With the help of* (3.8) *or of* (3.12), *using Problem* 3.2 *or Problem* 3.3, *show that in this case,* $u \in C^2(\Pi_T)$ *where* $[0,T)$ *is the maximal interval of existence of a classical solution.*

Now let us exploit the following consideration, which is purely mathematical: we try to reveal such properties of the smooth (for $t < T$) solutions that do not weaken (or which are conserved) while time approaches the critical value $t = T$. Such properties will therefore characterize the naturally arising singularities of a solution u. Denote $p = u_x(t,x)$ and differentiate the equation (5.1) in x. We have

$$0 = p_t + f'(u)\cdot p_x + f''(u)\cdot p^2 \geqslant p_t + f'(u)p_x\,.$$

Along any characteristics $x = x(t)$, $\dot{x} = f'\left(u(t,x(t))\right)$ (recall that the characteristics fill the whole domain Π_T of existence of a smooth solution), the latter inequality reads as

$$0 \geqslant p_t + \frac{dx}{dt}p_x = \frac{dp(t,x(t))}{dt}\,,$$

that is, the function p does not increase along the characteristics $x = x(t)$. Thus,

$$p(t,x(t)) \leqslant p(0,x(0)) = u_x(0,x(0)) \leqslant \sup_{x\in\mathbb{R}} u_0'(x) =: K_0\,.$$

Consequently, at any point $(t,x) \in \Pi_T$ there holds

$$p(t,x) = u_x(t,x) \leqslant K_0\,. \tag{5.4}$$

As the derivative $u_x(T,x)$ is not defined for some values of x, we pass to the following equivalent form of the inequality (5.4):

$$\frac{u(t,x_2)-u(t,x_1)}{x_2-x_1} \leqslant K_0 \qquad \forall x_1, x_2 \,. \tag{5.5}$$

A similar inequality was introduced in the works of O. A. Oleĭnik (see [37]); the inequality played the role of the admissibility condition in the theory of generalized solutions. From (5.5) it follows that $u(t,x_2)-u(t,x_1) \leqslant K_0(x_2-x_1)$ for $x_1 < x_2$; thus at the limit as $x_2 \to x^* + 0$, $x_1 \to x^* - 0$, where x^* is a discontinuity point of $u(T,x)$, we have

$$u_+ = u(t,x^*+0) < u(t,x^*-0) = u_- \,. \tag{5.6}$$

(Rigorously speaking, passing to the limit implies $u_+ \leqslant u_-$, but $u_+ \neq u_-$ since we assumed that x^* is a discontinuity point.)

Let us require (5.6) to be satisfied at every point of discontinuity of a generalized solution $u = u(t,x)$ (the solution is assumed to be piecewise smooth). It is natural to interpret this condition as an *admissibility condition* on strong discontinuities (jumps) within the class of piecewise smooth solutions.

Remark 5.1. In the example of non-uniqueness exposed above (see Section 4.3) for the Cauchy problem (4.11)–(4.12), where we have $f''(u) = 2 > 0$, the solutions u_δ, $\delta > 0$, of the form (4.13) fail to verify the admissibility condition (5.6) on the discontinuity line $x = 0$. The unique admissible solution of this problem will be the function $u(t,x) \equiv 0$, which is the classical solution of the problem considered.

If $f''(u) \leqslant 0$, then substituting $u = -v$ into equation (5.1) we obtain the equation $v_t + (\tilde{f}(v))_x = 0$, where $\tilde{f}(v) \equiv -f(-v)$; notice that $\tilde{f}''(v) = -f''(-v) \geqslant 0$. For the solution $v = v(t,x)$ of the above equation, we should have $v_+ < v_-$, according to the admissibility condition (5.6). We conclude that in the case $f''(u) \leqslant 0$, the admissibility condition is the inequality $u_+ = -v_+ > -v_- = u_-$, converse to the inequality (5.6).

To summarize, for the case of a convex or a concave flux function $f = f(u)$, we have deduced the following condition for admissibility of discontinuities. Let u_-, respectively u_+, be the one-sided limit of a generalized solution $u = u(t,x)$ as the discontinuity curve is approached from the left, respectively from the right, along the x-axis. Then

- in the case of a convex function $f = f(u)$ (for instance, $f(u) = u^2/2, e^u, \ldots$), generalized solutions of equation (5.1) may have jumps from u_- to u_+ only when $u_- > u_+$;
- in the case of a concave function $f = f(u)$ ($f(u) = -u^2, \ln u, \ldots$), jumps from u_- to u_+ are only possible when $u_- < u_+$.

Let us provide a "physical" explanation of the admissibility condition obtained for the case where the monotonicity of f' is strict. At any point of an admissible discontinuity curve $x = x(t)$, consider the slopes $f'(u_+)$ and $f'(u_-)$ of the characteristics

$x = f'(u_\pm)t + C$ which impinge at this point from the two sides of the discontinuity. Consider also the slope $\omega = \frac{dx}{dt} = \frac{f(u_+)-f(u_-)}{u_+-u_-}$ of the discontinuity curve (more exactly, the slope of its tangent line); notice that ω is equal to the value $f'(\tilde{u})$ at some point $\tilde{u}$ which lies strictly between u_+ and u_-. These three slopes satisfy the so-called *Lax admissibility condition*

$$f'(u_+) < \omega = \frac{f(u_+) - f(u_-)}{u_+ - u_-} = f'(\tilde{u}) < f'(u_-). \tag{5.7}$$

Indeed, if f is strictly convex, then f' is a monotone increasing function, and the admissibility condition for this case of a convex flux function f ensures that $u_+ < \tilde{u} < u_-$. Similarly, if f is strictly concave, then the admissibility condition yields $u_+ > \tilde{u} > u_-$, so that we get (5.7) again, since f' is a monotone decreasing function in this case.

Condition (5.7) is a particular case of the admissibility condition which is fundamental for the theory of systems of conservation laws. It was first formulated by the American mathematician P. D. Lax (see [30]).

Therefore, we observe that, as t grows, the characteristics approach the discontinuity curve from both sides (see Fig. 10a); none of the two characteristics can move away from it (the case where the characteristics move away from the discontinuity curve as t grows is depicted in Fig. 10b). This means that those discontinuities are admissible which are due to the fact that characteristics of a smooth solution (smooth from each side of the discontinuity curve) tend to have intersections as t grows (the intersections eventually occur on the discontinuity curve). On the contrary, the situation when the discontinuity curve is "enforced", with some of the characteristics originating out of the discontinuity curve as time grows, is not admissible.

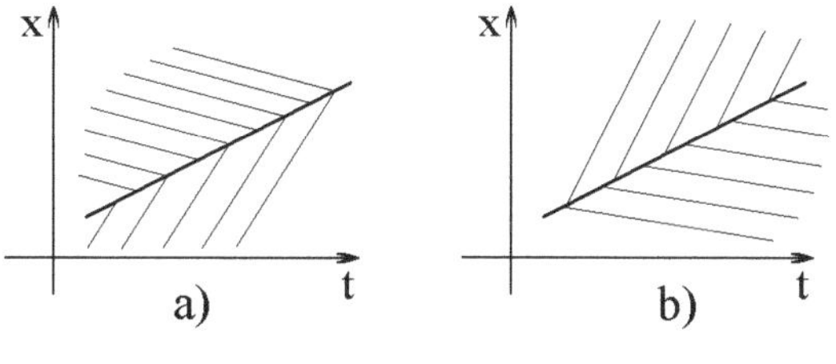

Figure 10. Lax condition: admissible and non-admissible discontinuity curves.

Example 5.2. Let us illustrate the above statement with the example of the Hopf equation (1.1), i.e., the equation (5.1) with $f(u) = u^2/2$. This equation describes the displacement of freely moving particles (see Section 1). Assume that the particles situated, at the initial instant of time, in a neighbourhood of $+\infty$ (i.e., particles with the x-coordinate larger than some sufficiently large value), move with a velocity u_+; assume that the particles initially located in a neighbourhood of $-\infty$ have a velocity u_-; and let $u_+ < u_-$. The latter constraint means that, as time passes, collisions are inevitable, and eventually, a shock wave will form. The velocity of propagation of this shock wave created by particle collisions will be equal to

$$\omega = \frac{f(u_+) - f(u_-)}{u_+ - u_-} = \frac{u_+^2/2 - u_-^2/2}{u_+ - u_-} = \frac{u_+ + u_-}{2}.$$

When the initial velocity profile is a monotone non-increasing function, it can be justified that for sufficiently large t, we obtain a generalized solution of the Hopf equation of the following form:

$$u(t,x) = \begin{cases} u_- & \text{for } x < \omega t + C, \\ u_+ & \text{for } x > \omega t + C. \end{cases} \tag{5.8}$$

This solution can be interpreted as follows. The particles with velocities u_- and u_+ collide when the quicker one (with the velocity u_-) overtakes the slower one (of velocity u_+); this collision is not elastic, and the two particles agglomerate into one single particle. After the collision, the particles continue to move with the velocity $(u_+ + u_-)/2$, creating a shock wave. The velocity of propagation of this wave is calculated with the help of the law of momentum conservation: this velocity is the arithmetic mean of the particles' velocities before the collision. Let us point out that such collisions induce a loss of the kinetic energy of the particles (we will further discuss this question later).

If, on the contrary, the speeds of the particles near $+\infty$ and near $-\infty$ were related by the inequality $u_+ > u_-$ and if the initial velocity distribution were a smooth monotone non-decreasing function, then no collision of particles would ever occur: at any time instant $t > 0$, the velocity distribution $u(t,\cdot)$ would be a smooth non-decreasing function as at the time $t = 0$, and no shock wave might form (see Section 3.1). Therefore, in the case $u_+ > u_-$, the function u given by (5.8), although it does satisfy the integral identity (5.3), is not a physically correct solution of the Hopf equation.

5.2 The vanishing viscosity method

In order to generalize the admissibility condition of the previous section to the case of a flux function $f = f(u)$ which is neither convex nor concave, we make the following observation and reformulate this condition in the terms of the respective location of the graph and the chords of convex or concave functions. We see that the jump between u_- and u_+ is admissible in the sense of the previous section if $u_- > u_+$ (respectively, $u_- < u_+$) and the graph of the flux function f is situated under the chord (respectively, above the chord) joining the points $(u_-, f(u_-))$ and $(u_+, f(u_+))$ (see Fig. 11).

It turns out that the above reformulation of the admissibility rule for convex/concave flux functions remains appropriate for the case of an arbitrary flux function f.

For a rather rigorous justification of this statement, let us use "physical" (more exactly, "fluid dynamics") considerations based on the concepts of an ideal gas and a viscous gas. If $x = x(t)$ is the trajectory of a particle of an ideal gas in a tube aligned with the x-axis, and if the function $u = u(t,x)$ represents the velocity of the particle that occupies the space location x at the time instant t, then (see Section 1)

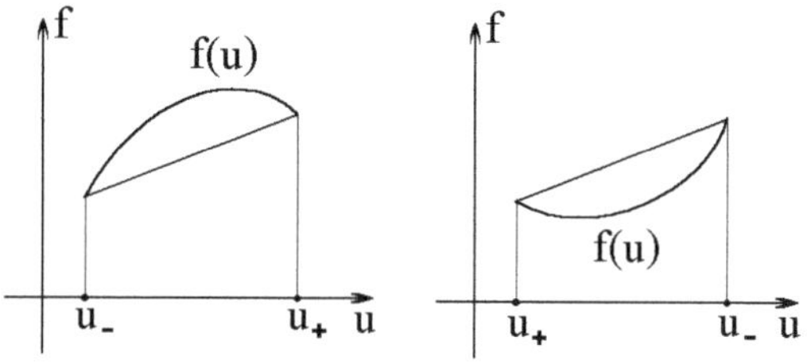

Figure 11. Visualization of admissible jumps, I.

$\dot{x}(t) = u(t, x(t))$, $\ddot{x}(t) = \frac{du}{dt} = 0$; this calculation previously led us to the Hopf equation (1.1). But, ideal gases "do not exist"; they only exist theoretically, as limits when the viscosity of a real gas is neglected because of its smallness.

If $\varepsilon > 0$ is the viscosity coefficient of a real gas, then (under certain assumptions) the force of viscous friction which acts on the particle $x(t)$ at time t and relative to the mass unit can be taken to be $\varepsilon u_{xx}(t, x(t))$. Then $\ddot{x} = \frac{du}{dt} = \varepsilon u_{xx}$, and instead of the Hopf equation we obtain the so-called Burgers equation[5]

$$u_t + uu_x = \varepsilon u_{xx}\,. \tag{5.9}$$

It is natural to admit that — this is what actually takes place — all admissible generalized solutions of the Hopf equation can be obtained as the limit of some solutions $u^\varepsilon = u^\varepsilon(t, x)$ of the equation (5.9) as the viscosity coefficient ε tends to 0. The procedure of introducing the term εu_{xx} into a first-order equation and the subsequent study of the limits of the solutions u^ε as $\varepsilon \to +0$ is called the *"vanishing viscosity" method.*

Before we continue with the application of the vanishing viscosity method to a justification of the general admissibility condition formulated above, let us point out an important method of "linearization" (in a sense) of the Burgers equation (5.9). Observe that we have $u_t = (\varepsilon u_x - u^2/2)_x$; thus we can introduce a potential $U = U(t, x)$, determined from the equality

$$dU = u\,dx + \left(\varepsilon u_x - u^2/2\right)\,dt.$$

In this case

$$U_x = u, \qquad U_t = \varepsilon u_x - u^2/2 = \varepsilon U_{xx} - (U_x)^2/2,$$

i.e., the function U satisfies the equation

$$U_t + \frac{1}{2}(U_x)^2 = \varepsilon U_{xx}\,. \tag{5.10}$$

In (5.10), let us make the substitution $U = -2\varepsilon \ln z$. Then

$$U_t = -2\varepsilon\frac{z_t}{z}, \quad U_x = -2\varepsilon\frac{z_x}{z}, \quad U_{xx} = -2\varepsilon\frac{z_{xx}}{z} + 2\varepsilon\frac{(z_x)^2}{z^2}.$$

[5] *NT* — In the western literature, it is customary to call this equation, "the Burgers equation with viscosity"; accordingly, the term "Burgers equation" then designs what is called the Hopf equation in our lectures.

Equation (5.10) then rewrites as

$$-2\varepsilon\frac{z_t}{z}+2\varepsilon^2\frac{(z_x)^2}{z^2}=-2\varepsilon^2\frac{z_{xx}}{z}+2\varepsilon^2\frac{(z_x)^2}{z^2},$$

so that we are reduced to a linear equation for the function $z=z(t,x)$, which is the classical heat equation:

$$z_t=\varepsilon z_{xx}\,. \tag{5.11}$$

Remark 5.3. The linearization method pointed out hereabove was first used by the Russian mechanicist V. A. Florin in 1948 in his investigation of a physical application. Later on, in the 1950th, this method was rediscovered by the American scholars E. Hopf and S. Cole; nowadays the transformation is often named after them (it would be more correct to speak about the Florin–Hopf–Cole transformation).

It follows from the above substitution that a solution of equation (5.9) has the form

$$u=U_x=-2\varepsilon\frac{z_x}{z}\,,$$

where $z=z(t,x)$ is a solution of the heat equation (5.11).

As is well known from the theory of second-order linear PDEs, solutions of the Cauchy problem for the heat equation (5.11), even with initial data that are only piecewise continuous, become infinitely differentiable for $t>0$. Hence, solutions of the Burgers equation (5.9) are also infinitely differentiable functions, and, consequently, they cannot include shock waves.

Now assume that the so-called "simple wave", given by

$$u(t,x)=u_-+\frac{u_+-u_-}{2}[1+\operatorname{sign}(x-\omega t)]=\begin{cases}u_- & \text{for } x<\omega t,\\ u_+ & \text{for } x>\omega t,\end{cases} \tag{5.12}$$

where $\omega=$ const, is a generalized solution of equation (5.1) in the sense of the integral identity (5.3). For this to hold, it is necessary and sufficient that the Rankine–Hugoniot condition

$$\omega\equiv\frac{dx}{dt}=\frac{f(u_+)-f(u_-)}{u_+-u_-} \tag{5.13}$$

holds on the discontinuity line $x(t)=\omega t$.

For this case, the idea of the vanishing viscosity method can be applied as follows. Let us consider a solution $u=u(t,x)$ of the form (5.12) as admissible, if it can be obtained as a pointwise limit (for $x\neq\omega t$) of solutions $u^\varepsilon=u^\varepsilon(t,x)$ of the equation

$$u_t^\varepsilon+\left(f(u^\varepsilon)\right)_x=\varepsilon u_{xx}^\varepsilon \tag{5.14}$$

as $\varepsilon\to+0$. (The approach developed below has been suggested by I. M. Gel'fand [18]).

Taking into account the special structure of the solution $u=u(t,x)$, let us seek a solution of (5.14) under the form

$$u^\varepsilon(t,x)=v(\xi),\qquad \xi=\frac{x-\omega t}{\varepsilon}. \tag{5.15}$$

Substituting this ansatz into equation (5.14), we infer that the function $v = v(\xi)$ satisfies the equation

$$-\omega v' + (f(v))' = v''. \tag{5.16}$$

On the other hand, it is clear that the function $u^\varepsilon = v\left(\frac{x-\omega t}{\varepsilon}\right)$ converges pointwise (for $x \neq \omega t$) to a function $u = u(t,x)$ of the form (5.12) as $\varepsilon \to +0$ if and only if the function $v = v(\xi)$ satisfies the boundary conditions

$$v(-\infty) = u_-, \qquad v(+\infty) = u_+. \tag{5.17}$$

Remark 5.4. One cannot hope for uniqueness of such a function $v = v(\xi)$. Indeed, if v is a solution of the problem (5.16)–(5.17), then the functions $\tilde{v} = v(\xi - \xi_0)$ are also solutions of this problem, for all $\xi_0 \in \mathbb{R}$.

Integrating (5.16), we obtain

$$v' = -\omega v + f(v) + C = F(v) + C, \qquad C = \text{const}. \tag{5.18}$$

The ODE (5.18) is autonomous, of first-order, and its right-hand side $F(v) + C$ is smooth; thus (5.18) admits a solution which tends to constant states u_- (as $\xi \to -\infty$) and u_+ (as $\xi \to +\infty$) if and only if the following conditions are satisfied:

(i) u_- and u_+ are stationary points of this equation, i.e., the right-hand side of equation (5.18) is zero at these points:

$$F(u_-) + C = F(u_+) + C = 0,$$

so that $C = -F(u_-) = -F(u_+)$. Upon rewriting the equality $F(u_-) = F(u_+)$ under the form $f(u_-) - \omega u_- = f(u_+) - \omega u_+$, we see that it coincides with the Rankine–Hugoniot condition (5.13).

(ii) There is no stationary point in the open interval between u_- and u_+; moreover, the right-hand side $F(v) - F(u_-) = F(v) - F(u_+)$ of (5.18) restricted to this interval should be

a) positive if $u_- < u_+$ (then the solution increases):

$$F(v) - F(u_-) > 0 \quad \forall v \in (u_-, u_+) \quad \text{if } u_- < u_+; \tag{5.19}$$

b) negative if $u_- > u_+$ ($v = v(\xi)$ decreases):

$$F(v) - F(u_+) < 0 \quad \forall v \in (u_+, u_-) \quad \text{if } u_+ < u_-. \tag{5.20}$$

When the above conditions are satisfied, the solutions of equation (5.16) with the desired boundary behaviour are given by the formula

$$\int_{v_0}^{v} \frac{dw}{F(w) - F(u_-)} = \xi - \xi_0, \qquad v_0 = \frac{u_+ + u_-}{2}.$$

Our point is that the relations (5.19)–(5.20) express analytically the admissibility condition.

Now let us interpret this condition geometrically. Substituting $F(v) = f(v) - \omega v$ into (5.19) and (5.20), we have

$$f(v) - f(u_-) > \omega(v - u_-) \quad \forall v \in (u_-, u_+) \quad \text{if} \quad u_- < u_+,$$
$$f(v) - f(u_+) < \omega(v - u_+) \quad \forall v \in (u_+, u_-) \quad \text{if} \quad u_+ < u_-,$$

which, in view of the Rankine–Hugoniot condition (5.13), amounts to

$$\frac{f(u) - f(u_-)}{u - u_-} > \omega = \frac{f(u_+) - f(u_-)}{u_+ - u_-} \quad \forall u \in (u_-, u_+) \quad \text{if} \quad u_- < u_+, \tag{5.19'}$$

$$\frac{f(u) - f(u_+)}{u - u_+} < \omega = \frac{f(u_+) - f(u_-)}{u_+ - u_-} \quad \forall u \in (u_+, u_-) \quad \text{if} \quad u_+ < u_-. \tag{5.20'}$$

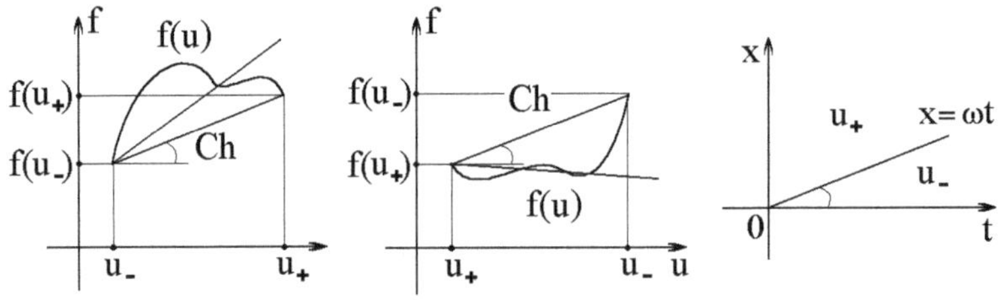

Figure 12. Visualization of admissible jumps, II.

Let us represent the graph of a flux function $f = f(u)$ (see Fig. 12). Condition (5.19′) means that the chord *Ch* with the endpoints $(u_-, f(u_-)), (u_+, f(u_+))$ has a smaller slope (the slope is measured as the inclination of the chord with respect to the positive direction of the u-axis) than the slope of the segment joining the point $(u_-, f(u_-))$ with the point $(u, f(u))$, where u runs over the interval (u_-, u_+)). Consequently, the point $(u, f(u))$ and thus the whole graph of $f = f(u)$ on the interval (u_-, u_+) lies above the chord *Ch*. In the same way, condition (5.20′) signifies that the graph of $f = f(u)$ for $u \in (u_+, u_-)$ is situated below the chord *Ch*.

Remark 5.5. Upon varying the values u_-, u_+ and also the function $f = f(u)$, one can construct different convergent sequences of admissible generalized solutions of the form (5.15). It is natural to consider as admissible also the pointwise limits of the admissible solutions. Therefore, it is clear that any situation where the graph of $f = f(u)$ touches the chord *Ch* should also be considered as admissible.

In conclusion, we obtain that a solution u of the equation (5.1) may have a jump from u_- to u_+ (a jump in the direction of increasing x) when the following **jump admissibility condition** holds:

- in the case $u_- < u_+$, the graph of the function $f = f(u)$ on the segment $[u_-, u_+]$ is situated **above** the chord (in the non-strict sense) with the endpoints $(u_-, f(u_-))$ and $(u_+, f(u_+))$;
- in the case $u_- > u_+$, the graph of the function $f = f(u)$ on the segment $[u_+, u_-]$ is situated **below** the chord (in the non-strict sense) with the endpoints $(u_-, f(u_-))$ and $(u_+, f(u_+))$.

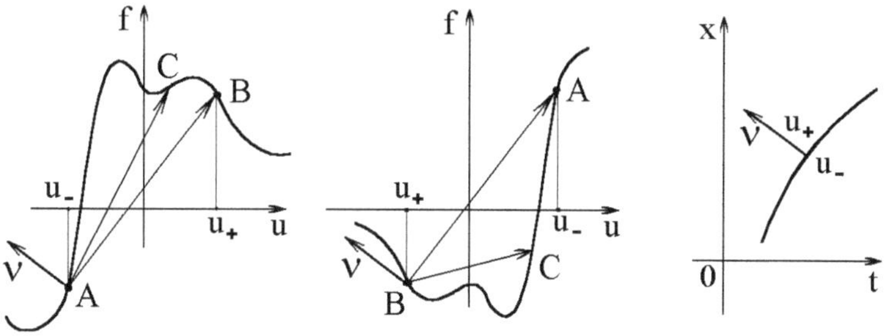

Figure 13. Visualization of admissible jumps, III.

Let us give another analytical expression of the condition obtained. Consider a curve on which the solution has a jump from u_- to u_+. In coordinates (u, f) we draw the graph of the function $f = f(u)$ on the interval between u_- and u_+ and the chord joining the endpoints of this graph. As in Fig. 8 and Fig. 9 (see Section 4.3), we mean that the axes (u, f) are aligned with the axes (t, x). Now on the same graph, let us situate the unit normal vector $\nu = (\cos(\nu, t), \cos(\nu, x))$ to the discontinuity curve (see Fig. 13). Introduce the points $A = (u_-, f(u_-))$, $B = (u_+, f(u_+))$, and let the point $C = (u, f(u))$ run along the graph. The vector ν is orthogonal to the vector $\overrightarrow{AB}$ (this is an expression of the Rankine–Hugoniot condition (5.13)) and is oriented "upwards", i.e., $\cos(\nu, x) > 0$ (this is because we have chosen the normal which forms an acute angle with the positive direction of the x-axis). The condition stating that the graph of the function $f = f(u)$ on the interval between u_- and u_+ is located over the chord ("over", in the non-strict sense) means exactly that the angle between the vectors $\overrightarrow{AC}$ (or, equivalently, $\overrightarrow{BC}$) and ν does not exceed $\pi/2$, that is, the scalar product $(\overrightarrow{AC}, \nu)$ of these vectors is nonnegative. Thus for the case $u_- < u_+$, we have

$$(u - u_-)\cos(\nu, t) + (f(u) - f(u_-))\cos(\nu, x) \geqslant 0 \qquad \forall u \in (u_-, u_+). \tag{5.21}$$

Similarly, the condition stating that the graph is located under the chord ("under", in the non-strict sense) means that the angle between the same vectors as before is greater than or equal to $\pi/2$, that is, the scalar product $(\overrightarrow{BC}, \nu)$ of these vectors is non-positive. Thus for the case $u_- > u_+$, we have

$$(u - u_+)\cos(\nu, t) + (f(u) - f(u_+))\cos(\nu, x) \leqslant 0 \qquad \forall u \in (u_+, u_-). \tag{5.22}$$

Remark 5.6. The admissibility conditions deduced with the vanishing viscosity approach agree perfectly with the conditions obtained in the previous section for the case of a convex/concave flux function $f = f(u)$. Indeed the convexity (respectively, the concavity) of a function means, by definition, that the chord joining two arbitrary points of the graph of the function lies above (respectively, lies below) the graph itself.

In the sequel of these lectures, unless an additional precision is given, by a *solution* of equation (5.1) we will tacitly mean a piecewise smooth function that satisfies the integral identity (5.3) and, in addition, the admissibility condition formulated in the present section.

Exercise 5.1. *Examine the question of admissibility of each of the jumps (jumps satisfying the Rankine–Hugoniot condition* (5.13)) *present in the solutions* $u = u(t, x)$ *to the corresponding equations of the form* (5.1):

(i) *for the generalized solutions* $u = u(t, x)$ *given in Exercise* 4.1;

(ii) *for the generalized solutions* $u = u(t, x)$ *constructed in Exercise* 4.2;

(iii) *for the generalized solutions* $u = u(t, x)$ *constructed in Exercise* 4.7.

5.3 The notion of entropy and irreversibility of processes

The jump admissibility conditions obtained in the previous sections are often called entropy-increase type conditions.[6] Where does this name come from? The reason is, the equations we study model nonlinear physical phenomena (called "processes" in the sequel) which are time-irreversible, and the function which characterizes this irreversibility is called "entropy".

The Hopf equation (1.1) is, certainly, the simplest model for the displacement of a gas in a tube; in more correct (more precise) models, also the pressure of the gas is present, moreover, the density of the gas enters the equations when the gas is compressible. The entropy function S is expressed with the help of the two latter quantities characterizing the gas, namely the pressure and the density. In the field of fluid dynamics, already in the 19th century it has been known that the entropy function does not decrease in time across the front of a shock wave Γ:

$$S_+ = S(t+0, x) \geqslant S_- = S(t-0, x), \qquad (t, x) \in \Gamma. \tag{5.23}$$

Therefore, all the inequalities that express irreversibility of processes in nature are called "inequalities of the entropy increase type". For the simplest gas dynamics equation, which is the Hopf equation, the role of entropy is played by the kinetic energy of the particle located at the point x at the time instant t:

$$S(t, x) \equiv \frac{1}{2} u^2(t, x).$$

[6] *NT* — In the literature on conservation laws, one often speaks of "entropy dissipation conditions". This term refers to the inequalities such as (5.28), (5.30) or (5.42) below. Each of these inequalities states the decrease (the *dissipation*) and not the increase of another quantity related to various functions called "entropies".

Let us show that inequality (5.23) for this "entropy" function S does hold across an admissible shock wave.

For the case of the Hopf equation (i.e., for $f(u) = u^2/2$), the Rankine–Hugoniot condition (5.13) has the form

$$\frac{u_- + u_+}{2} = \frac{dx}{dt}. \tag{5.24}$$

Since the flux function $f(u) = u^2/2$ is convex, the jump admissibility condition reduces to the inequality

$$u_- - u_+ > 0. \tag{5.25}$$

If $dx/dt > 0$, then (according to Fig. 14) we have $S_- = u_+^2/2$ and $S_+ = u_-^2/2$. Multiplying inequality (5.25) by the expression $(u_- + u_+)/2$ (this expression is positive thanks to (5.24)), we have $(u_-^2 - u_+^2)/2 > 0$, thus $S_- < S_+$.

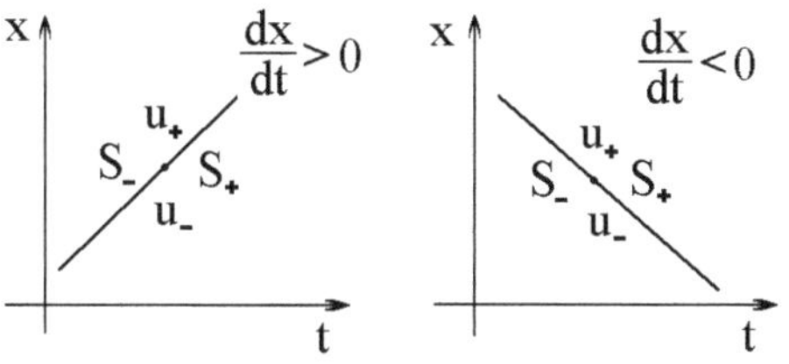

Figure 14. Increase of S for the Hopf equation.

Similarly, if $dx/dt < 0$, then (see Fig. 14)

$$S_- = \frac{1}{2}(u_-)^2 < \frac{1}{2}(u_+)^2 = S_+.$$

5.4 Energy estimates

Let us provide another characterization of irreversibility for equation (5.1), a characterization which has a clear physical meaning. Consider the full kinetic energy of the particle system under consideration:

$$E(t) = \int_{-\infty}^{+\infty} \frac{1}{2} u^2(t, x)\, dx. \tag{5.26}$$

For smooth (and, say, compactly supported) initial data, there exists a classical solution u of problem (5.1)–(5.2) on some time interval $[0, T)$, $T > 0$; moreover, for all fixed t, this solution has compact support in x. In the present section, we will only consider those solutions u of equation (5.1) for which the kinetic energy (5.26) is finite (this holds, e.g., in the above situation where $u = u(t, x)$ is of compact support in the variable x).

Proposition 5.7. *For classical solutions of equation* (5.1) *there holds*

$$E(t) \equiv \text{const},$$

i.e., the kinetic energy (5.26) *is a* first integral *of the equation* (5.1).

Proof. Since we have assumed that $u(t, \pm\infty) = 0$, we have

$$\begin{aligned} \frac{dE}{dt} &= \int_{-\infty}^{+\infty} u u_t \, dx = -\int_{-\infty}^{+\infty} u(f(u))_x \, dx \\ &= -uf(u)\Big|_{x=-\infty}^{x=+\infty} + \int_{-\infty}^{+\infty} f(u) u_x \, dx = \int_{u(t,-\infty)}^{u(t,+\infty)} f(u) \, du = 0. \end{aligned}$$

□

Now consider the corresponding equation with viscosity:

$$u_t^\varepsilon + (f(u^\varepsilon))_x = \varepsilon u_{xx}^\varepsilon \tag{5.27}$$

Proposition 5.8. *Let $u^\varepsilon \not\equiv 0$ be a solution of equation* (5.27) *such that, in addition, u^ε, u_x^ε, and u_{xx}^ε decay to zero as $x \to \pm\infty$ at a sufficiently high rate, and uniformly in t. Then the full kinetic energy $E = E(t)$ of this solution is a decreasing function of time.*

Proof. As in the proof of the previous proposition, we find

$$\begin{aligned} \frac{dE}{dt} &= \int_{-\infty}^{+\infty} u^\varepsilon u_t^\varepsilon \, dx \\ &= \int_{-\infty}^{+\infty} u^\varepsilon \left(\varepsilon u_{xx}^\varepsilon - (f(u^\varepsilon))_x\right) dx = -\varepsilon \int_{-\infty}^{+\infty} \left(u_x^\varepsilon\right)^2 \, dx \leqslant 0. \end{aligned} \tag{5.28}$$

Notice that we have the equality sign in (5.28) only in the case of a function u^ε that is constant in x. Since we assume that this function decays to zero as $x \to \infty$, we have $dE/dt < 0$ unless $u^\varepsilon \equiv 0$. □

Recall (see Section 5.2) that admissible generalized entropy solutions u of equation (5.1) were obtained as limits of solutions u^ε of equations (5.27); on the latter solutions, the kinetic energy is dissipated. Therefore, it can be expected that also on the limiting solutions u, the kinetic energy does not increase with time.

Proposition 5.9. *Assume that $u = u(t, x)$ is a piecewise smooth admissible generalized entropy solution of equation* (5.1) *with one curve of jump discontinuity $x = x(t)$. Then the speed of decrease of the kinetic energy $E = E(t)$ of this solution is equal, at any instant of time $t = t_0$, to the area $S(t_0)$ delimited by the graph of the flux function $f = f(u)$ on the segment $[u_-, u_+]$ (or on the segment $[u_+, u_-]$) and by the chord joining the endpoints $(u_-, f(u_-))$ and $(u_+, f(u_+))$ of this graph (see Fig. 15):*

$$\frac{dE}{dt}(t_0) = -S(t_0). \tag{5.29}$$

As previously, by $u_\pm = u_\pm(t_0)$ we denote the one-sided limits (as $x \to x(t_0)$) of the function $x \mapsto u(t_0, x)$ as the point approaches the discontinuity position $x(t_0)$.

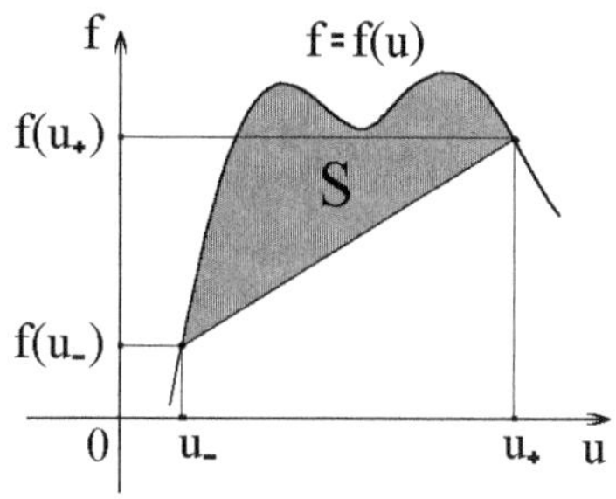

Figure 15. Area that determines the energy decrease rate.

Proof. To be specific, consider the case where $u_- < u_+$ and, consequently, the graph of the function $f = f(u)$ on the segment $[u_-, u_+]$ lies above the corresponding chord. Then

$$S = \int_{u_-}^{u_+} f(u)\,du - \frac{f(u_+) + f(u_-)}{2}(u_+ - u_-).$$

On the other hand,

$$\begin{aligned}
\frac{dE}{dt} &= \frac{d}{dt}\int_{-\infty}^{+\infty} \frac{1}{2}u^2(t,x)\,dx = \frac{d}{dt}\left(\int_{-\infty}^{x(t)} \frac{1}{2}u^2(t,x)\,dx + \int_{x(t)}^{+\infty} \frac{1}{2}u^2(t,x)\,dx\right) \\
&= \frac{1}{2}u_-^2 \cdot \dot{x}(t) + \int_{-\infty}^{x(t)} uu_t(t,x)\,dx - \frac{1}{2}u_+^2 \cdot \dot{x}(t) + \int_{x(t)}^{+\infty} uu_t(t,x)\,dx \\
&= \frac{u_-^2 - u_+^2}{2} \cdot \dot{x}(t) - \int_{-\infty}^{x(t)} u\,(f(u))_x\,dx - \int_{x(t)}^{+\infty} u\,(f(u))_x\,dx \\
&= \frac{u_-^2 - u_+^2}{2} \cdot \dot{x}(t) - uf(u)\Big|_{x=-\infty}^{x=x(t)} + \int_{-\infty}^{x(t)} f(u)u_x\,dx \\
&\qquad - uf(u)\Big|_{x=x(t)}^{x=+\infty} + \int_{x(t)}^{+\infty} f(u)u_x\,dx.
\end{aligned}$$

Thanks to the Rankine–Hugoniot condition (5.13) and taking into account the fact that $u(t, \pm\infty) = 0$, we have

$$\begin{aligned}
\frac{dE}{dt} &= \frac{u_-^2 - u_+^2}{2} \cdot \frac{f(u_+) - f(u_-)}{u_+ - u_-} - u_-f(u_-) + \int_0^{u_-} f(u)\,du + u_+f(u_+) + \int_{u_+}^{0} f(u)\,du \\
&= u_+f(u_+) - u_-f(u_-) - \frac{(u_+ + u_-)(f(u_+) - f(u_-))}{2} - \int_{u_-}^{u_+} f(u)\,du \\
&= \frac{(u_+ - u_-)(f(u_+) + f(u_-))}{2} - \int_{u_-}^{u_+} f(u)\,du = -S.
\end{aligned}$$

□

Remark 5.10. If the solution contains several shock waves (i.e., several jump discontinuities), then *on each of the discontinuity curves* the energy is lost (dissipated) according to the inequality (5.29). (The proof of this fact is left to the reader.)

Conclusion. We see that, according to Proposition 5.7, we have $E(t) = \text{const} = E(0)$ on smooth solutions $u = u(t,x)$ of the equation (5.1), up to the critical instant of time T (the instant when singularities arise in the solutions), i.e., up to the time T there is no dissipation of the kinetic energy; the kinetic energy stays constant on $[0,T)$.

However, when shock waves appear, according to (5.29), we have

$$\frac{dE}{dt} < 0,$$

so that the kinetic energy dissipates (on a shock wave, a part of it is transformed into heat). Consequently, the evolution of admissible generalized solutions with shock waves is related to the decrease of the kinetic energy; this is what makes the physical processes modelled by equation (5.1) irreversible.

The readers who sometimes spend vacations at the sea are probably acquainted with this phenomenon. Near the shore, if the sea is calm and the waves are temperate, the sea temperature near the surface is almost the same as the air temperature above. When the wind becomes stronger, waves become foamy, turbulent structures occur; these "broken waves" can be seen as shock waves on the sea surface. In this case, after some time, one can observe that the temperature of the surface layer of the sea has become higher than the air temperature. This heating phenomenon is conditioned by the heat production that occurs on the shock waves.

From the purely mathematical point of view, this situation stems from the fact that equation (5.1) does not change under the simultaneous change of t into $-t$ and of x into $-x$ (similarly, any of the shift transformations along the axes, namely $x \to x - x_0$ or $t \to t - T$, does not change the equation); in this case, it is said that the equation remains invariant under the corresponding transformation. Consequently, along with any *smooth*, as $t < T$, solution $u = u(t,x)$ of equation (5.1), the transformed function $\tilde{u}(t,x) \equiv u(T-t,-x)$ will also be a *smooth* solution of the same equation.

The same property holds for generalized solutions (in the sense of integral equality (5.3); the admissibility condition is not required), because the identity (5.3) is invariant under the same transformations.

If, on the contrary, $u = u(t,x)$ is an *admissible discontinuous generalized* solution of equation (5.1), then the corresponding function $\tilde{u}$ **will not be** an *admissible generalized ("entropy")* solution of the equation considered. This is because the entropy increase condition is not invariant under the transformation which includes the time reversal (the entropy increase condition is then replaced by the converse entropy decrease condition). Therefore, the simultaneous change of t into $T-t$ and of x into $-x$ is not allowed in the presence of discontinuous solutions. Hence, an admissible discontinuous generalized solution $u = u(t,x)$ is transformed into the non-admissible ("wrong") discontinuous generalized solution $\tilde{u}(t,x) \equiv u(T-t,-x)$.

5.5 Kruzhkov's definition of a generalized solution

In the preceding sections we discussed the requirements which one should impose on jumps (i.e., discontinuities of the first kind occurring along smooth curves) of generalized solutions (in the sense of the integral identity (5.3)) of equation (5.1). However, this kind of restrictions is only meaningful for piecewise smooth functions; in this case the notion of a jump, i.e., a discontinuity curve with one-sided limits of a solution on this curve, is meaningful. In contrast, while defining a generalized solution $u = u(t, x)$ of this equation in the sense of the integral identity (5.3), we only need that the integrals in (5.3) make sense. Clearly, the latter assumption is by far less restrictive compared with the assumption of piecewise smoothness of the function $u = u(t, x)$. Therefore, a natural question arises, namely, how could one define an admissible generalized solution to the Cauchy problem (5.1)–(5.2), so that the new notion includes both the integral identity and a condition of the entropy increase type (we need some generalization of the entropy increase conditions stated above as we want to extend them to solutions which may not be piecewise smooth). The answer to this question was given by S. N. Kruzhkov (see [25, 26]), and the answer applies not only to the problem we consider in these lectures but also to a wider class of equations and systems. In the same works of S. N. Kruzhkov, the existence and uniqueness of an admissible generalized solution, in the sense of the new definition, was proved.

Let us now give the aforementioned definition. One of the widest spaces of functions in which generalized solutions of our problem can be searched is the space of bounded measurable functions $u = u(t, x)$ defined in the strip $\Pi_T = [0, T) \times \mathbb{R}_x$.

Definition 5.11. A bounded measurable function $u = u(t, x) : \Pi_T \to \mathbb{R}$ is called a *generalized entropy solution*[7] (in the sense of Kruzhkov) of the problem (5.1)–(5.2) if

(i) for any constant $k \in \mathbb{R}$ and any nonnegative test function $\varphi = \varphi(t, x) \in C_0^\infty(\Pi_T)$ there holds the inequality

$$\int_{\Pi_T} \big[|u - k|\varphi_t + \operatorname{sign}(u - k)\, \big(f(u) - f(k)\big)\varphi_x\big]\, dx\, dt \geqslant 0; \tag{5.30}$$

(ii) there holds $u(t, \cdot) \to u_0$ as $t \to +0$ in the topology of $L_{1,\mathrm{loc}}(\mathbb{R})$, i.e.,

$$\forall\ [a, b] \subset \mathbb{R}, \quad \lim_{t \to +0} \int_a^b |u(t, x) - u_0(x)|\, dx = 0. \tag{5.31}$$

Proposition 5.12. *If a function $u = u(t, x)$ is a generalized entropy solution in the sense of Definition* 5.11 *of problem* (5.1)–(5.2)*, then it is also a generalized solution of equation* (5.1) *in the sense of the integral identity* (5.3).

Proof. Note that the function taking everywhere a constant value k is a classical solution and, therefore, it is also a generalized solution of equation (5.1). It follows that for

[7] *NT* — The western literature refers to "Kruzhkov entropy solutions" or merely to "entropy solutions".

any test function $\varphi \in C_0^\infty(\Pi_T)$, there holds

$$\int_{\Pi_T} [k\varphi_t + f(k)\varphi_x]\, dx\, dt = 0. \tag{5.32}$$

This identity can also be checked by a direct calculation.

Choose a value $k > \text{ess-sup}_{(t,x)\in\Pi_T} u(t,x)$ in (5.30) . We have

$$\int_{\Pi_T} \big[(k-u)\varphi_t + (f(k) - f(u))\varphi_x\big]\, dx\, dt \geqslant 0$$

for any function $\varphi \in C_0^\infty(\Pi_T)$, $\varphi(t,x) \geqslant 0$. Taking into account (5.32), we conclude that

$$-\int_{\Pi_T} \big[u\varphi_t + f(u)\varphi_x\big]\, dx\, dt \geqslant 0. \tag{5.33}$$

Then taking $k < \text{ess-inf}_{(t,x)\in\Pi_T} u(t,x)$, we obtain in the same way

$$\int_{\Pi_T} [u\varphi_t + f(u)\varphi_x]\, dx\, dt \geqslant 0. \tag{5.34}$$

Comparing the inequalities (5.33) and (5.34), we arrive at the equality

$$\int_{\Pi_T} [u\varphi_t + f(u)\varphi_x]\, dx\, dt = 0 \qquad \forall \varphi(t,x) \in C_0^\infty(\Pi_T),\ \varphi(t,x) \geqslant 0.$$

This is the integral identity we were aiming at, except that we need it for an arbitrary (not necessarily nonnegative) function $\phi \in C_0^\infty(\Pi_T)$. Therefore, in order to conclude the proof, it remains to notice that any function $\varphi \in C_0^\infty(\Pi_T)$ can be represented as the difference $\varphi = \varphi_1 - \varphi_2$ of two *nonnegative* test functions φ_1 and φ_2. It is sufficient to take a nonnegative function $\varphi_1 \in C_0^\infty(\Pi_T)$ with $\varphi_1 \equiv \sup_{\Pi_T} \varphi$ on the support of φ. Since the relation (5.3) holds for both φ_1 and φ_2, it also holds true for φ. □

Proposition 5.13. *Let $u = u(t,x)$ be a piecewise smooth function that is a generalized entropy solution of equation* (5.1) *in the sense of Definition* 5.11. *Then on each discontinuity curve Γ (given by the equation $x = x(t)$) the adequate admissibility condition,* (5.21) *or* (5.22)*, holds.*

Proof. Fix a point $(t_0, x_0) \in \Gamma$, $x_0 = x(t_0)$, on the discontinuity curve Γ. As usual, denote by $u_\pm(t_0,x_0)$ the one-sided limits of $u(t_0,x)$ on Γ as x approaches x_0. To be specific, assume that $u_-(t_0,x_0) < u_+(t_0,x_0)$. Let us fix an arbitrary number $k \in (u_-, u_+)$ and choose a small neighbourhood $O \subset \Pi_T$ of the point (t_0,x_0) such that

$$u(t,x) < k \quad \text{for} \quad (t,x) \in O_- \equiv \{(t,x) \in O \mid x < x(t)\}, \tag{5.35}$$

$$u(t,x) > k \quad \text{for} \quad (t,x) \in O_+ \equiv \{(t,x) \in O \mid x > x(t)\}. \tag{5.36}$$

This is always possible since we consider a piecewise smooth solution. Moreover, without loss of generality, we can assume that u is smooth in each of the subdomains O_+ and O_-.

From (5.30) it follows that for any test function $\varphi \in C_0^\infty(O)$, $\varphi(t,x) \geqslant 0$, there holds

$$\int_O \left[|u-k|\varphi_t + \operatorname{sign}(u-k)\,(f(u)-f(k))\,\varphi_x\right] dx\,dt \geqslant 0. \tag{5.37}$$

Let us split the latter integral over the domain O into the sum of integrals over O_- and O_+. Taking into account (5.35)–(5.36), we obtain

$$-\int_{O_-} \left[(u-k)\varphi_t + (f(u)-f(k))\,\varphi_x\right] dx\,dt \\ + \int_{O_+} \left[(u-k)\varphi_t + (f(u)-f(k))\,\varphi_x\right] dx\,dt \geqslant 0.$$

Now let us transfer the t and x derivatives according to the integration-by-parts formula (4.1). In addition to the integrals over the domains O_- and O_+, also integrals over their boundaries will arise, that is, we will get integrals over ∂O and over $\Gamma \cap O$. As φ is compactly supported in O, the integral over ∂O is zero. Consequently, we obtain

$$\int_{O_-} [u_t + (f(u))_x]\varphi\,dx\,dt \\ - \int_{\Gamma\cap O} ((u_- - k)\cos(\nu,t) + (f(u_-)-f(k))\cos(\nu,x))\varphi\,dS \\ - \int_{O_+} [u_t + (f(u))_x]\varphi\,dx\,dt \\ - \int_{\Gamma\cap O} ((u_+ - k)\cos(\nu,t) + (f(u_+)-f(k))\cos(\nu,x))\varphi\,dS \geqslant 0.$$

Here ν is the normal vector to the curve Γ pointing from O_- to O_+ (i.e., the outward normal vector to the boundary of O_- and, at the same time, the interior normal vector for O_+). According to Proposition 5.12, the function $u = u(t,x)$ is a generalized (in the sense of the integral identity (5.3)) solution of equation (5.1). Since u is smooth in $O_\pm$, it is also a classical solution of the equation in each of the subdomains O_- and O_+. Consequently, we have in both O_- and O_+ the pointwise identity $u_t + (f(u))_x = 0$. Thus for any nonnegative test function $\varphi \in C_0^\infty(O)$, there holds

$$\int_{\Gamma\cap O} ((2k - u_- - u_+)\cos(\nu,t) + (2f(k) - f(u_-) - f(u_+))\cos(\nu,x))\varphi\,dS \geqslant 0.$$

This means that for all $k \in (u_-, u_+)$, we have

$$(2k - u_- - u_+)\cos(\nu,t) + (2f(k) - f(u_-) - f(u_+))\cos(\nu,x) \geqslant 0. \tag{5.38}$$

As already mentioned, $u = u(t,x)$ is a generalized solution of equation (5.1). This means, in particular, that the Rankine–Hugoniot condition (5.13) is satisfied along the discontinuity curve Γ (here we take this condition in the equivalent form (4.6)):

$$(u_+ - u_-)\cos(\nu,t) + (f(u_+) - f(u_-))\cos(\nu,x) = 0. \tag{5.39}$$

Taking into account (5.39), we can rewrite inequality (5.38) under the form

$$\begin{aligned} 2\big[(k-u_-)\cos(\nu,t) + (f(k)-f(u_-))\cos(\nu,x)\big] \\ - \big[(u_+-u_-)\cos(\nu,t) + (f(u_+)-f(u_-))\cos(\nu,x)\big] \\ = 2\big[(k-u_-)\cos(\nu,t) + (f(k)-f(u_-))\cos(\nu,x)\big] \geqslant 0 \end{aligned}$$

for all $k \in (u_-, u_+)$. This is exactly the jump admissibility condition (5.21).

As to the case $u_+ < u_-$, transforming the term $\operatorname{sign}(u-k)$ and the term with the absolute value in equality (5.37) in the same vein as before, we obtain the minus signs in front of the same expressions. Accordingly, in place of the relation (5.38), we get

$$(2k - u_- - u_+)\cos(\nu,t) + (2f(k) - f(u_-) - f(u_+))\cos(\nu,x) \leqslant 0$$

for all $k \in (u_+, u_-)$. With the help of (5.39), we obtain the inequality

$$\begin{aligned} 2\big[(k-u_+)\cos(\nu,t) + (f(k)-f(u_+))\cos(\nu,x)\big] \\ + \big[(u_+-u_-)\cos(\nu,t) + (f(u_+)-f(u_-))\cos(\nu,x)\big] \\ = 2\big[(k-u_+)\cos(\nu,t) + (f(k)-f(u_+))\cos(\nu,x)\big] \leqslant 0, \end{aligned}$$

which holds for all $k \in (u_+, u_-)$. This statement coincides with (5.22). □

Finally, let us show that inequality (5.30) can be derived from the vanishing viscosity approach. Indeed, let $u = u(t,x)$ be a limit in the topology of $L_{1,\mathrm{loc}}(\Pi_T)$, as $\varepsilon \to +0$, of classical solutions $u^\varepsilon = u^\varepsilon(t,x)$ to the Cauchy problem consisting of the equation

$$u_t + f'(u)u_x = \varepsilon u_{xx} \tag{5.40}$$

and the initial datum $u(0,x) = u_0(x)$.

Take any convex function $E = E(u) \in C^2(\mathbb{R})$ and multiply equation (5.40) by $E'(u)$. The equalities

$$E'(u)u_t = \frac{\partial E(u(t,x))}{\partial t}, \qquad f'(u)E'(u)u_x = \frac{\partial}{\partial x}\Big(\int_k^{u(t,x)} f'(\xi)\,E'(\xi)\,d\xi\Big),$$
$$E'(u)u_{xx} = (E(u))_{xx} - E''(u)u_x^2,$$

imply

$$E_t + \Big(\int_k^u f'(\xi)E'(\xi)\,d\xi\Big)_x = \varepsilon\,(E(u))_{xx} - \varepsilon E''(u)u_x^2 \leqslant \varepsilon\,(E(u))_{xx} \tag{5.41}$$

since $E''(u) \geqslant 0$ and $\varepsilon > 0$. Now let us multiply inequality (5.41) by a test function $\varphi = \varphi(t,x) \geqslant 0$ from Definition 5.11 and integrate it over Π_T. Using the integration-by-parts formula, we transfer all the derivatives to the test function φ:

$$-\int_{\Pi_T}\Big[\varphi_t E(u) + \varphi_x \int_k^u f'(\xi)E'(\xi)\,d\xi\Big]\,dx\,dt \leqslant \varepsilon \int_{\Pi_T} \varphi_{xx} E(u)\,dx\,dt$$

Passing to the limit as $\varepsilon \to +0$, we get

$$\int_{\Pi_T} \left[\varphi_t E(u) + \varphi_x \int_k^u f'(\xi)E'(\xi)\,d\xi\right] dx\,dt \geqslant 0. \tag{5.42}$$

Let $\{E_m\}$ be a sequence of C^2-functions approximating the function $u \mapsto |u-k|$ uniformly on $\mathbb{R}$. Substitute $E = E_m(u)$ in the inequality (5.42) and pass to the limit as $m \to \infty$. We can choose E_m in such a way that E'_m is bounded and $E'_m(\xi) \to \operatorname{sign}(\xi - k)$ for all $\xi \in \mathbb{R}$, $\xi \neq k$. Thus, we have

$$\begin{aligned}\int_k^u f'(\xi)E'_m(\xi)\,d\xi \longrightarrow &\int_k^u f'(\xi)\operatorname{sign}(\xi-k)\,d\xi\\ &= \operatorname{sign}(u-k)\int_k^u f'(\xi)\,d\xi = \operatorname{sign}(u-k)\,(f(u)-f(k)).\end{aligned}$$

In this way, we deduce (5.30) from (5.42).

Problem 5.2. *Justify in detail the last passage to the limit in the above proof.*

Remark 5.14. In the case of a convex flux function $f = f(u)$, we can replace the integral inequality (5.30) in the definition of a generalized entropy solution by, first, the integral identity (5.3), and, second, the additional admissibility requirement that the inequality (5.42) holds for one fixed strictly convex function $E = E(u)$. Uniqueness of the so defined solution is shown in [39].

In the context of the inequality (5.42), a convex function $E = E(u)$ is called an "entropy" of the equation (5.1); indeed, inequality (5.42) is another variant of the "entropy increase-type conditions" in the sense of Section 5.3.

Remark 5.15. The definition of a generalized entropy solution on the basis of the inequality (5.30) extends to the multi-dimensional analogue of the problem (5.1)–(5.2). In this case, we have $x \in \mathbb{R}^n$,

$$f : \mathbb{R} \to \mathbb{R}^n, \qquad (f(u))_x \equiv \nabla_x f(u(t,x)), \qquad \varphi_x = \nabla_x \varphi,$$

and $(f(u) - f(k))\,\varphi_x$ is the scalar product of the vector $(f(u) - f(k))$ with the gradient of φ with respect to the space variable x. This way to define the notion of a solution $u = u(t,x)$, and also the family of entropies $|u-k|$, $k \in \mathbb{R}$, is often named after S. N. Kruzhkov (Kruzhkov's solutions, the Kruzhkov entropies). These notions were introduced in the works [25, 26]. Also the techniques of existence and uniqueness proofs, techniques deeply rooted in the physical context of the problem, were set up in these papers.

6 The Riemann problem (evolution of a primitive jump)

In this section, we consider the so-called Riemann problem for equation (4.2), which is the problem of evolution from a simplest piecewise constant initial datum. That is, we will construct admissible generalized solutions $u = u(t,x)$ of the following problem in a strip $\Pi_T = \{-\infty < x < +\infty, 0 < t < T\}$:

$$u_t + (f(u))_x = 0, \qquad u\big|_{t=0} = u_0(x) = \begin{cases} u_- & \text{for } x < 0, \\ u_+ & \text{for } x > 0, \end{cases} \tag{6.1}$$

where u_- and u_+ are two arbitrary constant states. The solutions we want to construct will be piecewise smooth in Π_T. This means that, first, they will satisfy the equation in the classical pointwise sense on all smoothness components of the solution; and second, they will satisfy both the Rankine–Hugoniot condition (4.5) and the entropy increase condition on each curve of jump discontinuity. These solutions will converge to the function u_0 as $t \to +0$ at all points, except for the point $x = 0$.

The proof of the uniqueness of an admissible generalized solution (in the sense of the integral identity and entropy increase condition) of the problem (6.1) can be found in [27, Lectures 4–6]; its existence is demonstrated below with an explicit construction.

First of all, let us notice that the equation we consider is invariant under the change $x \to kx$, $t \to kt$; moreover, the initial datum also remains unchanged under the action of homotheties $x \to kx$, $k > 0$. Furthermore, the entropy increase condition is also invariant under the above transformations. Admitting the uniqueness of an admissible generalized solution of the above problem, we conclude that any change of variables $x \to kx$, $t \to kt$ with $k > 0$ transforms the unique solution $u = u(t,x)$ of the problem into itself, i.e.,

$$u(kt, kx) \equiv u(t,x) \quad \forall k > 0.$$

This exactly means that the function $u = u(t,x)$ remains constant on each ray $x = \xi t$, $t > 0$, issued from the origin $(0,0)$, so that $u(t,x)$ depends only on the variable $\xi = x/t$:

$$u(t,x) = u(x/t), \quad t > 0. \tag{6.2}$$

Solutions that only depend on x/t are called *self-similar*. In particular, jump discontinuity curves of self-similar solutions can only be straight rays emanating from the origin $(0,0)$.

Exercise 6.1. *Find all the self-similar solutions of the equations from Exercise* 4.2 *such that the solutions are smooth in the whole half-plane* $t > 0$.

6.1 The Hopf equation

To start with, consider the Riemann problem (6.1) in the case $f(u) = u^2/2$:

$$u_t + uu_x = 0, \qquad u\big|_{t=0} = u_0(x) = \begin{cases} u_- & \text{for } x < 0, \\ u_+ & \text{for } x > 0. \end{cases} \tag{6.3}$$

First of all, we describe all the smooth self-similar solutions of the Hopf equation. Substituting (6.2) into the equation (6.3), we find

$$-\frac{x}{t^2}u'\left(\frac{x}{t}\right)+\frac{1}{t}u\left(\frac{x}{t}\right)u'\left(\frac{x}{t}\right)=\frac{1}{t}u'\left(\frac{x}{t}\right)\left(u\left(\frac{x}{t}\right)-\frac{x}{t}\right)=0,$$

i.e., either $u'=0$, so that we have $u\equiv C$ where C is a constant, or $u=x/t$. Consequently, the set of all smooth self-similar solutions of the Hopf equation reduces to the constant solutions and to the function x/t.

Now our task is to juxtapose pieces of the above smooth self-similar solutions in a correct way (i.e., respecting the Rankine–Hugoniot and the entropy increase condition on the discontinuity rays), with the goal to comply with the initial datum $u_0=u_0(x)$.

First, let us see which rays can separate two smoothness components of such a solution: two adjacent components may correspond either to two different constant states, or to a constant state and to the restriction of the function x/t on some cone with the vertex $(0,0)$.

It follows from the Rankine–Hugoniot condition (4.5) that two constant functions $u(t,x)\equiv u_1$ and $u(t,x)\equiv u_2,\ u_i=\text{const}$, can only be juxtaposed along the ray

$$x=\frac{f(u_2)-f(u_1)}{u_2-u_1}t=\frac{1}{2}\frac{u_2^2-u_1^2}{u_2-u_1}t=\frac{u_2+u_1}{2}t,$$

and because of the entropy increase condition, the jump is admissible only when u jumps from a greater to a smaller value (we mean that the direction of the jump is such that x grows). Consequently, if we specify, e.g., that $u_2>u_1$, then we should have

$$u(t,x)=u_2 \quad \text{for } x<\frac{u_2+u_1}{2}t, \quad \text{and} \quad u(t,x)=u_1 \quad \text{for } x>\frac{u_2+u_1}{2}t.$$

As to the juxtaposition of a constant $u(t,x)\equiv u_3=\text{const}$ and the function $u(t,x)=x/t$, we have the following. If the two functions juxtapose along a ray $x=\xi t$, then the limit of the function x/t on this ray equals ξ, and (4.5) yields

$$\xi=\frac{dx}{dt}=\frac{f(u_3)-f(\xi)}{u_3-\xi}=\frac{1}{2}\frac{u_3^2-\xi^2}{u_3-\xi}=\frac{u_3+\xi}{2},$$

so that $\xi=u_3$. The latter means that the function obtained by the juxtaposition turns out to be continuous on the border ray $x=\xi t=u_3t,\ t>0$. Consequently, here the discontinuity is a weak, not a strong one.

Now we can solve completely the Riemann problem for the Hopf equation. Here, two substantially different situations should be considered:

(i) When $u_->u_+$, we can construct a *shock wave* solution, where the two constants u_- and u_+ are joined across the ray $x=\frac{u_2+u_1}{2}t$, according to the Rankine–Hugoniot condition (see Fig. 16):

$$u(t,x)=\begin{cases} u_- & \text{for } x<\frac{u_-+u_+}{2}t, \\ u_+ & \text{for } x>\frac{u_-+u_+}{2}t. \end{cases} \tag{6.4}$$

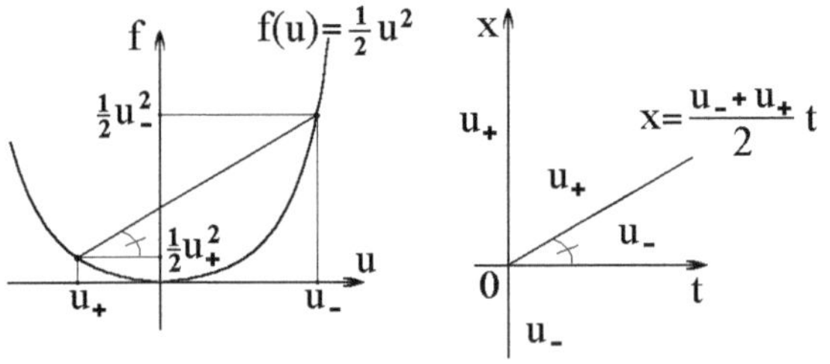

Figure 16. Shock-wave solution to the Riemann problem.

As has already been mentioned, the jump discontinuity in the desired solution is compatible with the admissibility condition of increase of entropy.

(ii) If $u_- < u_+$, we cannot take the shock wave solution analogous to the previous case, because the jump discontinuity would not satisfy the entropy increase condition. Here the function x/t is helpful; it can be combined continuously with the constant states u_- and u_+ (see Fig. 17):

$$u(t,x) = \begin{cases} u_- & \text{for } x \leqslant u_- t, \\ x/t & \text{for } u_- t < x < u_+ t, \\ u_+ & \text{for } x \geqslant u_+ t. \end{cases} \tag{6.5}$$

The so defined solution is indeed continuous in the whole half-plane $t > 0$. The cone determined by the inequalities $u_- t < x < u_+ t$, $t > 0$, in which the smoothing of the initially discontinuous function takes place, is called the *region of rarefaction* of the solution, and the solution (6.5) itself is called a *centered rarefaction wave*.

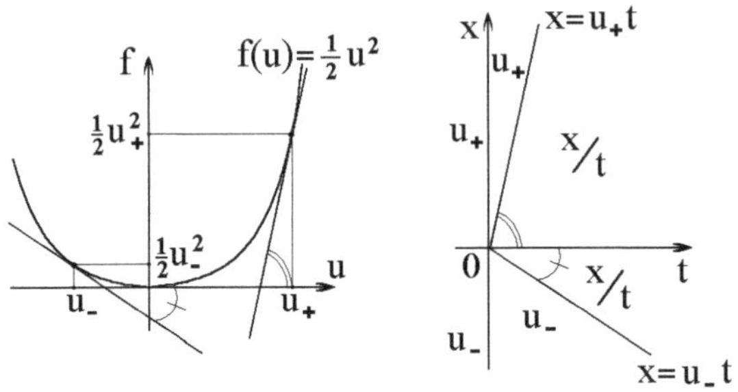

Figure 17. Rarefaction-wave solution to the Riemann problem.

Let us give a comment of geometrical nature to the solutions obtained. Drawing the graph of the function $f(u) = u^2/2$ relative to the axes (u, f), parallel to the axes

(t, x), let us mark the points $(u_-, u_-^2/2)$ and $(u_+, u_+^2/2)$ on the graph. Then, as it has already been mentioned, the discontinuity ray in solution (6.4) is parallel to the segment joining these two points (see Fig. 16). Also observe the following fact (in the sequel, we will see that this is by no means incidental): the lines of weak discontinuity of the solution $u = u(t, x)$ given by (6.5), namely the two rays $x = u_- t$ and $x = u_+ t$, are parallel to the tangent directions to the graph of the function $f(u) = u^2/2$ at the points $(u_-, f(u_-))$ and $(u_+, f(u_+))$, respectively.

Remark 6.1. When $u_- > u_+$, formula (6.5) is meaningless: no function in the upper half-plane $t > 0$ is determined by this formula.

Problem 6.1. *Show that the solution constructed above, given by* (6.4) *or by* (6.5), *according to the sign of* $(u_- - u_+)$, *is the unique admissible generalized solution of the Riemann problem* (6.3) *within the class of all self-similar piecewise-smooth functions.*

6.2 The case of a convex flux function

In the case where $f = f(u)$ is a smooth strictly convex function, the solution of the Riemann problem (6.1) is almost the same as for the case of the Hopf equation (i.e., as for the case $f(u) = u^2/2$). The only difference is that the non-constant smooth self-similar solution $u(t, x) = x/t$ of the Hopf equation is replaced by the appropriate smooth function $\psi = \psi(x/t)$. Let us find this function ψ. As above, we substitute (6.2) into (6.1) and obtain

$$-\frac{x}{t^2}u' + \frac{1}{t}f'(u)u' = \frac{1}{t}u'(x/t)\left(f'\left(u\left(x/t\right)\right) - x/t\right) = 0.$$

Therefore, besides the constants obtained from the equation $u' = 0$, there exists one more function $u(\xi) = \psi(\xi)$ (here $\xi = x/t$) defined as the solution of the equation

$$f'(\psi) = \xi.$$

That is, ψ is the function inverse to f': we have $\psi = (f')^{-1}$. The inverse function does exist since f is strictly convex, so that f' is a strictly monotone function. The solution $u(t, x) = \psi(x/t)$, which is discontinuous at $(0, 0)$ and continuous for $t > 0$, is a *centered rarefaction wave.*

Remark 6.2. In the previous section, for the particular case of the Hopf equation, we had $f'(u) = u$, so that $\psi(\xi) = (f')^{-1}(\xi) = \xi$.

In the case of a general strictly convex flux function $f = f(u)$, we construct the solution of the Riemann problem (6.1) similar to the case of the Hopf equation, namely:

(i) When $u_- > u_+$, then we can use the shock wave again, juxtaposing the two constant states u_- and u_+ separated by the ray $\frac{x}{t} = \frac{f(u_+) - f(u_-)}{u_+ - u_-}$, $t > 0$, the slope of the ray being found from the Rankine–Hugoniot condition:

$$u(t, x) = \begin{cases} u_- & \text{for } x < \frac{f(u_+) - f(u_-)}{u_+ - u_-}t, \\ u_+ & \text{for } x > \frac{f(u_+) - f(u_-)}{u_+ - u_-}t. \end{cases} \tag{6.6}$$

(Compare with (6.4) and Fig. 16.) The jump in the solution obtained is admissible according to the entropy increase condition.

(ii) When $u_- < u_+$, then the function given by (6.6) is a generalized solution but it does not satisfy the entropy increase condition. Then, similar to the construction of (6.5), we combine the constant states u_- and u_+ with the non-trivial smooth solution $\psi = \psi(x/t)$. The rays $x = \xi_- t$ and $x = \xi_+ t$, where the transition occurs, are determined by the requirement of continuity of the solution: $u_\pm = \psi(\xi_\pm)$, i.e., $\xi_\pm = f'(u_\pm)$, so that

$$u(t,x) = \begin{cases} u_- & \text{for } x \leqslant f'(u_-)t, \\ \psi(x/t) & \text{for } f'(u_-)t < x < f'(u_+)t, \\ u_+ & \text{for } x \geqslant f'(u_+)t. \end{cases} \tag{6.7}$$

The function given by (6.7) is well-defined in the upper half-plane $t > 0$; indeed, the flux function $f = f(u)$ is strictly convex, thus f' is an increasing function, so that $f'(u_-) < f'(u_+)$ whenever $u_- < u_+$.

The rarefaction wave $\psi = \psi(x/t)$, being continuous for $t > 0$, takes all the intermediate values between u_- and u_+. As ψ is defined as the inverse function of f', the condition $\psi(x/t) = \hat{u}$ is equivalent to the equality $x = f'(\hat{u})t$ valid for all $\hat{u} \in [u_-, u_+]$. This means that the rarefaction wave $\psi = \psi(x/t)$ takes a given intermediate value $\hat{u}$ on the ray $x = f'(\hat{u})t$, $t > 0$. We can see that this ray is parallel to the direction tangent to the graph $f = f(u)$ at the point $(\hat{u}, f(\hat{u}))$ of the graph. Thus in particular, we have justified the statement already noted in the previous section: the rays of weak discontinuity of the solution $u = u(t,x)$ given by formula (6.7) (i.e., the rays $x = f'(u_\pm)t$) are aligned with the directions tangent to the graph $f = f(u)$ at the endpoints $(u_\pm, f(u_\pm))$ (see Fig. 17). (As always, we assume that the axes (u, f) are aligned with the axes (t, x).)

Remark 6.3. Note that the convexity of $f = f(u)$ is only needed on the segment $[u_-, u_+]$ (or $[u_+, u_-]$, if $u_+ < u_-$).

Concerning the case of a strictly concave and smooth (on the segment between u_- and u_+) flux function $f = f(u)$, the unique self-similar admissible generalized solution to the Riemann problem is constructed by exchanging, in a sense, the two situations described above. Namely: for the case $u_- < u_+$, we obtain the shock wave (6.6); if $u_- > u_+$, then the solution is given by (6.7) (in this case f' is a decreasing function, consequently, here we have $f'(u_-) < f'(u_+)$). The careful derivation of the formulas is left to the reader:

Problem 6.2. *Solve the Riemann problem* (6.1) *in the case of a general smooth strictly concave flux function* $f = f(u)$*; represent the piecewise smooth solution graphically (as in Fig.* 16 *and* 17*); check the validity of the Rankine–Hugoniot condition, and of the entropy increase inequality on the jumps.*

Exercise 6.2. *Solve the following Riemann problems:*

(i) $u_t - (u^2)_x = 0$,

$$u\big|_{t=0} = \begin{cases} -1 & \text{for } x < 0, \\ 1 & \text{for } x > 0 \end{cases} \quad \text{and} \quad u\big|_{t=0} = \begin{cases} 1 & \text{for } x < 0, \\ -1 & \text{for } x > 0; \end{cases}$$

(ii) $u_t + u^2 \cdot u_x = 0$,

$$u\big|_{t=0} = \begin{cases} 0 & \text{for } x < 0, \\ 2 & \text{for } x > 0 \end{cases} \quad \text{and} \quad u\big|_{t=0} = \begin{cases} 2 & \text{for } x < 0, \\ 0 & \text{for } x > 0; \end{cases}$$

(iii) $u_t + \cos u \cdot u_x = 0,\ u\big|_{t=0} = \begin{cases} 0 & \text{for } x < 0, \\ \pi & \text{for } x > 0, \end{cases}$

$$u\big|_{t=0} = \begin{cases} \pi & \text{for } x < 0, \\ 0 & \text{for } x > 0 \end{cases} \quad \text{and} \quad u\big|_{t=0} = \begin{cases} \pi & \text{for } x < 0, \\ 2\pi & \text{for } x > 0; \end{cases}$$

(iv) $u_t + e^u \cdot u_x = 0$,

$$u\big|_{t=0} = \begin{cases} 0 & \text{for } x < 0, \\ 1 & \text{for } x > 0 \end{cases} \quad \text{and} \quad u\big|_{t=0} = \begin{cases} 1 & \text{for } x < 0, \\ 0 & \text{for } x > 0; \end{cases}$$

(v) $u_t + (\ln u)_x = 0$,

$$u\big|_{t=0} = \begin{cases} e & \text{for } x < 0, \\ 1 & \text{for } x > 0 \end{cases} \quad \text{and} \quad u\big|_{t=0} = \begin{cases} 1 & \text{for } x < 0, \\ e & \text{for } x > 0. \end{cases}$$

6.3 The case of a flux function with inflexion point

In order to treat the Riemann problem in the case where $f = f(u)$ is neither convex nor concave, let us first give two definitions.

Definition 6.4. The *concave hull* of a function $f = f(u)$ on a segment $[\alpha, \beta]$ is the function

$$\hat{f}(u) = \inf_{\tilde{f} \in \hat{F}} \tilde{f}(u), \quad u \in [\alpha, \beta],$$

where $\hat{F}$ is the family of all concave functions $\tilde{f} = \tilde{f}(u)$ defined on $[\alpha, \beta]$ such that $\tilde{f}(u) \geqslant f(u)$ for all $u \in [\alpha, \beta]$.

Definition 6.5. The *convex hull* of a function $f(u)$ on a segment $[\alpha, \beta]$ is the function

$$\check{f}(u) = \sup_{\tilde{f} \in \check{F}} \tilde{f}(u), \quad u \in [\alpha, \beta],$$

where $\check{F}$ is the family of all convex functions $\tilde{f} = \tilde{f}(u)$ defined on $[\alpha, \beta]$ such that $\tilde{f}(u) \leqslant f(u)$ for all $u \in [\alpha, \beta]$.

Remark 6.6. If f is a concave (respectively, convex) function on $[\alpha, \beta]$, then the function itself is its concave (respectively, convex) hull: $\hat{f} = f$ (respectively, $\check{f} = f$); furthermore, the graph of its convex (respectively, concave) hull is the straight line segment joining the endpoints $(\alpha, f(\alpha))$ and $(\beta, f(\beta))$ of the graph.

Exercise 6.3. *Construct the concave and the convex hulls for the function* $f(u) = u^3$ *on the segment* $[-1, 1]$ *as well as for the function* $f(u) = \sin u$ *on the segment* $[0, 3\pi]$.

To solve the Riemann problem (6.1) for a given smooth flux function $f = f(u)$ in the case $u_- < u_+$, we first construct the convex hull of f on the segment $[u_-, u_+]$. In the case $u_- > u_+$, we construct the concave hull of f on the segment $[u_+, u_-]$.

The graph of any of the hulls consists of some parts of the graph of f, where the graph has the right convexity/concavity direction, and of straight line segments connecting these pieces of the graph of f (see the above exercise). Each of the straight line segments will correspond to a jump ray (thus, to a shock wave) in the solution of the Riemann problem; each of such rays will separate two components of smoothness of the solution. Each of these components can either be a constant state (u_- or u_+), or a smooth self-similar solution of the form $u(t, x) = \psi(x/t)$ (i.e., a centered rarefaction wave). Here $\psi = \psi(\xi)$ is the function (locally) inverse to f', so that $\xi = f'(u)$ (see Section 6.2). Notice that on each segment of strict convexity/concavity of $f = f(u)$ the function f' is indeed invertible.

Example 6.7. Let us construct the solution (i.e., the self-similar admissible generalized solution) of the following Riemann problem:

$$u_t + (u^3)_x = 0, \quad u\big|_{t=0} = \begin{cases} 1 & \text{for } x < 0, \\ -1 & \text{for } x > 0. \end{cases} \tag{6.8}$$

First, because of $u_- = 1 > -1 = u_+$, we construct the concave hull of the flux function $f(u) = u^3$ on the segment $[-1, 1]$. To perform the construction, we draw the tangent line to the graph at the right endpoint $(1, 1)$ of this graph. The tangency point, denoted by $(\hat{u}, \hat{u}^3)$ can be determined from the condition

$$\frac{1 - \hat{u}^3}{1 - \hat{u}} = f'(\hat{u}) = 3\hat{u}^2, \quad \hat{u} \neq 1,$$

i.e., $1 + \hat{u} + \hat{u}^2 = 3\hat{u}^2$, whence $\hat{u} = -1/2$. Notice that the piece of the graph of $f(u) = u^3$ between the left endpoint $(-1, -1)$ of the graph and the tangency point $(-1/2, (-1/2)^3)$ is concave. Thus we see that[8] the graph of the concave hull $\hat{f}$ of the function $f(u) = u^3$ on the segment $[-1, 1]$ consists of: first, the piece of the "cubic parabola" $f = f(u) = u^3$ on the segment $[-1, -1/2]$; and second, the straight line segment that joins the points $(-1/2, -1/8)$ and $(1, 1)$ (see Fig. 18). Therefore, the solution of the Riemann problem under consideration has one and only one ray $x = \xi t$, $t > 0$, on which the solution has a jump. This ray is parallel to the straight line segment in the graph of $\hat{f} = \hat{f}(u)$ (as usual, for the sake of convenient graphical representation, the axes (t, x) are aligned with the axes (u, f)); expressing analytically the slope of the strong discontinuity ray, we have

$$\xi = \frac{1 + 1/8}{1 + 1/2} = \frac{3}{4}.$$

[8] *NT* — This conclusion requires some thinking; it is based on several easy-to-justify properties of the concave hull. In particular, one always has $\hat{f}(\alpha) = f(\alpha) = \check{f}(\alpha)$, $\hat{f}(\beta) = f(\beta) = \check{f}(\beta)$, with the notation of the definitions. The reader who analyzed the examples of Exercise 6.3 has already performed this construction.

This ray separates the constant state $u_- = 1$ (taken from the side $x < \frac{3}{4}t$) and a piece of rarefaction $\psi(x/t)$. Here $\psi = \psi(\xi)$ is the function inverse to $\xi = f'(u) = 3u^2$ on the segment $[-1, -1/2]$, so that we have

$$u = \psi(\xi) = -\sqrt{\xi/3}, \qquad 3/4 \leqslant \xi \leqslant 3.$$

The limit of the solution $u = u(t,x)$ from the side $x > \frac{3}{4}t$ on the jump ray $x = \frac{3}{4}t$ equals $\psi(\frac{3}{4}) = -\frac{1}{2}$ (this stems from the fact that $f'(-\frac{1}{2}) = 3(-\frac{1}{2})^2 = \frac{3}{4}$).

As for the case of a convex flux function (see Section 6.2), the juxtaposition of the rarefaction wave $\psi = \psi(x/t)$ and the constant state $u_+ = -1$ occurs continuously, that is, these two smoothness components are separated by the weak discontinuity ray $x = 3t$, $t > 0$. Once more, this ray is aligned with the tangent direction at the point $(u_+, f(u_+)) = (u_+, u_+^3) = (-1, -1)$ of the graph of the flux function $f(u) = u^3$.

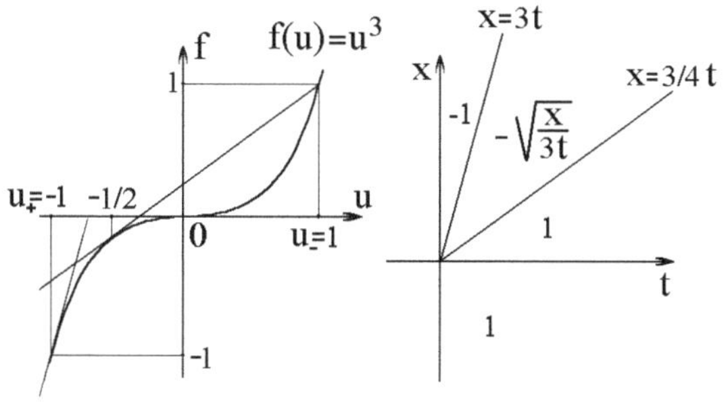

Figure 18. Solution for Example 6.7.

Thus we obtain the following solution of problem (6.8):

$$u(t,x) = \begin{cases} 1 & \text{for } x < \frac{3}{4}t, \\ -\sqrt{\frac{x}{3t}} & \text{for } \frac{3}{4}t < x < 3t, \\ -1 & \text{for } x \geqslant 3t. \end{cases}$$

Exercise 6.4. *Construct the solution of the Riemann problem*

$$u_t + u^2 \cdot u_x = 0, \quad u\big|_{t=0} = \begin{cases} -2 & \text{for } x < 0, \\ 2 & \text{for } x > 0. \end{cases}$$

Example 6.8. Let us solve the Riemann problem

$$u_t + (\sin u)_x = 0, \quad u\big|_{t=0} = \begin{cases} 3\pi & \text{for } x < 0, \\ 0 & \text{for } x > 0. \end{cases}$$

As we have $u_- = 3\pi > 0 = u_+$, we have to construct the concave hull $\hat{f} = \hat{f}(u)$ of the graph of $f(u) = \sin u$ on the segment $[0, 3\pi]$. The graph of $\hat{f}$ (see Fig. 19) consists of two pieces of concavity of the graph of $f(u) = \sin u$, those on the segments $[0, \pi/2]$ and $[5\pi/2, 3\pi]$, and of the horizontal segment joining the points $(\pi/2, 1)$ and $(5\pi/2, 1)$ of the sine curve. We conclude that the solution $u = u(t, x)$ should have one strong discontinuity (jump) along the ray $x = 0$, separating the one-sided limit states

$$\frac{5\pi}{2} = \lim_{x\to -0} u(t,x) \quad \text{and} \quad \frac{\pi}{2} = \lim_{x\to +0} u(t,x).$$

We also see that

$$u(t,x) = \begin{cases} 3\pi & \text{for } x < f'(3\pi)\cdot t = \cos 3\pi \cdot t = -t, \\ 0 & \text{for } x > f'(0)\cdot t = t. \end{cases}$$

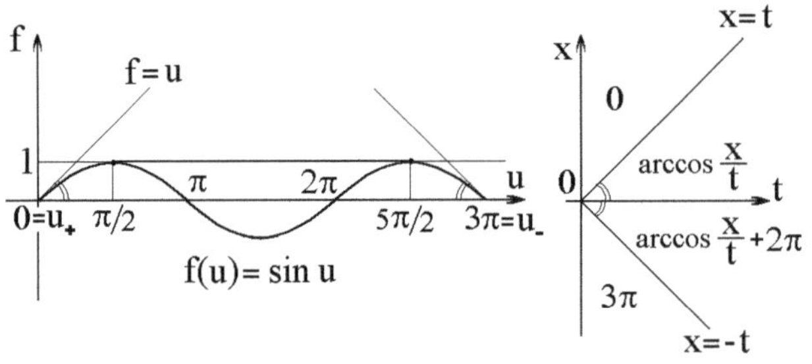

Figure 19. Solution for Example 6.8.

It remains to express u from the equation

$$f'(u) = \cos u = \xi = x/t$$

on the segments $[0, \pi/2]$ and $[5\pi/2, 3\pi]$. By construction, it is not surprising that the function $f'(u) = \cos u$ is monotone on these segments. Solutions of the equation $\cos u = \xi$, $-1 \leqslant \xi \leqslant 1$, are well known: we have $u = \pm\arccos\xi + 2\pi n, n \in \mathbb{Z}$. On the segment $[0, \pi/2]$, the solution specifies to $u = \arccos\xi$, while on the segment $[5\pi/2, 3\pi]$ we get $u = \arccos\xi + 2\pi$. Recapitulating, the solution we have constructed looks as follows (see Fig. 19):

$$u(t,x) = \begin{cases} 3\pi & \text{for } x \leqslant -t, \\ \arccos x/t + 2\pi & \text{for } -t < x < 0, \\ \arccos x/t & \text{for } 0 < x < t, \\ 0 & \text{for } x \geqslant t. \end{cases}$$

The solution of the Riemann problem will change drastically if we exchange the values u_+ and u_-.

Example 6.9. Construct the solution of the Riemann problem

$$u_t + (\sin u)_x = 0, \quad u\big|_{t=0} = \begin{cases} 0 & \text{for } x < 0, \\ 3\pi & \text{for } x > 0. \end{cases}$$

Now we have to start by constructing the convex hull $\check{f} = \check{f}(u)$ of the function $f(u) = \sin u$ on the segment $[0, 3\pi]$ (see Fig. 20). It consists of two segments of the lines issued from the graph's endpoints $(0,0)$ and $(3\pi, 0)$, the lines being tangent to the sine graph at some points contained within $[\pi, 2\pi]$, each of the segments being taken between the endpoint and the tangency point, and of the convex piece of the sine curve between the two tangency points $(u_1, \sin u_1)$ and $(u_2, \sin u_2)$. Symmetry considerations readily yield the equalities $u_1 + u_2 = 3\pi$, $\sin u_1 = \sin u_2$; also the slopes of the two tangent segments constructed above only differ by their sign. Denote by

$$-k = \frac{f(u_1) - f(0)}{u_1 - 0} = \frac{\sin u_1}{u_1} = f'(u_1) = \cos u_1$$

the slope of the tangent segment passing through the endpoint $(0,0)$. Then $+k$ is the slope of the other tangent segment. We cannot find explicitly the exact values of u_1, u_2 and k, but we can say that u_1 is the smallest strictly positive solution of the equation $\tan u_1 = u_1$, that $u_2 = 3\pi - u_1$, and that $k = -\cos u_1 = \cos u_2$.

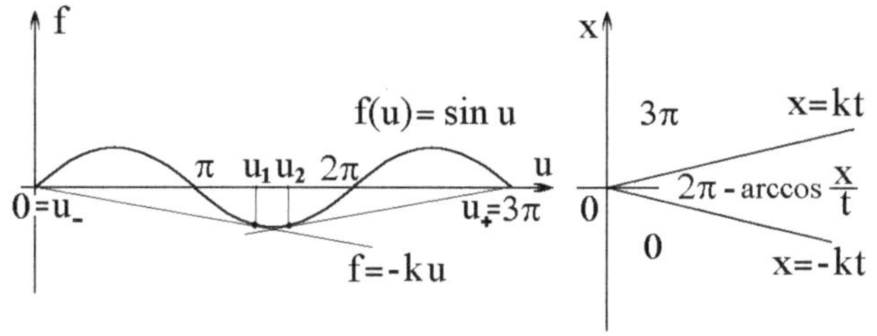

Figure 20. Solution for Example 6.9.

On the segment $[u_1, u_2] \subset [\pi, 2\pi]$, we can invert the function $f'(u) = \cos u$. In this case, $u = (f')^{-1}(\xi) = 2\pi - \arccos \xi$, $-k \leqslant \xi \leqslant k$. Now we can write down the "almost explicit" solution (depicted in Fig. 20):

$$u(t,x) = \begin{cases} 0 & \text{for } x \leqslant -kt, \\ 2\pi - \arccos x/t & \text{for } -kt < x < kt, \\ 3\pi & \text{for } x \geqslant kt. \end{cases}$$

The solution above has two strong discontinuities: the one across the line $x = -kt$ with the jump from 0 to u_1, and the one across the line $x = kt$ with the jump from u_2 to 3π.

Exercise 6.5. *Construct the solution of the Riemann problem*

$$u_t + \sin(2u) \cdot u_x = 0, \quad u\big|_{t=0} = \begin{cases} -5\pi/4 & \text{for } x < 0, \\ 5\pi/4 & \text{for } x > 0. \end{cases}$$

Afterword

In the present lecture course, we have introduced the reader to the notions and tools which underly the nonlocal theory of the Cauchy problem for the one-dimensional (in the space variable) quasilinear conservation law of the form

$$u_t + (f(u))_x = 0. \tag{6.9}$$

As to the nonlocal theory for the multidimensional scalar equation

$$u_t + \operatorname{div}_x f(u) = 0, \qquad x \in \mathbb{R}^n, \tag{6.10}$$

where f is an n-dimensional vector-function, it appeared in a rather complete form at the end of the 1960s (see [25, 26]), for the case where the components $f_i = f_i(u)$ of the flux function vector $f = f(u)$ satisfy a Lipschitz continuity condition. This assumption of Lipschitz continuity results in the effects of finite speed of propagation of perturbations and of finite domain of dependence (at a fixed point (t, x)) on the initial data for the solutions of equation (6.10).

A further challenge in the nonlocal theory of equations (6.9) and (6.10) lies in its generalization to the case where the flux function $f = f(u)$ is merely continuous, i.e., it is not necessarily differentiable. In this case, one expects that purely "parabolic", "diffusive" effects should appear: namely, the effects of infinite speed of propagation of perturbations and of infinite domain of dependence of entropy solutions on the initial data.

Indeed, let us look at the construction of the admissible generalized entropy solution of the Cauchy problem

$$u_t + \left(\frac{|u|^\alpha}{\alpha}\right)_x = 0, \qquad x \in \mathbb{R}, \qquad \alpha \in (0,1), \tag{6.11}$$

$$u\big|_{t=0} = u_0(x) \equiv \frac{[\operatorname{sign}(x+1) - \operatorname{sign} x]}{2} = \begin{cases} 1, & x \in (-1,0), \\ 0, & x \notin (-1,0). \end{cases} \tag{6.12}$$

As we have initially $u_0(x) \geqslant 0$, it can be deduced from Definition 5.11 that the generalized entropy solution $u = u(t,x)$ of the problem (6.11)–(6.12) is also nonnegative. Consequently, in (6.9) the flux function $f(u) \equiv u^\alpha/\alpha$ is concave on the interval of all values that could be possibly taken by the solution $u = u(t,x)$. On the other hand, because of the special ("single-step") structure of the initial function, it can be expected that, for a sufficiently small time interval $0 \leqslant t \leqslant \delta$, the admissible generalized solution of our problem will be determined by the solutions of the two Riemann problems with the initial functions $\operatorname{sign}(x+1)$ and $\operatorname{sign} x$, respectively.

Problem 6.3. *Check that the function*

$$u(t,x) = \begin{cases} 0 & \text{for } x < \frac{t}{\alpha} - 1, \\ 1 & \text{for } \frac{t}{\alpha} - 1 < x \leqslant t, \\ \left(\frac{t}{x}\right)^{\frac{1}{1-\alpha}} & \text{for } x > t \end{cases}$$

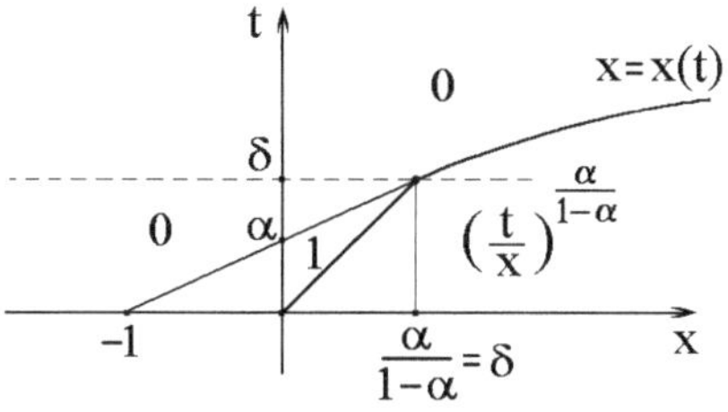

Figure 21. Solution of problem (6.11)–(6.12).

(see Fig. 21 *for a graphic representation of this function) defines a piecewise smooth admissible generalized solution of the problem* (6.11)–(6.12) *in the time interval* $0 < t < \frac{\alpha}{1-\alpha} = \delta$.

Problem 6.4. *Extend the above solution* $u = u(t,x)$ *of the problem* (6.11)–(6.12) *to the half-space* $t > \delta = \frac{\alpha}{1-\alpha}$. *More exactly, find the equation of the discontinuity curve* $x = x(t)$, *using for* $t > \delta$ *the ansatz (see Fig.* 21*)*

$$u(t,x) = \begin{cases} 0 & \text{for } x < x(t), \\ \left(\frac{t}{x}\right)^{\frac{1}{1-\alpha}} & \text{for } x > x(t). \end{cases}$$

Consequently, for the compactly supported initial function (6.12), the generalized entropy solution $u = u(t,x)$ of the Cauchy problem for equation (6.11) has in x a non-compact (unbounded) support, for all time $t > 0$ (thus, for an instant of time as small as desired!). It is known that, in the theory of parabolic PDEs (modeling diffusive processes in nature), such effect of infinite speed of propagation leads to non-uniqueness of a solution of the Cauchy problem. What would be the influence of this effect on the theory of nonlocal solvability of the Cauchy problem for equation (6.10), within the class of all essentially bounded measurable functions in the upper half-plane? It turns out that, without any further restriction on the continuous components $f_i = f_i(u)$ of the flux function, there exists at least one generalized entropy solution of the Cauchy problem. Contrarily (as it has been observed for the first time in the work [28]), the property of uniqueness of a generalized entropy solution of this problem can be connected with the product of the moduli of continuity ω_i of the functions f_i. If for all $u, v \in \mathbb{R}$

$$|f_i(u) - f_i(v)| \leqslant \omega_i(|u - v|), \tag{6.13}$$

where ω_i is a concave, strictly increasing and continuous function on $[0, +\infty)$ with $\omega_i(0) = 0$, then it is sufficient that for small ρ

$$\Omega(\rho) \equiv \prod_{i=1}^{n} \omega_i(\rho) \leqslant \text{const}\ \rho^{n-1}; \tag{6.14}$$

i.e., the restriction (6.13)–(6.14) ensures the uniqueness of a generalized entropy solution to a Cauchy problem for equation (6.11).

Further, let us stress that for the equation

$$u_t + \Big(\frac{|u|^\alpha}{\alpha}\Big)_x + \Big(\frac{|u|^\beta}{\beta}\Big)_y = 0, \qquad 0 < \alpha < \beta < 1,$$

the restriction (6.14) (which, for this concrete case, takes the form $\alpha + \beta \geqslant 1$), is both necessary and sufficient for the uniqueness of a generalized entropy solution to the Cauchy problem with general initial datum. The corresponding counterexample was constructed by E. Yu. Panov (see, e.g., [28]).

Notice that in the case $n = 1$ the condition (6.14) imposes no restriction at all on the merely continuous flux function $f = f(u)$: in the one-dimensional situation, a generalized entropy solution to the Cauchy problem is always unique.

Also notice that in the work [28] a rather simple proof of the uniqueness of a generalized entropy solutions is given under the assumption $\Omega(\rho)/\rho^{n-1} \to 0$ as $\rho \to 0$ that is slightly stronger than (6.14).

In conclusion, let us say that the nonlocal theory of first-order quasilinear conservation laws, whose rigorous mathematical treatment started in the 1950th, is yet actively developing. Many interesting problems remain unsolved, even for the one-dimensional equation (6.10). But most topical and interesting are the problems of conservation laws in the vector case, even for the simplest situations. Indeed, let us consider the well-known "wave equation" system

$$\begin{cases} u_t - v_x &= 0, \\ v_t - u_x &= 0. \end{cases}$$

This system was the very first object of research in PDEs (then called "mathematical physics"), in the works of D'Alembert and Euler. In order to take into account certain nonlinear dependencies in the process of wave propagation considered, one replaces the linear expression v_x in the first equation by the nonlinear expression $(p(v))_x$, where p is a function with $p'(v) > 0$. In this case, there arises the so-called "p-system", which is well known in the theory of hyperbolic systems of conservation laws:

$$\begin{cases} u_t - (p(v))_x &= 0, \\ v_t - u_x &= 0. \end{cases}$$

This system is another simple (although more complex than the Hopf equation (1.1)) but important model in the field of gas dynamics. Alas, nowadays, whatever be the non-linearity $p = p(v)$, nobody in the entire world knows how to define the "correct" entropy solution of this problem.

Thus a slightest nonlinear perturbation of a simple linear system results in an extremely difficult unsolved problem[9] in the field of nonlinear analysis.

[9] *NT* — These are words of S. N. Kruzhkov, spoken out in 1997 shortly before his passing away. Since then,

Hopefully, the topical, simple-to-formulate, both "natural" and difficult field of nonlocal theory of quasilinear conservation laws will yet attract the attention of young, deep-thinking researchers, able to invent new approaches away from the traditional guidelines.

Acknowledgments

The authors would like to express their gratitude to the editors Etienne Emmrich and Petra Wittbold for having provided the possibility to publish the S. N. Kruzhkov lectures and thus make the lectures available to a wide audience. The manuscript benefited much from their careful reading and many helpful remarks.

The present publication would not have been possible without the active involvement by Boris Andreianov. He modestly positioned himself as the translator of the Russian manuscript; in fact, he initiated the present publication and moreover could be considered as its rightful co-author. Along with the translation he revised our original manuscript to the current western mathematical presentation standards. On top of that he has provided comments on the present state of the subject of conservation laws, that

further advance was made in the research on first-order quasilinear equations. New important approaches became standard, which shed more light on the fundamental theory and on advanced qualitative properties of admissible generalized solutions. We refer the interested reader to the monographs and textbooks [11, 14, 22, 32, 33, 35, 47, 48] which appeared since 1996, and to the references therein.

In particular, the problem mentioned here has been, at least partially, solved. A definition of a generalized solution which leads to a complete well-posedness theory, and which applies in particular to the p-system in one space dimension, has been given in the works of A. Bressan and collaborators (see [11]). These results, and the methods developed to achieve the results, represent a breakthrough in the theory of systems of conservation laws, a breakthrough that occurred more than thirty-five years after the pioneering works of S. N. Kruzhkov establishing the notion of entropy solution for the case of one scalar equation.

Yet the most important case for the applications, the one of multi-dimensional systems of conservation laws, remains very far from being solved. We can simply repeat S. N. Kruzhkov's words, saying that nowadays, in 2008, nobody in the entire world knows how to define the "correct" notion of solution for this problem!

For the case (also discussed in the above Afterword) of a general merely continuous (but not necessarily Lipschitz, nor Hölder continuous) vector flux function $f = f(u)$, in spite of some further progress (see [2, 5, 42]), a difficult open question persists: whether or not there is uniqueness of a generalized entropy solution in $L^\infty(0, T; L^1(\mathbb{R}^n)) \cap L^\infty(\Pi_T)$ without any additional restriction (such as (6.13)–(6.14)) on the flux function $f = f(u)$.

Let us mention, without any tentative of exhaustivity, that in the last fifteen years progress has been achieved: on the study of boundary-value problems for conservation laws (see, e.g., [35, 38]), on the numerical approximation of entropy solutions (see, e.g., [8, 22]), on the study of fine properties of general (not necessarily piecewise smooth, see, e.g., [24])) entropy solutions using methods of geometric measure theory (see, e.g., [15]) and the new tools of kinetic solutions (see, e.g., [9, 10, 19, 34, 43, 45, 47, 51]) and parameterized families of H-measures (see [40, 41, 46]), on the study of linear problems with irregular coefficients (see, e.g., [1]), on the convergence of the vanishing viscosity method (see [7]), on the study of stability of shock waves, on various generalizations of conservation law (6.10) including nonlocal problems, problems with oscillating or discontinuous in (t, x) coefficients, stochastic problems, problems on manifolds, on the related degenerated diffusion problems (see, e.g., [12, 13, 4]), on the study of singular solutions (see, e.g., [44]), of unbounded solutions (see, e.g., [21, 42]), and on the related new notion of renormalized solution (see [6]). Even a theory of "non-Kruzhkov" solutions to conservation laws was constructed (see [33]), stimulated by physical models with a specific notion of admissibility. Much of the above progress was inspired by "physical" considerations and by the investigation of applied problems.

Thus, although the above Afterword does not reflect the most recent challenges in the theory of first-order quasilinear PDEs, S. N. Kruzhkov's words sound as topical as ever. And it is certain that, after ten more years, the present footnote will look somewhat obsolete with respect to the new front of research.

the authors would simply be unable to survey. The authors do not belong to the scientific school of S. N. Kruzhkov, but are rather his colleagues who have collaborated with S. N. Kruzhkov on the task of creating and promoting the present course of lectures as a new element of the mathematical education at the Moscow Lomonosov State University. Their scientific interests lie in connected, but yet different branches of PDEs with respect to the subject of the lectures. On the contrary, Boris Andreianov learned the subject directly from S. N. Kruzhkov as a student and as a Ph.D. student. Now he continues to work in the field of the first-order quasilinear PDEs. His contribution to the preparation of the present edition is extremely valuable.

Translated from the Russian manuscript
by Boris P. Andreianov (Besançon)

References

[1] L. Ambrosio. The flow associated to weakly differentiable vector fields: recent results and open problems, *Boll. Unione Mat. Ital. Sez. B Artic. Ric. Mat.*, **(8) 10**:1 (2007), 25–41.

[2] B. Andreianov, Ph. Bénilan and S. N. Kruzhkov. L^1-theory of scalar conservation law with continuous flux function, J. Funct. Anal., **171**:1 (2000), 15–33.

[3] V. I. Arnol'd. *Supplementary chapters to the theory of ordinary differential equations.* (Russian), Nauka eds, Moscow, 1982. French transl. in *Chapitres supplémentaires de la théorie des équations différentielles ordinaires.* Éditions Mir, Moscow, 1984.

[4] M. Bendahmane and K. H. Karlsen. Renormalized solutions for quasilinear anisotropic degenerate parabolic equations, *SIAM J. Math. Anal.*, **36**:2 (2004), 405–422.

[5] Ph. Bénilan and S. N. Kruzhkov. Conservation laws with continuous flux functions, *NoDEA Nonlin. Diff. Eq. Appl.*, **3**:4 (1996), 395–419.

[6] Ph. Bénilan, J. Carrillo and P. Wittbold. Renormalized entropy solutions of scalar conservation laws, *Ann. Scuola Norm. Sup. Pisa Cl. Sci., (4)* **29**:2 (2000), 313–327.

[7] S. Bianchini and A. Bressan. Vanishing viscosity solutions of nonlinear hyperbolic systems, *Ann. Math.*, (2) **161**:1 (2005), 223–342.

[8] F. Bouchut and B. Perthame. Kruzhkov's estimates for scalar conservation laws revisited, *Trans. Amer. Math. Soc.*, **350**:7 (1998), 2847–2870.

[9] Y. Brenier. Résolution d'équations d'évolution quasilinéaires en dimension N d'espace à l'aide d'équations linéaires en dimension $N + 1$. *J. Diff. Eq.*, **50**:3 (1983), 375–390.

[10] Y. Brenier. L^2 formulation of multidimensional scalar conservation laws, to appear in *Arch. Ration. Mech. Anal.*

[11] A. Bressan. *Hyperbolic systems of conservation laws. The one-dimensional Cauchy problem*, Oxford Univ. Press, Oxford, 2000.

[12] J. Carrillo. Entropy solutions for nonlinear degenerate problems, *Arch. Ration. Mech. Anal.*, **147**:4 (1999), 269–361.

[13] G.-Q. Chen and B. Perthame. Well-posedness for anisotropic degenerate parabolic-hyperbolic equations, *Ann. Inst. H. Poincaré Anal. Non Lin.*, **20**:4 (2003), 645–668.

[14] C. M. Dafermos. *Hyperbolic conservation laws in continuum physics*, 2nd ed. Springer, Berlin, 2005.

[15] C. De Lellis, F. Otto and M. Westdickenberg. Structure of entropy solutions for multidimensional scalar conservation laws, Arch. Ration. Mech. Anal., **170**:2 (2003), 137–184.

[16] L. C. Evans. *Partial differential equations*, American Math. Soc., Providence, RI, 1998.

[17] A. F. Filippov. *Problems book on differential equations*, Nauka eds., Moscow, 1985. French transl. in A. Philippov. *Recueil de problèmes d'équations différentielles.* Éditions Mir, Moscow, 1976.

[18] I. M. Gel'fand. Some problems in the theory of quasi-linear equations. (Russian), *Uspehi Mat. Nauk*, **14**:2 (1959), 87–158. English transl. in *Russian Mathematical Surveys*, **2** (1959).

[19] Y. Giga and T. Miyakawa. A kinetic construction of global solutions of first order quasilinear equations, *Duke Math. J.*, **50**:2 (1983), 505–515.

[20] E. Godlewski and P.-A. Raviart. *Hyperbolic systems of conservation laws*, Ellipses, Paris, 1991.

[21] A. Yu. Goritskiĭ and E. Yu. Panov. On locally bounded generalized entropy solutions of the Cauchy problem for a first-order quasilinear equation. (Russian), *Tr. Mat. Inst. Steklova* **236**:1 (2002), *Differ. Uravn. i Din. Sist.*, 120–133. English transl. in *Proc. Steklov Inst. Math.*, **236**:1 (2002), 110–123.

[22] H. Holden and N. H. Risebro. *Front tracking for hyperbolic conservation laws*, Springer, New York, 2002.

[23] E. Hopf. The partial differential equation $u_t + uu_x = \mu u_{xx}$, *Comm. Pure Appl. Math.*, **3** (1950), 201–230.

[24] M. V. Korobkov and E. Yu. Panov. Isentropic solutions of quasilinear equations of the first order, *Mat. Sb.*, **197**:5 (2006), 99-124. English transl. in *Sb. Math.*, **197**:5 (2006), 727-752.

[25] S. N. Kružkov. Generalized solutions of the Cauchy problem in the large for first order nonlinear equations. (Russian) , *Dokl. Akad. Nauk. SSSR*, **187**:1 (1969), 29–32. English transl. in Soviet Math. Dokl. **10** (1969), 785–788.

[26] S. N. Kružkov. First order quasilinear equations with several independent variables. (Russian), *Mat. Sb.*, **81**:123 (1970), 228–255. English transl. in *Math. USSR Sb.*, **10** (1970), 217–243.

[27] S. N. Kružkov. *Nonlinear Partial Differential Equations (Lectures). Part II. First-order equations*, Moscow State Lomonosov Univ. eds, Moscow, 1970.

[28] S. N. Kruzhkov and E. Yu. Panov. First-order conservative quasilinear laws with an infinite domain of dependence on the initial data. (Russian), Dokl. Akad. Nauk SSSR, **314**:1 (1990), 79–84. English transl. in *Soviet Math. Dokl.*, **42**:2 (1991), 316–321.

[29] O. A. Ladyženskaya. On the construction of discontinuous solutions of quasi-linear hyperbolic equations as limits of solutions of the corresponding parabolic equations when the "coefficient of viscosity" tends towards zero. (Russian), *Dokl. Akad. Nauk SSSR*, **111**:2 (1956), 291–294.

[30] P. D. Lax. Hyperbolic systems of conservation laws. II, *Comm. Pure Appl. Math.*, **10**:1 (1957), 537–566.

[31] P. D. Lax. Weak solutions of nonlinear hyperbolic equations and their numerical computation. *Comm. Pure Appl. Math.*, **7**:1 (1954), 159–193.

[32] P. D. Lax. *Hyperbolic partial differential equations.* Courant Lect. Notes in Math., 14. American Math. Soc., Providence, RI, 2006.

[33] P. G. LeFloch. *Hyperbolic systems of conservation laws. The theory of classical and nonclassical shock waves*, Lect. Math. ETH Zürich. Birkhäuser, Basel, 2002.

[34] P.-L. Lions, B. Perthame and E. Tadmor. A kinetic formulation of multidimensional scalar conservation laws and related equations, *J. Amer. Math. Soc.*, **7**:1 (1994), 169–191.

[35] J. Málek, J. Nečas, M. Rokyta and M. Růžička. *Weak and measure-valued solutions to evolutionary PDEs*, Chapman & Hall, London, 1996.

[36] O. A. Oleĭnik. On Cauchy's problem for nonlinear equations in a class of discontinuous functions. (Russian), *Doklady Akad. Nauk SSSR*, **95**:3 (1954), 451–454.

[37] O. A. Oleĭnik. Discontinuous solutions of non-linear differential equations. (Russian), *Uspehi Mat. Nauk*, **12**:3 (1957), 3–73. English transl. in *Russian Mathematical Surveys*, **3** (1957).

[38] F. Otto. Initial-boundary value problem for a scalar conservation law, C. R. Acad. Sci. Paris Sér. I Math., **322**:8 (1996), 729–734.

[39] E. Yu. Panov. Uniqueness of the solution of the Cauchy problem for a first-order quasilinear equation with an admissible strictly convex entropy. (Russian), *Mat. Zametki*, **55**:55 (1994), 116–129. English transl. in *Math. Notes*, **55**:5-6 (1994), 517–525.

[40] E. Yu. Panov. On sequences of measure valued solutions for a first order quasilinear equation. (Russian), *Mat. Sb.*, **185**:2 (1994), 87–106; English transl. in *Russian Acad. Sci. Sb. Math.*, **81**:1 (1995), 211–227.

[41] E. Yu. Panov. Property of strong precompactness for bounded sets of measure valued solutions of a first-order quasilinear equation. (Russian), *Mat. Sb.*, **190**:3 (1999), 109–128; English transl. in Sb. Math., **190**:3 (1999), 427–446.

[42] E. Yu. Panov. On the theory of generalized entropy solutions of the Cauchy problem for a first-order quasilinear equation in the class of locally integrable functions. (Russian), *Izvest. RAN*, **66**:6 (2002), 91–136; English transl. in *Izvestiya: Mathematics*, **66**:6 (2002), 1171–1218.

[43] E. Yu. Panov. On kinetic formulation of first-order hyperbolic quasilinear systems, Ukr. Math. Bull., **1**:4 (2004), 553–568.

[44] E. Yu. Panov and V. M. Shelkovich. δ'-Shock waves as a new type of solutions to systems of conservation laws, *J. Diff. Eq.*, **228**:1 (2006), 49–86.

[45] E. Yu. Panov. On kinetic formulation of measure valued solutions for hyperbolic first-order quasilinear equations with several space variables, in: *Analytical Approaches to Multidimensional Balance Laws*, O. Rozanova ed., Nova Press, 2007.

[46] E. Yu. Panov. Existence of strong traces for quasi-solutions of multidimensional conservation laws, J. Hyperbolic Differ. Equ., **4**:4 (2007), 729–770.

[47] B. Perthame. *Kinetic formulation of conservation laws*, Oxford Univ. Press, Oxford, 2002.

[48] D. Serre. *Systems of conservation laws. 1. Hyperbolicity, entropies, shock waves.* and *Systems of conservation laws. 2. Geometric structures, oscillations, and initial-boundary value problems.* Transl. from the 1996 French original. Cambridge Univ. Press, Cambridge, 1999, and 2000.

[49] J. Smoller. *Shock waves and reaction-diffusion equations*, 2nd ed., Springer, New York, 1994.

[50] A. N. Tihonov and A. A. Samarskiĭ. On discontinuous solutions of a quasilinear equation of first order. (Russian), *Dokl. Akad. Nauk SSSR*, **99**:1 (1954), 27–30.

[51] A. Vasseur. Strong traces for solutions of multidimensional scalar conservation laws, *Arch. Ration. Mech. Anal.*, **160**:3 (2001), 181–193.

[52] A. I. Vol'pert. Spaces BV and quasilinear equations, *Mat. Sb.*, **73** (115) (1967), 255–302. English transl. in *Math. USSR Sb.*, **2** (1967), 225–267.

Author information

Gregory A. Chechkin, Moscow Lomonosov State University, Russia.
E-mail: chechkin@mech.math.msu.su

Andrey Yu. Goritsky, Moscow Lomonosov State University, Russia.
E-mail: goritsky@mech.math.msu.su

Adaptive semi-Lagrangian schemes for Vlasov equations

Martin Campos Pinto

Abstract. This lecture presents a new class of adaptive semi-Lagrangian schemes – based on performing a semi-Lagrangian method on adaptive interpolation grids – in the context of solving Vlasov equations with underlying "smooth" flows, such as the one-dimensional Vlasov–Poisson system. After recalling the main features of the semi-Lagrangian method and its error analysis in a uniform setting, we describe two frameworks for implementing adaptive interpolations, namely multilevel meshes and interpolatory wavelets. For both discretizations, we introduce a notion of good adaptivity to a given function and show that it is preserved by a low-cost prediction algorithm which transports multilevel grids along any "smooth" flow. As a consequence, error estimates are established for the resulting *predict and readapt* schemes under the essential assumption that the flow underlying the transport equation, as well as its approximation, is a stable diffeomorphism. Some complexity results are stated in addition, together with a conjecture of the convergence rate for the overall adaptive scheme. As for the wavelet case, these results are new and also apply to high-order interpolation.

Keywords. Fully adaptive scheme, semi-Lagrangian method, Vlasov equation, smooth characteristic flows, adaptive mesh prediction, error estimates, interpolatory wavelets.

AMS classification. 65M12, 65M50, 82D10.

1 Introduction

In this lecture, we shall describe adaptive numerical methods for approximating Vlasov equations, i.e., kinetic equations which model in statistical terms the nonlinear evolution of a collisionless plasma. In order to give the reader a specific example, we shall somehow focus our presentation on the one-dimensional Vlasov–Poisson system

$$\partial_t f(t,x,v) + v \cdot \partial_x f(t,x,v) + E(t,x) \cdot \partial_v f(t,x,v) = 0, \quad t > 0, \quad x, v \in \mathbb{R},$$

$$\partial_x E(t,x) = \int_{\mathbb{R}} f(t,x,v)\,\mathrm{d}v - 1, \quad f(0,\cdot,\cdot) = f^0,$$

but we emphasize that our results actually apply to any nonlinear transport problem associated with a smooth characteristic flow, "smooth" meaning here that the flow is a Lipschitz diffeomorphism, see in particular Assumptions 3.2 and 3.3 below.

In order to save computational resources while approximating the complex and thin structures that may appear in the solutions as time evolves, several adaptive schemes have been proposed in the past few years, see, e.g., [3, 6, 7, 20], all based on the semi-Lagrangian method originally introduced by Chio-Zong Cheng and Georg Knorr [9] and later revisited by Eric Sonnendrücker, Jean Rodolphe Roche, Pierre Bertrand and Alain Ghizzo [23]. A common feature of these new schemes lies in the multilevel, tree-structured discretization of the phase space, and a central issue appears to be the *prediction strategy* for generating adaptive meshes that are optimal, i.e., that only retain the "necessary" grid points.

Based on the regularity analysis of the numerical solution and how it gets transported by the numerical flow, it has recently been shown in [7] that a low-cost prediction strategy could achieve an accurate evolution of the adaptive multilevel meshes from one time step to the other, in the sense that the overall accuracy of the scheme was monitored by a prescribed tolerance parameter ε representing the local interpolation error at each time step. In this lecture, we shall follow the same approach and propose new algorithms for high-order wavelet-based schemes, together with error estimates.

The lecture is organized as follows. We shall start by presenting the Vlasov–Poisson system in Section 2 and recall some important properties satisfied by its classical solutions. The backward semi-Lagrangian method is then described in Section 3, together with the main points of its error analysis, which yields a first convergence rate in the case where the discretization is uniform. We detail next two distinct frameworks for generating adaptive multilevel discretizations, namely adaptive meshes in Section 4 and interpolatory wavelets in Section 5. Finally, Section 6 is devoted to describing algorithms that build multilevel grids which either are well-adapted to given functions or *remain* well-adapted to transported functions. More precisely, for both discretizations we introduce a notion of ε-adaptivity to a given function and show that it is preserved by a low-cost prediction algorithm which transports multilevel grids along any smooth flow. We end by proving new error estimates for the resulting semi-Lagrangian schemes (with arbitrary interpolation order, under a central assumption on the approximate flow) and state some partial complexity results.

In these notes, we shall often write $A \lesssim B$ to express that $A \le CB$ holds with a constant C independent of the parameters involved in the inequality, $A \sim B$ meaning that both $A \lesssim B$ and $B \lesssim A$ hold. As often as possible, we will nevertheless indicate in the text the parameters on which these "invisible" constants may depend.

2 The Vlasov–Poisson system

The Vlasov system, as introduced by Anatoliĭ Aleksandrovich Vlasov [24] in the late 1930s, describes in statistical terms the time evolution of rarefied plasmas, which are gases of charged particles such as ions and electrons. In this section, we recall the form of the system as well as its interpretation and describe some of its properties. Although we shall mention the general equations in $d = 3$ physical dimensions, this lecture will essentially focus on a simplified model in $d = 1$ physical dimension, leading to a two-dimensional transport problem in the phase space. Most of our algorithms and proofs, however, can be generalized to higher dimensions with no particular difficulty.

2.1 Description of the model

In the Vlasov model, the state of every species $\mathcal{E}$ of charged particles in the plasma is represented at time t by a *density function* $f_{\mathcal{E}}(t)$ defined in the phase space, i.e., the subset of $\mathbb{R}^d \times \mathbb{R}^d$ which contains every possible position x and speed v. In particular, the number of $\mathcal{E}$-type particles that are located at time t in a physical domain $\omega_x \subset \mathbb{R}^d$, with speed $v \in \omega_v \subset \mathbb{R}^d$, reads

$$Q_{\mathcal{E}}(t,\omega) = \iint_{\omega_x \times \omega_v} f_{\mathcal{E}}(t,x,v)\,\mathrm{d}x\,\mathrm{d}v. \tag{2.1}$$

For the sake of simplicity, we shall assume that the effect of the magnetic field can be neglected, which corresponds to an electrostatic approximation, and consider only two species, namely:

- positive (and heavy) ions, assumed to be uniformly distributed in space and time; their density function will be denoted by $f_p(t,x,v) = f_p(v)$, and we shall assume that it is normalized, i.e., $\int f_p(v)\,\mathrm{d}v = 1$, and
- much lighter electrons; their density function f is the main unknown of the model.

In dimension $d = 1$, the Vlasov–Poisson system reads as

$$\partial_t f(t,x,v) + v \cdot \partial_x f(t,x,v) + E(t,x) \cdot \partial_v f(t,x,v) = 0, \tag{2.2}$$

$$\partial_x E(t,x) = \int_{\mathbb{R}} f(t,x,v)\,\mathrm{d}v - 1, \tag{2.3}$$

where E represents the normalized, self-consistent electric field. In order to consider the corresponding Cauchy problem, we supplement (2.2)–(2.3) with an initial condition

$$f(0,\cdot,\cdot) = f^0 \tag{2.4}$$

assumed to be smooth (at least continuous) and compactly supported in $\mathbb{R}^2$.

2.2 Physical interpretation and characteristic flows

Since the end of the 18th century and the works of Charles-Augustin Coulomb, it is known that a closed system of charged particles $\{q_i \in \mathbb{R}, x_i(t) \in \mathbb{R}^3\}_{i \in \mathcal{I}}$ is subject to binary interactions – the so-called Coulomb interactions – yielding an N-body problem. A less expensive model follows by considering the electromagnetic field which, according to the theory that James Clerk Maxwell developed one century later, is created by the particles and simultaneously influences their motion through the Lorentz force. More precisely, if the associated charge and current densities are denoted by

$$\rho(t,x) := \sum_{i \in \mathcal{I}} q_i \delta_{\{x_i(t)\}}(x) \quad \text{and} \quad \jmath(t,x) := \sum_{i \in \mathcal{I}} v_i(t) q_i \delta_{\{x_i(t)\}}(x),$$

respectively, where $\delta_{x_i(t)}$ stands for the Dirac mass located at $x_i(t)$ and $v_i(t) := x_i'(t)$ is the particle's speed, then the electromagnetic field $(E,B)(t,x)$ created by the particles satisfies the following *Maxwell system*:

$$\nabla_x \cdot E = \rho/\varepsilon_0, \tag{2.5}$$

$$\nabla_x \times E = -\partial_t B, \tag{2.6}$$

$$\nabla_x \cdot B = 0, \tag{2.7}$$

$$\nabla_x \times B = \mu_0(\jmath + \varepsilon_0 \partial_t E). \tag{2.8}$$

Here ∇_x is the differential operator $(\partial_{x_1}, \partial_{x_2}, \partial_{x_3})$, $\times$ is the vector product in $\mathbb{R}^3$, and ε_0, μ_0 denote the permittivity and permeability constants. In turn, every particle is subject to the *Lorentz force*

$$F_i(t) = q_i[E(x_i(t)) + v_i(t) \times B(t, x_i(t))]. \tag{2.9}$$

According to Isaac Newton's fundamental law of dynamics $m_i v_i' = F_i$, it follows that every particle has a phase space trajectory that is a solution to the ordinary differential equation

$$x_i'(t) = v_i(t), \quad v_i'(t) = \frac{q_i}{m_i}[E(x_i(t)) + v_i(t) \times B(t, x_i(t))]. \tag{2.10}$$

The kinetic Vlasov model corresponds to a *continuous limit* of the above mean field model when the particles are so many that each species can be represented by a smooth, e.g., continuous, density function $f_{\mathcal{E}}$. In such a case, charge and current densities are defined as

$$\rho(t,x) := \sum_{\mathcal{E}} q_{\mathcal{E}} \int_{v \in \mathbb{R}^3} f_{\mathcal{E}}(t,x,v) \mathrm{d}v \quad \text{and} \quad \jmath(t,x) := \sum_{\mathcal{E}} q_{\mathcal{E}} \int_{v \in \mathbb{R}^3} v f_{\mathcal{E}}(t,x,v) \mathrm{d}v,$$

and the electromagnetic field is again given by Maxwell's system (2.5)–(2.8). As was written above, we shall only consider two species $\mathcal{E}$, namely positive ions and electrons. The so-called Vlasov–Poisson system, which corresponds to an electrostatic approximation, i.e., the effects of the magnetic field are neglected, can then be obtained by writing the pointwise conservation of the electron density f along the *characteristic curves*, which are defined as the solutions

$$t \mapsto (X(t), V(t)) = (X(t;s,x,v), V(t;s,x,v)) \tag{2.11}$$

to the ordinary differential system

$$\partial_t X(t) = V(t), \quad \partial_t V(t) = E(t, X(t)), \quad (X,V)(s) = (x,v). \tag{2.12}$$

Note that this is a natural extension of the trajectories (2.10) in the continuous limit. If the electron density f satisfies

$$\partial_t f(t, X(t;0,x,v), V(t;0,x,v)) = 0 \tag{2.13}$$

for any (x,v) in the phase space, one indeed finds

$$\partial_t f(t,x,v) + v \cdot \nabla_x f(t,x,v) + E(t,x) \cdot \nabla_v f(t,x,v) = 0, \tag{2.14}$$

with an obvious definition for ∇_v. Since E has then a vanishing curl, it derives from an electric potential ϕ, i.e.,

$$E = -\nabla_x \phi, \tag{2.15}$$

and equation (2.5) is equivalent to the Poisson equation

$$\Delta_x \phi = -\frac{\rho}{\varepsilon_0}, \tag{2.16}$$

where $\Delta_x \equiv \nabla_x \cdot \nabla_x$ denotes the Laplace operator. Equations (2.14)–(2.16) form the Vlasov–Poisson system in three physical dimensions, from which the simplified model (2.2)–(2.3) easily derives (with normalized constants) in one dimension.

Now, as long as E is bounded and continuously differentiable with respect to x, it is known that the associated *characteristic flow*

$$\mathcal{F}_{t,s} : (x, v) \mapsto (X, V)(t; s, x, v) \tag{2.17}$$

is a measure preserving $\mathcal{C}^1$-diffeomorphism with inverse $\mathcal{F}_{t,s}^{-1} = \mathcal{F}_{s,t}$, see, e.g., [22]. Let us then notice that equation (2.13), which expresses the pointwise transport of f along the characteristic curves, also yields a local transport property for the charge density, in the sense that

$$\iint_\omega f(0, x, v)\, \mathrm{d}x\, \mathrm{d}v = \iint_{\mathcal{F}_{t,s}(\omega)} f(t, x, v)\, \mathrm{d}x\, \mathrm{d}v \quad \text{for any } \omega \subset \mathbb{R}^{2d}, \tag{2.18}$$

which is equivalent to saying that the flow is measure preserving. This property can be seen as a consequence of the vanishing divergence of the field $(x, v) \mapsto (v, E(t, x))$. Indeed, it first allows to write (2.2) in a conservative form, i.e.,

$$\partial_t f(t, x, v) + \nabla_{x,v} \cdot [(v, E(t, x)) f(t, x, v)] = 0. \tag{2.19}$$

Second, introducing the divergence-free field $\Phi(t, x, v) := (1, v, E(t, x))$ defined on $\mathcal{D}_{\tau,\omega} = \{(t, x, v) : t \in [0, \tau], \mathcal{F}_{t,s}^{-1}(x, v) \in \omega\} \subset [0, \tau] \times \mathbb{R}^{2d}$, employing the Stokes formula gives

$$\iint_{\mathcal{F}_{\tau,s}(\omega)} \mathrm{d}x\, \mathrm{d}v - \iint_\omega \mathrm{d}x\, \mathrm{d}v = \iint_{\partial \mathcal{D}_{\tau,\omega}} \Phi \cdot n\, \mathrm{d}\sigma = \iiint_{\mathcal{D}_{\tau,\omega}} \nabla_{t,x,v} \cdot \Phi\, \mathrm{d}t\, \mathrm{d}x\, \mathrm{d}v = 0,$$

where n denotes the outward unit vector normal to $\partial \mathcal{D}_{\tau,\omega}$. Here the first equality comes from the fact that the boundary of $\mathcal{D}_{\tau,\omega}$ is parallel to the field lines of Φ outside the "faces" $\mathcal{F}_{\tau,s}(\omega)$ and ω. As the latter equality precisely means that $\mathcal{F}_{t,s}$ preserves the Lebesgue measure, we see that its Jacobian determinant is equal to one. In particular, property (2.18) readily follows from equation (2.13).

2.3 Existence of smooth solutions

According to the previous section, we shall say that (f, E) is a classical solution of the Vlasov–Poisson system (2.2)–(2.4) if

(i) f is continuous,

(ii) E is continuously differentiable,

(iii) the Vlasov equation (2.2) is satisfied in the sense of distributions.

Now, because the characteristic curves are defined on any point (x, v) of the phase space, condition (iii) is equivalent to

(iii)' f is constant along the characteristic curves defined by (2.12).

Remark 2.1. If f is only continuous, the derivatives appearing in (2.2) must be understood in the (weak) sense of distributions. Nevertheless, the solution is said to be classical (or strong), because the characteristic curves are well-defined, and hence equation (2.13) is satisfied in a classical sense.

One of the first results is due to Sergei Iordanskii (see [21]), who has proven in the early 1960s global existence in time and uniqueness of classical solutions under certain conditions. First, the initial datum f^0 must be continuous, and it must be integrable in the following sense:

$$\rho^0(x) = \int_{\mathbb{R}} f^0(x,v)\,\mathrm{d}v - 1 < \infty \quad \text{and} \quad \int_{\mathbb{R}} v^2\theta(v)\,\mathrm{d}v < \infty, \tag{2.20}$$

where θ is a non-increasing function of $|v|$ that dominates f^0 and f_p (the density of the positive ions). Second, the electric field must satisfy the limit condition

$$\lim_{x \to -\infty} E(t,x) = 0, \quad t > 0, \tag{2.21}$$

which is a reasonable condition since E is only defined up to a constant. Note that the assumptions (2.20) are rather natural, since they allow to define the current and kinetic energy densities,

$$\jmath(t,x) := \int_{\mathbb{R}} v[f(t,x,v) - f_p(v)]\,\mathrm{d}v \quad \text{and} \quad \varepsilon_k(t,x) := \int_{\mathbb{R}} v^2[f(t,x,v) + f_p(v)]\,\mathrm{d}v,$$

respectively, which are two fundamental physical quantities. Finally, Iordanskii shows that f and E satisfy an additional equation, namely Ampère's equation

$$\partial_t E(t,x) = \int_{\mathbb{R}} v[f_p(v) - f(t,x,v)]\,\mathrm{d}v. \tag{2.22}$$

In 1980, Jeffrey Cooper and Alexander Klimas [13] have extended these results to more general limit conditions than (2.21), in particular they have addressed the periodic case where $x \in \mathbb{R}/\mathbb{Z}$.

From now on, we shall only consider this case, moreover our initial data will always be assumed to have a compact support in the phase space

$$\Omega_\infty := (\mathbb{R}/\mathbb{Z}) \times \mathbb{R}. \tag{2.23}$$

Their result is the following.

Theorem 2.2. *If f^0 is continuous on Ω_∞ (hence 1-periodic with respect to x), if it satisfies*

$$\rho^0(x) := \int_{\mathbb{R}} f^0(x,v)\,\mathrm{d}v - 1 < \infty \quad \textit{and} \quad \int_{\mathbb{R}} |v|\theta(v)\,\mathrm{d}v < \infty, \tag{2.24}$$

where θ is defined as in (2.20), *and if the plasma is globally neutral, i.e.,*

$$\int_0^1 \rho^0(x)\,\mathrm{d}x = \int_0^1 \int_{\mathbb{R}} f^0(x,v)\,\mathrm{d}v\,\mathrm{d}x - 1 = 0, \tag{2.25}$$

then there exists a unique classical solution to (2.2)–(2.4) *such that $\int_0^1 E(0,x)\,\mathrm{d}x = 0$. Moreover, this solution is 1-periodic with respect to x.*

Notice that the global neutrality, namely (2.25) at time $t = 0$, and

$$\iint_{\Omega_\infty} f(t,x,v)\,\mathrm{d}x\,\mathrm{d}v = \iint_{\Omega_\infty} f^0(x,v)\,\mathrm{d}x\,\mathrm{d}v = 1 \tag{2.26}$$

for positive times (which follows from the transport properties of f) is equivalent to the continuity of the 1-periodic field $E(t,\cdot)$. Now, it is possible to give an analytical expression for E: Indeed, if we denote by $-G(x,y)$ the Green function associated with the one-dimensional Poisson equation (2.3) defined in such a way that

$$\partial^2_{xx} G(\cdot,y) = \delta(\cdot - y) \quad \text{on} \quad (0,1) \tag{2.27}$$

holds for any $y \in (0,1)$ with periodic boundary conditions $G(0,y) = G(1,y)$, then E reads

$$E(t,x) = \int_0^1 K(x,y) \left(\int_{\mathbb{R}} f(t,y,v)\,\mathrm{d}v - 1 \right) \mathrm{d}y \tag{2.28}$$

with

$$K(x,y) = \partial_x G(x,y) = \begin{cases} y - 1 & \text{if } 0 < x < y, \\ y & \text{if } y \le x < 1. \end{cases} \tag{2.29}$$

In order to study later the accuracy of the numerical schemes, we now state some smoothness estimates for f and E. In general, it is known that any initial order of smoothness is preserved by the equation, see, e.g., [14]. Since the analysis is simple in our case, we give a detailed proof for the following estimates, inspired by the techniques presented in the book of Robert Glassey [19]. Here and below, we shall rely on the usual notations for Sobolev spaces, see, e.g., [1].

Lemma 2.3. *If f^0 belongs to $W^{1,\infty}(\Omega_\infty)$ and satisfies the conditions* (2.24)–(2.25), *then for any final time $T < \infty$, the solution f is compactly supported in the v-variable, i.e.,*

$$\Sigma_v(t) := \sup\{|v| : \exists x \in \mathbb{R}/\mathbb{Z}, f(t,x,v) > 0\} \le \Sigma_v(0) + 2T, \quad t \le T, \tag{2.30}$$

and satisfies the following smoothness estimates:

$$\|f(t,\cdot,\cdot)\|_{W^{1,\infty}(\Omega_\infty)} \le C, \tag{2.31}$$

$$\|\partial_t f(t,\cdot,\cdot)\|_{L^\infty(\Omega_\infty)} \le C, \tag{2.32}$$

$$\|E(t,\cdot)\|_{W^{2,\infty}(0,1)} \le C, \tag{2.33}$$

$$\|\partial_t E(t,\cdot)\|_{W^{1,\infty}(0,1)} \le C, \tag{2.34}$$

$$\|\partial^2_{tt} E(t,\cdot)\|_{L^\infty(0,1)} \le C, \tag{2.35}$$

for all $t \in (0,T)$*, with a constant* $C > 0$ *depending on* f^0 *and* T *only.*

Proof. Let us first show the weaker assertion that

$$\sup_{t\in(0,T)} \|E(t,\cdot)\|_{W^{1,\infty}(0,1)} \le C \quad \text{and} \quad \sup_{t\in(0,T)} \|\partial_t E(t,\cdot)\|_{L^\infty(0,1)} \le C \tag{2.36}$$

hold as long as f^0 is continuous: Indeed, the conservation of f along the characteristic curves (2.11) yields a maximum principle

$$0 \le f \le \|f^0\|_{L^\infty(\Omega_\infty)}, \tag{2.37}$$

and a bounded support in the v-direction, i.e., for all $t \in [0,T]$,

$$\Sigma_v(t) - \Sigma_v(0) \le \sup_{(x,v)\in\Omega_\infty} \int_0^T |\partial_t V(\tau;0,x,v)|\,\mathrm{d}\tau \le T\|E\|_{L^\infty((0,T)\times(0,1))}. \tag{2.38}$$

By using (2.28), (2.37) and (2.26), it follows that for all t, we have

$$\|E(t,\cdot)\|_{L^\infty(0,1)} \le \|K\|_{L^\infty}\left(\iint_{\Omega_\infty} |f(t,x,v)|\,\mathrm{d}x\,\mathrm{d}v + 1\right) \le 2. \tag{2.39}$$

According to (2.38), the above inequality yields (2.30). We also have, for all t,

$$\|\partial_x E(t,\cdot)\|_{L^\infty(0,1)} \le \Sigma_v(t)\|f^0\|_{L^\infty(\Omega_\infty)} + 1 \tag{2.40}$$

by using the Poisson equation (2.3), and

$$\|\partial_t E(t,\cdot)\|_{L^\infty(0,1)} \le \Sigma_v(t)^2\|f^0\|_{L^\infty(\Omega_\infty)} + \int_{\mathbb{R}} v f_p(v)\,\mathrm{d}v$$

by using the Ampère equation (2.22), which establishes both inequalities in (2.36). If we now assume $f^0 \in W^{1,\infty}(\Omega_\infty)$, we can write

$$\begin{aligned} |f(t,x,v) - f(t,\tilde{x},\tilde{v})| &= |f^0(X_0(t),V_0(t)) - f^0(\tilde{X}_0(t),\tilde{V}_0(t))| \\ &\le |f^0|_{W^{1,\infty}(\Omega_\infty)}\big(|e_x(t)| + |e_v(t)|\big) \end{aligned}$$

where for any s with $0 \le s \le t \le T$, we have set

$$\begin{cases} (X_0, V_0)(s) := (X, V)(t-s; t, x, v) \\ (\tilde{X}_0, \tilde{V}_0)(s) := (X, V)(t-s; t, \tilde{x}, \tilde{v}) \end{cases} \text{and} \quad \begin{cases} e_x(s) := X_0(s) - \tilde{X}_0(s) \\ e_v(s) := V_0(s) - \tilde{V}_0(s) \end{cases}. \quad (2.41)$$

By using the equations (2.12), we see that these quantities satisfy $e_x'(s) = e_v(s)$ and $e_v'(s) = E(t-s, X_0(s)) - E(t-s, \tilde{X}_0(s))$. The first bound in (2.36) thus yields

$$|e_x'(s)| + |e_v'(s)| \lesssim \big(|e_x(s)| + |e_v(s)|\big). \quad (2.42)$$

In particular, the function $\psi(s) := |e_x(s)| + |e_v(s)|$ satisfies

$$\psi(t) = \psi(0) + \Big|\int_0^t e_x'(s)\,\mathrm{d}s\Big| + \Big|\int_0^t e_v'(s)\,\mathrm{d}s\Big| \le \psi(0) + C(T)\int_0^t \psi(s)\,\mathrm{d}s, \quad (2.43)$$

and by applying the Gronwall lemma we find

$$\big(|e_x(t)| + |e_v(t)|\big) \lesssim \big(|e_x(0)| + |e_v(0)|\big) \lesssim \big(|x - \tilde{x}| + |v - \tilde{v}|\big) \quad (2.44)$$

with constants depending only on T, f^0. This shows that $\sup_{t\in(0,T)} \|f(t,\cdot,\cdot)\|_{W^{1,\infty}(\Omega_\infty)}$ is bounded by a constant that only depends on T and f^0, and we note that for all $t \in (0,T)$, the bound

$$\|\partial_t f(t,\cdot,\cdot)\|_{L^\infty(\Omega_\infty)} \le \big(Q(T) + \|E(t,\cdot)\|_{L^\infty(0,1)}\big)\|f(t,\cdot,\cdot)\|_{W^{1,\infty}(\Omega_\infty)}$$

with $Q(T) := \sup_{t\in(0,T)} \Sigma_v(t)$ follows from the Vlasov equation (2.2). Let us now turn to the electric field: By differentiating the Poisson equation (2.3) with respect to x and t, we find

$$\|\partial^2_{xx} E\|_{L^\infty((0,T)\times(0,1))} \le Q(T)\|\partial_x f\|_{L^\infty((0,T)\times\Omega_\infty)}, \quad (2.45)$$

$$\|\partial^2_{tx} E\|_{L^\infty((0,T)\times(0,1))} \le Q(T)\|\partial_t f\|_{L^\infty((0,T)\times\Omega_\infty)}, \quad (2.46)$$

and the seminorm $\|\partial^2_{tt} E\|_{L^\infty((0,T)\times(0,1))}$ is easily bounded by differentiating the Ampère equation (2.22) with respect to t. □

3 The backward semi-Lagrangian method

Based on the pointwise transport property (2.13), the semi-Lagrangian approach consists in combining a transport and a projection operator within every time step as in

$$f_{n+1} := PTf_n,$$

where $f_n \approx f(t_n)$ denotes the numerical solution. More precisely, the schemes that we will consider in this lecture decompose as follows (see, e.g., [9] or [23]):

(i) given f_n, approach the exact backward flow $\mathcal{F}_{t_n,t_{n+1}}$, see (2.17), by some computable diffeomorphism $\mathcal{B}[f_n]$,

(ii) define an intermediate solution by transporting f_n along this approximate flow

$$\mathcal{T} f_n := f_n \circ \mathcal{B}[f_n], \tag{3.1}$$

(iii) obtain f_{n+1} by interpolating the intermediate solution $\mathcal{T} f_n$, for instance on the nodes of some triangulation $\mathcal{K}$.

Following the same principle, the *adaptive* semi-Lagrangian approach consists in interpolating $\mathcal{T} f_n$ on an adaptive grid of the phase space. A major issue resides then in transporting the grid along the flow in such a way that both its *sparsity* and its *accuracy* are guaranteed.

From now on, we shall assume that every solution is supported in $\Omega := (0,1)^2$. According to Lemma 2.3, we know that this holds true for sufficiently small supports of the initial datum f^0. Moreover, we shall consider a uniform discretization involving N time steps and set $\Delta t := T/N$ and $t_n := n\Delta t$ for $n = 0, \dots, N$.

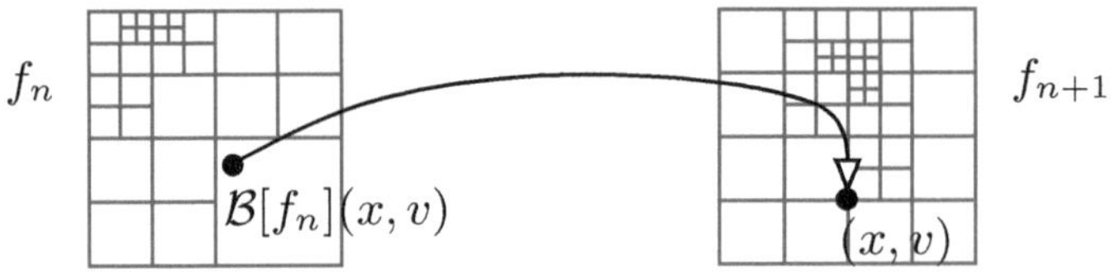

Figure 1. The backward semi-Lagrangian method (here with adaptive meshes).

Remark 3.1 (Computational cost). Like the numerical flow, the intermediate solution $\mathcal{T} f_n$ is computable everywhere, but only is *computed* on the interpolation grid corresponding to f_{n+1}. Hence if the latter is interpolated on a triangulation $\mathcal{K}$, the computational cost of one iteration is of the order $C\#(\mathcal{K})$, where C is the cost of applying the approximate flow $\mathcal{B}[f_n]$ to one point $(x, v) \in \Omega$ and $\#(\mathcal{K})$ denotes the cardinality of $\mathcal{K}$.

3.1 Approximation of smooth flows

We mentioned before that our main results apply to any transport problem with an underlying smooth flow. Besides, we will need that the approximate flow is also smooth and, moreover, stable and accurate in order to describe and further analyze our adaptive schemes. Let us formulate these properties as assumptions, for later reference.

Assumption 3.2. Let $s < t$ be arbitrary instants in $[0, T]$. The characteristic (backward) flow underlying the Vlasov equation, i.e., the mapping $\mathcal{F}_{s,t}$ for which we have

$$f(t, x, v) = f\big(s, \mathcal{F}_{s,t}(x, v)\big), \qquad (x, v) \in \Omega,$$

is a diffeomorphism from Ω into itself, i.e., it satisfies (with $\mathcal{B} = \mathcal{F}_{s,t}$)

$$\|\mathcal{B}(z + h) - \mathcal{B}(z)\| \le L_{\mathcal{B}}\|h\| \quad \text{and} \quad \|\mathcal{B}^{-1}(z + h) - \mathcal{B}^{-1}(z)\| \le L_{\mathcal{B}^{-1}}\|h\| \tag{3.2}$$

for all z, $z + h \in \Omega$ and with constants $L_{\mathcal{B}}$, $L_{\mathcal{B}^{-1}}$ independent of s, t (here and below, $\|\cdot\|$ denotes the Euclidean norm on $\mathbb{R}^2$).

Assumption 3.3. We are given a scheme $\mathcal{B}[\cdot] = \mathcal{B}_{\Delta t}[\cdot]$ which maps any Lipschitz continuous function $g \in W^{1,\infty}(\Omega)$ to a mapping $\mathcal{B}[g] : \Omega \to \Omega$ such that

- for any numerical solution f_n, $n = 0, \dots, N-1$, the approximate backward flow $\mathcal{B} = \mathcal{B}[f_n]$ is a diffeomorphism and (3.2) holds with constants independent of n,
- the mapping $\mathcal{B}[\cdot]$ is stable in the sense that there exists a constant independent of Δt such that
$$\big\|\mathcal{B}[g] - \mathcal{B}[\tilde{g}]\big\|_{L^\infty(\Omega)} \lesssim \Delta t \|g - \tilde{g}\|_{L^\infty(\Omega)} \tag{3.3}$$
holds for any pair g, $\tilde{g}$ of Lipschitz functions, and
- the approximation is locally r-th order accurate with $r > 1$ in the sense that
$$\big\|\mathcal{F}_{t_n,t_{n+1}} - \mathcal{B}[f(t_n)]\big\|_{L^\infty(\Omega)} \lesssim (\Delta t)^r \tag{3.4}$$
holds for all $n = 0, \dots, N-1$ with a constant depending on f_0 and $T = N\Delta t$ only.

We observe that a smooth flow preserves the local regularity of the solutions, measured in terms of the first order modulus of smoothness ω_1,
$$\omega_1(g, \tau, A)_\infty := \sup_{\|h\| \le \tau} \|\Delta_h^1 g\|_{L^\infty(\{A\}_h)}, \tag{3.5}$$
based on the finite difference $\Delta_h^1 g(z) := g(z+h) - g(z)$, and where we have set
$$\{A\}_h := \{z \in A : z + h \in A\} \tag{3.6}$$
for any open domain $A \subset \Omega$. Moduli of smoothness are classical functionals which enter for instance the definition of Besov spaces, see, e.g., [11]. Clearly the quantity (3.5) is monotone with respect to τ, and it is easily seen that if m is any positive integer, writing $\Delta_{mh}^1 g(z) = \Delta_h^1 g(z + (m-1)h) + \Delta_h^1 g(z + (m-2)h) + \dots + \Delta_h^1 g(z)$ yields
$$\omega_1(g, m\tau, A)_\infty \le m\omega_1(g, \tau, A)_\infty. \tag{3.7}$$
We are now ready to prove the following result:

Lemma 3.4. *If the flow $\mathcal{B}$ satisfies* (3.2)*, then for any $g \in L^\infty(\Omega)$, we have*
$$\omega_1(g \circ \mathcal{B}, \tau, A)_\infty \le \lceil L_\mathcal{B} \rceil \omega_1(g, \tau, A^{\mathcal{B},\tau})_\infty, \tag{3.8}$$
where $A^{\mathcal{B},\tau} := \mathcal{B}(\tilde{A}^\tau)$, $\tilde{A}^\tau := A + B_{\ell^2}(0, L_\mathcal{B} L_{\mathcal{B}^{-1}}\tau)$ (here and below, we use the standard notation $B_{\ell^p}(z, r)$ for the open ℓ^p ball of $\mathbb{R}^2$ with center z and radius r) and where $\lceil L_\mathcal{B} \rceil$ denotes the smallest integer greater or equal to $L_\mathcal{B}$.

Proof. Since $\mathcal{B}$ is Lipschitz, we have for any h
$$\begin{aligned}\|f(B(\cdot + h)) - f(B(\cdot))\|_{L^\infty(\{A\}_h)} &\le \sup_{z \in \{A\}_h} \sup_{\|h'\| \le L_\mathcal{B}\|h\|} |f(B(z) + h') - f(B(z))| \\ &\le \sup_{\|h'\| \le L_\mathcal{B}\|h\|} \|f(\cdot + h') - f\|_{L^\infty(\mathcal{B}(\{A\}_h))}.\end{aligned}$$

Now observe that $\mathcal{B}(\{A\}_h) \subset \mathcal{B}(A) \subset \{\mathcal{B}(A+B_{\ell^2}(0, L_{\mathcal{B}^{-1}}s))\}_{h'}$ holds for any h' with $\|h'\| \le s$, see (3.6). This yields

$$\|f(B(\cdot+h))-f(B(\cdot))\|_{L^\infty(\{A\}_h)} \le \sup_{\|h'\|\le L_{\mathcal{B}}\tau} \|f(\cdot+h')-f\|_{L^\infty(\{\mathcal{B}(A+B_{\ell^2}(0,L_{\mathcal{B}^{-1}}L_{\mathcal{B}}\tau))\}_{h'})}$$

for any h with $\|h\| \le \tau$, i.e., $\omega_1(g\circ\mathcal{B},\tau,A)_\infty \le \omega_1(g, L_{\mathcal{B}}\tau, A^{\mathcal{B},\tau})_\infty$, and (3.8) follows by applying (3.7). □

Remark 3.5. By observing that $|g|_{W^{1,\infty}(\Omega)} = \sup_{\tau>0}\tau^{-1}\omega_1(g,\tau,\Omega)_\infty$, Lemma 3.4 yields

$$|f(t,\cdot,\cdot)|_{W^{1,\infty}(\Omega)} \le C|f^0|_{W^{1,\infty}(\Omega)} \qquad \forall t\in[0,T]$$

under Assumption 3.2, with a constant $C>0$ independent of t.

Remark 3.6. If the flow $\mathcal{B}$ has more smoothness, it is possible to establish high-order estimates for the associated transport, involving moduli of (positive, integer) order ν,

$$\omega_\nu(g,t,A)_\infty := \sup_{\|h\|\le t} \|\Delta_h^\nu g\|_{L^\infty(\{A\}_{h,\nu})}, \tag{3.9}$$

based on the finite differences defined recursively by $\Delta_h^\nu g := \Delta_h^1(\Delta_h^{\nu-1}g)$, and now writing $\{A\}_{h,\nu} := \{z\in A : z+h\in A,\dots,z+\nu h\in A\}$.

For the sake of completeness, we now describe one approximation scheme for the flow that is based on a Strang splitting in time, and which is standard in the context of semi-Lagrangian methods, see, e.g., [9] or [23]. Denoting by

$$E[g] := \int_0^1 K(x,y)\Big(\int_{\mathbb{R}} g(y,v)\,\mathrm{d}v - 1\Big)\,\mathrm{d}y \tag{3.10}$$

the electric field associated with some arbitrary phase space density g, see (2.28), the scheme consists in defining one-directional flows

$$\mathcal{B}_x^{\frac12}(x,v) := \Big(x - v\frac{\Delta t}{2}, v\Big), \qquad \mathcal{B}_v[g](x,v) := (x, v - E[g]\Delta t), \tag{3.11}$$

and corresponding transport operators

$$\mathcal{T}_x^{\frac12} : g\mapsto g\circ\mathcal{B}_x^{\frac12}, \qquad \mathcal{T}_v : g\mapsto g\circ\mathcal{B}_v[g].$$

The full operator $\mathcal{T} := \mathcal{T}_x^{1/2}\mathcal{T}_v\mathcal{T}_x^{1/2} : g\mapsto g\circ\mathcal{B}[g]$ corresponds to the explicit flow

$$\mathcal{B}[g] : (x,v)\mapsto(\tilde{x},\tilde{v}), \qquad \begin{cases} \tilde{x} := x - v\Delta t + \dfrac{(\Delta t)^2}{2} E\Big[\mathcal{T}_x^{\frac12} g\Big]\Big(x - v\dfrac{\Delta t}{2}\Big), \\ \tilde{v} := v - \Delta t E\Big[\mathcal{T}_x^{\frac12} g\Big]\Big(x - v\dfrac{\Delta t}{2}\Big). \end{cases} \tag{3.12}$$

The following lemma states that this scheme is (locally) third order accurate in time, i.e., (3.4) is satisfied with $r=3$. Readers who are mostly interested in the analysis of adaptive schemes might skip the technical details of the proof.

Lemma 3.7. *If the initial datum f^0 is in $W^{1,\infty}(\Omega)$, we have, for all $n = 0, \dots, N-1$,*

$$\sup_{(x,v)\in\Omega} \big\|\mathcal{F}_{t_n,t_{n+1}}(x,v) - \mathcal{B}[f(t_n)](x,v)\big\| \lesssim (\Delta t)^3 \tag{3.13}$$

with a constant that only depends on f^0 and the final time $T = N\Delta t$.

Proof. For (x, v) fixed in $\mathbb{R}^2$, we set

$$(X,V)(s) := \mathcal{F}_{s,t_{n+1}}(x,v) \quad \text{and} \quad (X^n,V^n) := \mathcal{B}[f(t_n)](x,v).$$

Thus we need to prove that $\max\big(|X^n - X(t_n)|, |V^n - V(t_n)|\big) \le C(\Delta t)^3$. Denoting by $E_X(t) := E(t, X(t))$ the exact field along the characteristic curve, we use Lemma 2.3 together with the characteristic equation (2.12) to bound the following time derivatives (for conciseness, here $\|\cdot\|_\infty$ stands for $\|\cdot\|_{L^\infty((0,T)\times(0,1))}$):

$$\|E_X\|_{L^\infty(0,T)} \le C, \tag{3.14}$$

$$\|\dot{E}_X\|_{L^\infty(0,T)} \le \|\partial_t E\|_\infty + \|V\|_{L^\infty(0,T)}\|\partial_x E\|_\infty \lesssim \|\partial_t E\|_\infty + \|\partial_x E\|_\infty \le C, \tag{3.15}$$

$$\begin{aligned}\|\ddot{E}_X\|_{L^\infty(0,T)} \le{}& \|\partial^2_{tt}E\|_\infty + 2\|V\|_{L^\infty(0,T)}\|\partial^2_{tx}E\|_\infty \\ &+ \|V^2\|_{L^\infty(0,T)}\|\partial^2_{xx}E\|_\infty + \|E\|_\infty\|\partial_x E\|_\infty \le C.\end{aligned} \tag{3.16}$$

We next decompose

$$\begin{aligned} X^n - X(t_n) &= X(t_{n+1}) - X(t_n) - v\Delta t + \frac{(\Delta t)^2}{2}\, E\Big[\mathcal{T}_x^{\frac12} f(t_n)\Big]\Big(x - v\,\frac{\Delta t}{2}\Big) \\ &= \mathcal{E}_1 + \frac{(\Delta t)^2}{2}(\mathcal{E}_2 + \mathcal{E}_3) \end{aligned}$$

with auxiliary terms defined by

$$\begin{aligned} \mathcal{E}_1 &:= X(t_{n+1}) - X(t_n) - v\Delta t + \frac{(\Delta t)^2}{2}\, E_X(t_{n+\frac12}) \quad \text{with} \quad t_{n+\frac12} = \Big(n + \frac12\Big)\Delta t, \\ \mathcal{E}_2 &:= E\Big(t_{n+\frac12}, x - v\,\frac{\Delta t}{2}\Big) - E_X\Big(t_{n+\frac12}\Big), \\ \mathcal{E}_3 &:= E\Big[\mathcal{T}_x^{\frac12} f(t_n)\Big]\Big(x - v\,\frac{\Delta t}{2}\Big) - E\Big(t_{n+\frac12}, x - v\,\frac{\Delta t}{2}\Big). \end{aligned}$$

Similarly, we decompose

$$V^n - V(t_n) = V(t_{n+1}) - V(t_n) - \Delta t\, E\Big[\mathcal{T}_x^{\frac12} f(t_n)\Big]\Big(x - v\,\frac{\Delta t}{2}\Big) = \mathcal{E}_4 + \Delta t(\mathcal{E}_2 + \mathcal{E}_3) \tag{3.17}$$

with $\mathcal{E}_1, \mathcal{E}_2, \mathcal{E}_3$ defined as above and

$$\mathcal{E}_4 := V(t_{n+1}) - V(t_n) - \Delta t E_X(t_{n+\frac12}).$$

It thus remains to establish that $|\mathcal{E}_1|$, $|\mathcal{E}_4| \lesssim (\Delta t)^3$ and $|\mathcal{E}_2|$, $|\mathcal{E}_3| \lesssim (\Delta t)^2$. For the first term, we calculate

$$\mathcal{E}_1 = \int_{t_n}^{t_{n+1}} (V(t) - v)\,\mathrm{d}t + \frac{(\Delta t)^2}{2} E_X(t_{n+\frac{1}{2}}) = \int_{t_n}^{t_{n+1}} \int_t^{t_{n+1}} (E_X(t_{n+\frac{1}{2}}) - E_X(s))\,\mathrm{d}s\,\mathrm{d}t$$

and from (3.15) we have $|E_X(t_{n+1/2}) - E_X(s)| \le |\dot{E}_X|_{L^\infty(0,T)} |t_{n+1/2} - s| \lesssim \Delta t$, hence $|\mathcal{E}_1| \lesssim (\Delta t)^3$. For the second term $\mathcal{E}_2$, we find

$$|\mathcal{E}_2| = \left|E(t_{n+\frac{1}{2}}, x - v\frac{\Delta t}{2}) - E(t_{n+\frac{1}{2}}, X(t_{n+\frac{1}{2}}))\right| \le \|\partial_x E\|_\infty \left|X(t_{n+\frac{1}{2}}) - x + v\frac{\Delta t}{2}\right|$$
$$\lesssim |X(t_{n+\frac{1}{2}}) - X(t_{n+1}) + v\frac{\Delta t}{2}| \lesssim \int_{t_{n+\frac{1}{2}}}^{t_{n+1}} |v - V(t)|\,\mathrm{d}t \lesssim \|E_X\|_{L^\infty(0,T)} (\Delta t)^2 \lesssim (\Delta t)^2,$$

where the last inequality comes from (3.14). According to the definitions of E and $E[\mathcal{T}_x^{1/2} f(t_n)]$, see (2.28) and (3.10), respectively, we next bound the third term $\mathcal{E}_3$ according to

$$|\mathcal{E}_3| = \left| \int_0^1 K\Big(x - v\frac{\Delta t}{2}, y\Big) \int_{\mathbb{R}} \Big[f\Big(t_n, y - \tilde{v}\frac{\Delta t}{2}, \tilde{v}\Big) - f(t_{n+\frac{1}{2}}, y, \tilde{v}) \Big]\,\mathrm{d}\tilde{v}\,\mathrm{d}y \right|$$
$$\le \int_0^1 \left| \int_{\mathbb{R}} \Big[f\Big(t_n, y - \tilde{v}\frac{\Delta t}{2}, \tilde{v}\Big) - f(t_{n+\frac{1}{2}}, y, \tilde{v}) \Big]\,\mathrm{d}\tilde{v} \right| \mathrm{d}y = \int_0^1 \left| \int_{\mathbb{R}} \Phi(y, \tilde{v})\,\mathrm{d}\tilde{v} \right| \mathrm{d}y \tag{3.18}$$

with $\Phi(y, \tilde{v}) := f(t_n, y - \tilde{v}\Delta t/2, \tilde{v}) - f(t_{n+1/2}, y, \tilde{v})$. Setting $t_s := t_n + \Delta t/2 - s$ and $y_s(\tilde{v}) := y - \tilde{v}s$ for concise notations, we then observe that

$$\Phi(y, \tilde{v}) = \int_0^{\frac{\Delta t}{2}} \frac{\mathrm{d}}{\mathrm{d}s} f(t_s, y_s(\tilde{v}), \tilde{v})\,\mathrm{d}s = - \int_0^{\frac{\Delta t}{2}} (\partial_t f + \tilde{v}\partial_x f)(t_s, y_s(\tilde{v}), \tilde{v})\,\mathrm{d}s,$$

hence it follows from the Vlasov equation that $\Phi(y, \tilde{v}) = - \int_0^{\Delta t/2} \Theta(s, y, \tilde{v})\,\mathrm{d}s$ with $\Theta(s, y, \tilde{v}) := E(t_s, y_s(\tilde{v}))\partial_v f(t_s, y_s(\tilde{v}), \tilde{v})$. Now, instead of writing a straightforward bound $|\mathcal{E}_3| \lesssim \Delta t$, which is not sufficient, we integrate by parts by using $(y_s)' = -s$,

$$\int_{\mathbb{R}} s\partial_x E(t_s, y_s(\tilde{v})) f(t_s, y_s(\tilde{v}), \tilde{v})\,\mathrm{d}\tilde{v} = - \int_{\mathbb{R}} \frac{\mathrm{d}}{\mathrm{d}\tilde{v}} [E(t_s, y_s(\tilde{v}))] f(t_s, y_s(\tilde{v}), \tilde{v})\,\mathrm{d}\tilde{v}$$
$$= \int_{\mathbb{R}} E(t_s, y_s(\tilde{v})) \frac{\mathrm{d}}{\mathrm{d}\tilde{v}} [f(t_s, y_s(\tilde{v}), \tilde{v})]\,\mathrm{d}\tilde{v}$$
$$= \int_{\mathbb{R}} E(t_s, y_s(\tilde{v})) \big[(-s\partial_x f + \partial_v f)(t_s, y_s(\tilde{v}), \tilde{v})\big]\,\mathrm{d}\tilde{v},$$

which yields

$$\int_{\mathbb{R}} \Theta(s, y, \tilde{v})\,\mathrm{d}\tilde{v} = s \int_{\mathbb{R}} \Big[\partial_x E(t_s, y_s(\tilde{v})) f(t_s, y_s(\tilde{v}), \tilde{v})$$
$$+ E(t_s, y_s(\tilde{v})) \partial_x f(t_s, y_s(\tilde{v}), \tilde{v}) \Big]\,\mathrm{d}\tilde{v}.$$

It is then possible to get a satisfactory bound for Φ. Indeed, by using Lemma 2.3, we know that the characteristic curves have a bounded support with respect to the velocity, and that E, f are Lipschitz continuous. It follows that

$$\left|\int_{\mathbb{R}} \Phi(y,\tilde{v})\,\mathrm{d}\tilde{v}\right| = \left|\int_{\mathbb{R}}\int_0^{\frac{\Delta t}{2}} \Theta(s,y,\tilde{v})\,\mathrm{d}s\,\mathrm{d}\tilde{v}\right|$$
$$\leq \frac{\Delta t}{2} \sup_{|s|\leq \Delta t/2} \left|\int_{-Q(T)}^{Q(T)} \Theta(s,y,\tilde{v})\,\mathrm{d}\tilde{v}\right| \lesssim (\Delta t)^2,$$

which, together with (3.18), yields $|\mathcal{E}_3| \lesssim (\Delta t)^2$. For the fourth term $\mathcal{E}_4$ in (3.17), we finally obtain, by using (2.12) and the fact that E_X is the field along exact curves,

$$\begin{aligned}
\mathcal{E}_4 &= \left(V(t_{n+1}) - V(t_{n+\frac{1}{2}})\right) + \left(V(t_{n+\frac{1}{2}}) - V(t_n)\right) - \Delta t E_X(t_{n+\frac{1}{2}}) \\
&= \int_0^{\frac{\Delta t}{2}} \left[E_X(t_{n+1}-\tau) + E_X(t_n+\tau)\right]\mathrm{d}\tau - \Delta t E_X(t_{n+\frac{1}{2}}) \\
&= \int_0^{\frac{\Delta t}{2}} \left[E_X(t_{n+1}-\tau) - E_X(t_{n+\frac{1}{2}}) + E_X(t_n+\tau) - E_X(t_{n+\frac{1}{2}})\right]\mathrm{d}\tau \\
&= \int_0^{\frac{\Delta t}{2}} \int_\tau^{\frac{\Delta t}{2}} \left[\dot{E}_X(t_{n+1}-s) - \dot{E}_X(t_n+s)\right]\mathrm{d}s\,\mathrm{d}\tau,
\end{aligned}$$

which yields $|\mathcal{E}_4| \leq (\Delta t)^3 \|\ddot{E}_X\|_{L^\infty(0,T)} \lesssim (\Delta t)^3$ according to (3.16). □

3.2 Uniform interpolation and error analysis

In [2], Nicolas Besse established an a priori error estimate for the first order semi-Lagrangian method in the case where $\mathcal{K} = \mathcal{K}_h$ is a quasi-uniform (say, fixed) mesh of maximal diameter h and where the initial datum f^0 is in $W^{2,\infty}(\Omega)$. The scheme reads then:

$$f_0 := P_h f^0 \qquad \text{and} \qquad f_n := P_h \mathcal{T} f_{n-1} \qquad \text{for} \qquad n = 1,\dots,N,$$

with P_h denoting the associated (continuous, piecewise affine) $\mathcal{P}_1$-interpolation. Higher order schemes were also analyzed by Nicolas Besse and Michel Mehrenberger in [4], but we shall not describe their arguments here. The analysis consists in decomposing the error

$$e_{n+1} := \|f(t_{n+1}) - f_{n+1}\|_{L^\infty(\Omega)}$$

into three parts as follows: a first part

$$e_{n+1,1} := \|f(t_{n+1}) - \mathcal{T} f(t_n)\|_{L^\infty(\Omega)},$$

which corresponds to the approximation of the characteristics by the numerical transport operator $\mathcal{T}$, a second part

$$e_{n+1,2} := \|(I - P_h)\mathcal{T} f(t_n)\|_{L^\infty(\Omega)},$$

which corresponds to the interpolation error, and a third part

$$e_{n+1,3} := \|P_h(\mathcal{T}f(t_n) - \mathcal{T}f_n)\|_{L^\infty(\Omega)},$$

which can be seen as the propagation of the numerical errors from the previous time step. Estimating these three terms involves the properties (such as stability, smoothness and accuracy) of both the approximated characteristics and the interpolations. Let us detail the arguments.

In order to estimate the first term we can rely on the accuracy (3.4). Indeed, since the approximate and exact transport operators are characterized by the corresponding flows, according to $\mathcal{T}f(t_n) = f(t_n) \circ \mathcal{B}[f(t_n)]$ and $f(t_{n+1}) = f(t_n) \circ \mathcal{F}_{t_n,t_{n+1}}$, we have

$$e_{n+1,1} \le |f(t_n)|_{W^{1,\infty}(\Omega)} \sup_{(x,v)\in\Omega} \big\|\mathcal{F}_{t_n,t_{n+1}}(x,v) - \mathcal{B}[f(t_n)](x,v)\big\| \lesssim (\Delta t)^r$$

as long as $f^0 \in W^{1,\infty}(\Omega)$, by using (3.13) (with $r = 3$ if we use the Strang splitting scheme described above).

For the second term, we need interpolation error estimates, and for that purpose we recall a classical result of Jacques Deny and Jacques-Louis Lions, involving the space $\mathcal{P}_m := \mathrm{span}\{x^a y^b : a, b \in \{0, 1, \ldots, m\}, a + b \le m\}$ of polynomials of total degree less or equal to m (see [10, Th. 14.1]).

Theorem 3.8. *For $1 \le p \le \infty$, $m \in \mathbb{N}$ and $A \subset \mathbb{R}^2$ being a bounded connected domain with Lipschitz boundary, the estimate*

$$\inf_{q\in\mathcal{P}_m} \|g - q\|_{L^p(A)} \lesssim |g|_{W^{m+1,p}(A)} \tag{3.19}$$

holds for all $g \in W^{m+1,p}(A)$ with a constant independent of g.

In particular, it is possible to infer a local estimate by using a scaling argument: for any square or triangle K_h of diameter h, we have

$$\inf_{q\in\mathcal{P}_m} \|g - q\|_{L^p(K_h)} \lesssim h^{m+1}|g|_{W^{m+1,p}(K_h)} \tag{3.20}$$

with a constant that depends on the angles of K_h, but not on its diameter. In order to derive an estimate for the $\mathcal{P}_m$-interpolation error in the supremum norm, we next observe that for any $K \in \mathcal{K}_h$, we have $\|P_h g\|_{L^\infty(K)} \lesssim \|g\|_{L^\infty(K)}$ (with constant one in the $\mathcal{P}_1$ case). Hence, for any $p \in \mathcal{P}_m$, we have

$$\|(I - P_h)g\|_{L^\infty(K)} \le \|g - q\|_{L^\infty(K)} + \|P_h(p - g)\|_{L^\infty(K)} \lesssim \|g - q\|_{L^\infty(K)},$$

which, according to (3.20), yields

$$\|(I - P_h)g\|_{L^\infty(K)} \lesssim h^{m+1}|g|_{W^{m+1,\infty}(K)}. \tag{3.21}$$

By using high-order smoothness estimates for the exact and approximate transport operators, which can be established by arguments similar to those in Lemmas 2.3 and 3.4, the second term is then estimated by

$$e_{n+1,2} \lesssim h^2|\mathcal{T}f(t_n)|_{W^{2,\infty}(\Omega)} \lesssim h^2|f(t_n)|_{W^{2,\infty}(\Omega)} \lesssim h^2,$$

as long as $f^0 \in W^{2,\infty}(\Omega)$.

Finally, we observe that the third part $e_{n+1,3}$ is bounded by $\|\mathcal{T}f(t_n) - \mathcal{T}f_n\|_{L^\infty(\Omega)}$; indeed, piecewise affine interpolations never increase the L^∞-norm. Note that if $\mathcal{T}$ was linear, we would clearly have $\|\mathcal{T}f(t_n) - \mathcal{T}f_n\|_{L^\infty(\Omega)} = \|\mathcal{T}(f(t_n) - f_n)\|_{L^\infty(\Omega)} \le e_n$. However, the operator $\mathcal{T}$ is nonlinear but, according to (3.3), it is stable. Thus we have

$$\begin{aligned}
\|\mathcal{T}f(t_n) - \mathcal{T}f_n\|_{L^\infty(\Omega)} &\le \big\|f(t_n) \circ \mathcal{B}[f(t_n)] - f(t_n) \circ \mathcal{B}[f_n]\big\|_{L^\infty(\Omega)} \\
&\quad + \big\|(f(t_n) - f_n) \circ \mathcal{B}[f_n]\big\|_{L^\infty(\Omega)} \\
&\le |f(t_n)|_{W^{1,\infty}(\Omega)} \big\|\mathcal{B}[f(t_n)] - \mathcal{B}[f(t_n)]\big\|_{L^\infty(\Omega)} + e_n \\
&\le (1 + C\Delta t) e_n
\end{aligned}$$

with a constant that only depends on the initial datum f^0 and the final time T. By gathering the above estimates, we find that

$$e_{n+1} \le e_{n+1,1} + e_{n+1,2} + e_{n+1,3} \le (1 + C\Delta t)e_n + C((\Delta t)^r + h^2)$$

holds with a constant depending on f^0 and T only. Therefore, it follows

$$e_n \lesssim (\Delta t)^{r-1} + h^2/\Delta t \qquad \text{for} \quad n = 0, \dots, N,$$

from a discrete Gronwall argument. Balancing $(\Delta t)^r \sim h^2$, we finally observe the following convergence rate in terms of the cardinality $\#(\mathcal{K}_h)$ of the triangulation:

$$\sup_{n=0,\dots,N} \|f(t_n) - f_n\|_{L^\infty(\Omega)} \lesssim h^{2(1-\frac{1}{r})} \lesssim \#(\mathcal{K}_h)^{-(1-\frac{1}{r})}. \tag{3.22}$$

In other words, such a scheme is of global order $1 - 1/r$, at least. Let us recall that such an inequality somehow expresses a trade-off between the *accuracy* of the numerical approximations and their *complexity*, closely related to their computational cost. In particular, it allows

- to impose a maximal cardinality on the meshes, while guaranteeing the accuracy of the interpolations, or
- to prescribe a given accuracy on the interpolations, while giving a complexity bound for the associated meshes.

As we shall see, it is an essential purpose of adaptive strategies to improve the order of convergence. The remaining sections are devoted to describe and further analyze such adaptive variants of the above scheme. In particular, we shall describe in Sections 4 and 5 two distinct frameworks, namely adaptive meshes and interpolatory wavelets, that both have a multilevel tree structure and are suitable for adaptive interpolation. In Section 6, we will give the details for algorithms automatically adapting the interpolation grids to a given function, and accurately transporting, i.e., predicting, these grids along any given smooth flow.

4 Adaptive multilevel meshes

In this section, we describe a simple algorithmic setting for performing adaptive interpolations of finite element type. In order to motivate such constructions, we start by recalling how adaptive strategies can be proven to be more efficient than uniform ones when the unknown solution has a highly non-uniform smoothness (a precise meaning of this statement will be given in the text). Readers interested in more details about adaptive approximation and characterization of convergence rates in terms of smoothness spaces such as Sobolev or Besov spaces are strongly encouraged to read the excellent tutorial article of Ronald DeVore [17] and the book of Albert Cohen [11].

4.1 Why adaptive meshes?

We consider here the problem of interpolating some continuous function g known on the unit square $\Omega = (0,1)^2$. If we desire to use $\mathcal{P}_1$-finite elements, i.e., piecewise affine interpolations on conforming triangulations of Ω, we can think of two different approaches: The first one consists in using – as in the previous section – a sequence of uniform meshes $\mathcal{K}_h$ made of shape-regular triangles of diameter $\mathcal{O}(h)$, i.e., triangles K that contain and that are contained in balls of respective diameter d_K and d'_K such that

$$c_* h \lesssim d_K \leq d'_K \lesssim c^* h$$

with constants independent of h. If g belongs to the space $W^{2,\infty}(\Omega)$, i.e., if it is essentially bounded on Ω and if its first and second order derivatives are also essentially bounded on Ω, we have seen that estimate (3.20) allows to bound the global interpolation error on $\mathcal{K}_h$ by

$$\|(I - P_h)g\|_{L^\infty(\Omega)} \lesssim h^2 |g|_{W^{2,\infty}(\Omega)}. \tag{4.1}$$

Now, it is possible to write a convergence rate associated with this approximation, independently of the transport problem under consideration. Indeed, by using that the cardinality $\#(\mathcal{K}_h)$ is of order h^{-2}, we have

$$\|(I - P_h)g\|_{L^\infty(\Omega)} \lesssim \#(\mathcal{K}_h)^{-1} |g|_{W^{2,\infty}(\Omega)}, \tag{4.2}$$

which yields a convergence rate for uniform $\mathcal{P}_1$-interpolation (see the discussion in the previous section).

Now, to improve the trade-off between accuracy and complexity, a different approach consists in designing a mesh which is locally adapted to the target function g. A way of doing this could be, according to (3.20), to use bigger triangles (hence larger values of h) where g has a small $W^{2,\infty}$-seminorm, and smaller ones elsewhere. Intuitively, this should reduce the cardinality of the triangulation while not increasing much the global interpolation error. A more convenient setting, however, is given by the following local estimate that is substantially stronger than (3.20):

$$\|(I - P_K)g\|_{L^\infty(K)} \lesssim |g|_{W^{2,1}(K)}. \tag{4.3}$$

This estimate is valid with a constant that only depends on the angles of K. The foregoing estimate can be shown by using the continuous embedding of $W^{2,1}(K)$ into

$L^\infty(K)$, see, e.g., [1, Ch. 4], and a scaling argument. Note that the scale invariance, i.e., the fact that the constant does not depend on the diameter of K, corresponds to the fact that the Sobolev embedding is critical, indeed we have $\frac{1}{\infty} = \frac{1}{1} - \frac{2}{d}$ in dimension $d = 2$. According to this estimate, a natural desire is to find a triangulation $\mathcal{K}_\varepsilon$ that equilibrates the local seminorms $|g|_{W^{2,1}(K)}$, in the sense that it satisfies for all $K \in \mathcal{K}_\varepsilon$

$$\underline{c}\varepsilon \le |g|_{W^{2,1}(K)} \le \overline{c}\varepsilon \tag{4.4}$$

with constants $\underline{c}$, $\overline{c}$ independent of h. Clearly, the associated interpolation P_ε would satisfy

$$\|(I - P_\varepsilon)g\|_{L^\infty(\Omega)} \lesssim \varepsilon,$$

and because summing over the left inequalities in (4.4) yields

$$\#(\mathcal{K}_\varepsilon) \le (\underline{c}\varepsilon)^{-1}|g|_{W^{2,1}(\Omega)}, \tag{4.5}$$

the resulting adaptive approximation $(\mathcal{K}_\varepsilon, P_\varepsilon g)$ would achieve the estimate

$$\|(I - P_\varepsilon)g\|_{L^\infty(\Omega)} \lesssim \#(\mathcal{K}_\varepsilon)^{-1}|g|_{W^{2,1}(\Omega)}. \tag{4.6}$$

As this estimate holds for functions which are only in $W^{2,1}(\Omega)$, we see that it indicates better performances in the case where g is not very smooth. More generally, it reveals that such an adaptive approach should outperform a uniform one in the case where g is in $W^{2,\infty}(\Omega)$ but has a highly non-uniform smoothness, i.e., when $|g|_{W^{2,1}(\Omega)}$ is very small compared to $|g|_{W^{2,\infty}(\Omega)}$.

4.2 Multilevel FE-trees and associated quad-meshes

From the above arguments, it appears that an adaptive strategy is likely to yield better results than a uniform one when interpolating functions of non-uniform smoothness. What we did not mention is an algorithm to design a triangulation $\mathcal{K}_\varepsilon$ that fulfills (4.4), and in practice this might be a quite difficult task. For the sake of simplicity, we shall therefore restrict ourselves to a particular class of triangulations that are obtained by recursive splittings of dyadic quadrangles. The resulting multilevel meshes should then be seen as a *compromise* between uniform and pure adaptive triangulations. As is usual in compromises, we need to choose between the two inequalities in (4.4), and since we are first interested in the accuracy of the approximations, we shall choose the one giving an estimate from above. Nevertheless, we mention that such a choice still allows to derive complexity estimates, and we refer again to [17] for a survey on nonlinear (adaptive) and multilevel approximation.

Let us introduce first multilevel *quad-meshes*, and later on derive conforming triangulations. To this end, we consider at any level $j \in \mathbb{N}$ the uniform partitions

$$\mathcal{Q}_j := \{\Omega_\gamma : \gamma \in \mathcal{I}_j\} \quad \text{with} \quad \mathcal{I}_j := \{(j,k,k') : 0 \le k, k' \le 2^j - 1\}$$

consisting of all *dyadic quadrangles*, i.e., quadrangles of the form

$$\Omega_{(j,k,k')} := (2^{-j}k, 2^{-j}(k+1)) \times (2^{-j}k', 2^{-j}(k'+1))$$

that are included in $\Omega = (0,1)^2$. In the sequel we shall denote by $|\gamma| = j$ the level of any index $\gamma \in \mathcal{I}_j$. Since the meshes $\mathcal{Q}_j$ are nested, we can equip the associated index sets with a natural tree structure: we define the *children* of γ as the set

$$\mathcal{C}^*(\gamma) := \{\mu \in \mathcal{I}_{|\alpha|+1} : \Omega_\mu \subset \Omega_\gamma\}$$

(here the superscript * is in order to distinguish this children set from another set that will be introduced in Section 5 when defining trees of dyadic points), and we say that λ is a parent of γ whenever $\gamma \in \mathcal{C}^*(\lambda)$. Obviously, every γ has four children and (as long as $|\gamma| \geq 1$) a unique parent $\mathcal{P}(\gamma)$. In graph theory, a *tree* is a connected acyclic graph, which here corresponds to considering only index sets $\Lambda \subset \mathcal{I}^\infty := \bigcup_{j\geq 0} \mathcal{I}_j$ that contain the parent of any of their elements, i.e., that satisfy

$$\mathcal{P}(\gamma) \in \Lambda \qquad \forall \gamma \in \Lambda. \tag{4.7}$$

We also recall that the (inner) *leaves* of Λ are the nodes with no children in Λ,

$$\mathcal{L}_{\text{in}}(\Lambda) := \{\gamma \in \Lambda : \mathcal{C}^*(\gamma) \cap \Lambda = \emptyset\}. \tag{4.8}$$

In the framework of multilevel meshes associated with finite element type interpolation, we consider the following definition.

Definition 4.1. The set $\Lambda \subset \mathcal{I}^\infty := \bigcup_{j\geq 0} \mathcal{I}_j$ is said to be a *FE-tree* if all its nodes (except the leaves) have exactly four children in Λ, i.e., if

$$\mathcal{C}^*(\gamma) \subset \Lambda \text{ or } \mathcal{C}^*(\gamma) \cap \Lambda = \emptyset \qquad \forall \gamma \in \Lambda. \tag{4.9}$$

Its associated quad-mesh is then defined as

$$M(\Lambda) := \{\Omega_\gamma : \gamma \in \mathcal{L}_{\text{in}}(\Lambda)\}, \tag{4.10}$$

see (4.8).

For later purposes, we will also need that the levels of two adjacent cells in a quad-mesh do not differ too much. This yields the following definition.

Definition 4.2. The FE-tree Λ is said to be *graded* if it satisfies

$$\big||\gamma| - |\mu|\big| \leq 1 \qquad \forall\, \gamma, \mu \in \mathcal{L}_{\text{in}}(\Lambda) \quad \text{with} \quad \overline{\Omega}_\gamma \cap \overline{\Omega}_\mu \neq \emptyset. \tag{4.11}$$

Fundamentally, the tree structure should be seen as a convenient setting for algorithmic refinements: just as refining a cell in a mesh consists in replacing it by its four sub-cells, refining the corresponding node in Λ consists in adding its four children to Λ. To any quad-mesh (made of dyadic quadrangles), we can indeed associate the FE-tree

$$\Lambda(M) := \{\gamma \in \mathcal{I}^\infty : \exists \Omega_\lambda \in M,\ \Omega_\lambda \subset \Omega_\gamma\},$$

the leaves of which clearly coincide with M. Now, the simplest way to build a tree Λ is to recursively add new children to the root tree $\mathcal{I}_0 = \{(0,0,0)\}$, according to some growing criterion. Being interested in $\mathcal{P}_1$-interpolations, a natural criterion would be, according to (4.3), to check whether the local $W^{2,1}$-seminorm of the function g is larger than some prescribed tolerance ε. In the context of interpolating transported numerical solutions however, this is not well-posed since the second derivatives of a piecewise affine function g are not L^1-functions but only Radon measures supported on the edges of the underlying triangulation. Denoting by $\mathcal{M}(A) = (\mathcal{C}_c(A))'$ the set of Radon measures, that is the dual of the space of continuous functions with compact support on the open domain A, and by

$$\int_A \mu := \sup_{\substack{\varphi \in \mathcal{C}_c(A) \\ \|\varphi\|_{L^\infty(A)} \leq 1}} |\mu(\varphi)|$$

the total mass of the measure $\mu \in \mathcal{M}(A)$, we then relax the $W^{2,1}$-seminorm of g and use instead the total mass of its second derivatives,

$$|g|_{W^*(A)} := \int_A \Big(|\partial^2_{xx} g| + |\partial^2_{xv} g| + |\partial^2_{vv} g| \Big).$$

This defines a seminorm for any open domain A and can be extended to any measurable (Borel) set such as a closed polygonal set. Note that in this case it might include non-zero contributions from the edges. Accordingly, we denote by $W^*(\Omega)$ the space of any g such that $|g|_{W^*(\Omega)}$ is finite and we adopt the following definition.

Definition 4.3 (ε-adapted FE-trees and meshes). A mesh M consisting of quadrangles or triangles is said to be ε-*adapted* to g if it satisfies

$$\sup_{K \in M} |g|_{W^*(\overline{K})} \leq \varepsilon,$$

and the FE-tree Λ is said to be ε-adapted to g if its associated mesh $M(\Lambda)$ does so.

In [7], it is shown that the W^*-seminorm satisfies two interesting properties: first, it gives a generalization of the error estimate (4.3), i.e., for any triangle K, we have

$$\|(I - P_K)g\|_{L^\infty(K)} \lesssim |g|_{W^*(K)} \tag{4.12}$$

with a constant depending on the shape of K only, and, second, (up to another relaxation involving a weighted $W^{1,\infty}$-seminorm) the numerical transport operator resulting from the splitting scheme (3.12) can be assumed to be stable with respect to the W^*-seminorm, i.e., we will consider that

$$|g \circ \mathcal{B}[f_n]|_{W^*(A)} \lesssim |g|_{W^*(\mathcal{B}(A))} \tag{4.13}$$

holds with a constant independent of g. For details about the exact stability satisfied by the splitting scheme, we refer to [7].

4.3 Adaptive interpolations based on quad-meshes

In order to perform continuous interpolation (and later we shall focus on first order, i.e., piecewise affine elements), we now associate to any graded quad-mesh M a conforming triangulation $\mathcal{K}(M)$, which is in some sense equivalent to M, see (4.14) below. First, we build a nonconforming triangulation $\tilde{\mathcal{K}}(M)$ by simply splitting each quadrangle K of M into two triangles. More precisely, if K is an upper left or a lower right child (of its parent cell), it is split into its lower left and upper right halves, and the splitting is symmetric in the other two cases. We observe in Figure 2 that, unless M is uniform, the resulting triangulation $\tilde{\mathcal{K}}(M)$ is indeed nonconforming: when a quadrangle K shares an edge with two finer quadrangles K_1 and K_2, the splitting produces one big triangle (say K^-) that shares an edge with two smaller triangles (say K_1^- and K_2^+). Now, because M is graded, this is the only possible configuration where the triangles are nonconforming, and it is easily seen that a conforming triangulation $\mathcal{K}(M)$ is obtained by simply merging any such pair (K_1^-, K_2^+).

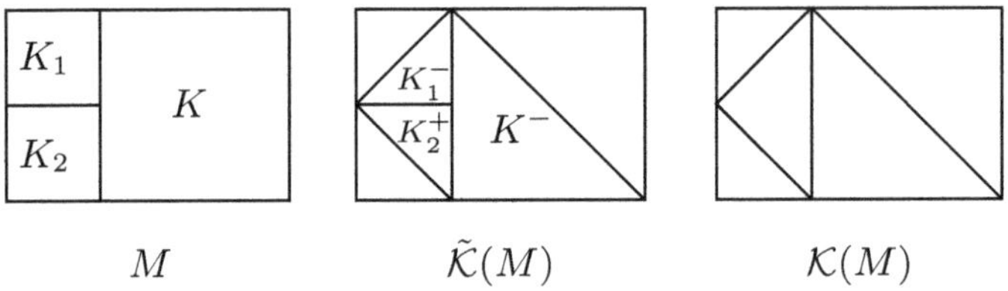

Figure 2. Deriving a conforming triangulation from a graded quad-mesh.

As every quadrangle in M (respectively, every triangle in $\mathcal{K}(M)$) intersects at most two triangles in $\mathcal{K}(M)$ (respectively, two quadrangles in M), we simultaneously have

$$\#(\mathcal{K}(M)) \sim \#(M) \qquad \text{and} \qquad \sup_{K\in\mathcal{K}(M)} |g|_{W^*(K)} \sim \sup_{K\in M} |g|_{W^*(K)} \tag{4.14}$$

for any g in $W^*(\Omega)$. It follows that the piecewise affine interpolation P_Λ associated with any graded FE-tree Λ via the conforming triangulation $\mathcal{K}(M(\Lambda))$ satisfies, in the case where Λ is ε-adapted to g,

$$\|(I - P_\Lambda)g\|_{L^\infty(\Omega)} \lesssim \sup_{K\in\mathcal{K}(M(\Lambda))} |g|_{W^*(K)} \lesssim \sup_{K\in M(\Lambda)} |g|_{W^*(K)} \lesssim \varepsilon. \tag{4.15}$$

5 Interpolatory wavelets

In this section, we shall recall the construction of interpolatory wavelets of arbitrary (even) order $2R$, which relies on a discrete interpolation scheme first introduced by Gilles Deslauriers and Serge Dubuc, see [16]. We also review the main properties of the associated hierarchical basis that will be used later in the analysis of our adaptive semi-Lagrangian scheme. For more information on wavelet constructions, we refer to the books of Ingrid Daubechies [15] and Albert Cohen [11].

For the sake of simplicity, we describe the construction of interpolatory wavelets in the entire $\mathbb{R}^2$.

5.1 A discrete multilevel framework: the iterative interpolation scheme

We first denote the two-dimensional uniform dyadic grids at every level $j \in \mathbb{N}$ by

$$\Gamma_j := \{(2^{-j}k, 2^{-j}k') : k, k' \in \mathbb{Z}\} \subset \Gamma_{j+1} \subset \cdots \subset \mathbb{R}^2,$$

and let

$$\nabla_{j+1} := \Gamma_{j+1} \setminus \Gamma_j$$

be the set of nodes of *level* $j+1$, i.e., appearing in the refinement of Γ_j into Γ_{j+1}. In the sequel, the level of a dyadic cell γ will be denoted by the short notation $|\gamma|$, and the set of all dyadic nodes will be denoted by

$$\Gamma_\infty := \bigcup_{j \geq 0} \Gamma_j.$$

Note that if we let $\nabla_0 := \Gamma_0$, the sets ∇_j, $j \geq 0$, form a partition of Γ_∞.

The next ingredients are inter-grid operators P_j^{j+1}, P_{j+1}^j acting on sequences and standing for restriction and reconstruction, respectively, the general idea being that if the sequences $\mathbf{g}^{[j]} := \{g(\gamma) : \gamma \in \Gamma_j\} \in \ell^\infty(\Gamma_j)$, $j \in \mathbb{N}$, correspond to samples of a given $g \in \mathcal{C}(\mathbb{R}^2)$, the restricted sequences always satisfy

$$P_j^{j+1}\mathbf{g}^{[j+1]} = \mathbf{g}^{[j]},$$

whereas the predicted sequences $P_{j+1}^j\mathbf{g}^{[j]}$ generally differ from $\mathbf{g}^{[j+1]}$ on the finer nodes $\gamma \in \nabla_{j+1}$. On the other hand, the discrepancy should be small in the regions where g is smooth.

In order to be more specific we introduce stencils $S_\gamma \subset \Gamma_j$, associated with nodes of level $|\gamma| = j+1$, as follows (see Figure 3 for an illustration).

- If $\gamma = (2^{-(j+1)}(2k+1), 2^{-j}k')$ corresponds to a refinement of Γ_j in the first dimension, we set

$$S_\gamma := \{(2^{-j}(k+l), 2^{-j}k') : -R+1 \leq l \leq R\}. \tag{5.1}$$

- Similarly, if $\gamma = (2^{-j}k, 2^{-(j+1)}(2k'+1))$ corresponds to a refinement in the second dimension, we set

$$S_\gamma := \{(2^{-j}k, 2^{-j}(k'+l)) : -R+1 \leq l \leq R\}. \tag{5.2}$$

- Finally, if $\gamma = (2^{-(j+1)}(2k+1), 2^{-(j+1)}(2k'+1))$ corresponds to a refinement of Γ_j in both dimensions, we set

$$S_\gamma := \{(2^{-j}(k+l), 2^{-j}(k'+l')) : -R+1 \leq l, l' \leq R\}. \tag{5.3}$$

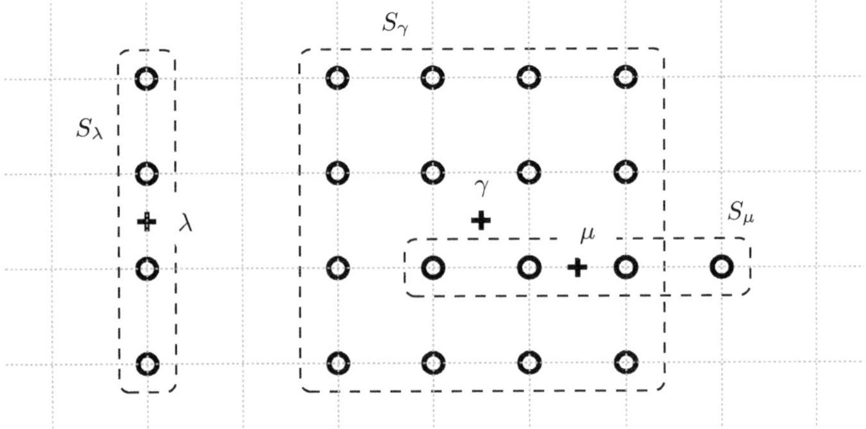

Figure 3. The three kinds of stencils corresponding to (5.1), (5.2) and (5.3), for $R = 2$.

Remark 5.1. Nodes of the latter type will play a particular role in the design of adaptive grids, and will be called **-nodes* in the sequel, see in particular Section 5.5.

The two inter-grid operators are then defined as follows:

- The restriction $P_j^{j+1} : \ell^\infty(\Gamma_{j+1}) \to \ell^\infty(\Gamma_j)$ is a simple decimation, i.e.,

$$(P_j^{j+1}\mathbf{g}^{[j+1]})(\gamma) := \mathbf{g}^{[j+1]}(\gamma) \qquad \forall \gamma \in \Gamma_j.$$

- The prediction $P_{j+1}^j : \ell^\infty(\Gamma_j) \to \ell^\infty(\Gamma_{j+1})$ is a reconstruction given by

$$(P_{j+1}^j\mathbf{g}^{[j]})(\gamma) := \begin{cases} \mathbf{g}^{[j]}(\gamma) & \text{if } \gamma \in \Gamma_j, \\ \sum_{\mu \in S_\gamma} \pi(\gamma,\mu)\mathbf{g}^{[j]}(\mu) & \text{if } \gamma \in \nabla_{j+1} := \Gamma_{j+1} \setminus \Gamma_j \end{cases}$$

for any $\gamma \in \Gamma_{j+1}$. Here the coefficients $\pi(\gamma,\mu)$, $\mu \in S_\gamma$, are defined in such a way that P_{j+1}^j corresponds to Lagrangian interpolation of maximal coordinate degree $2R-1$. More precisely, for any $\mu \in S_\gamma$, we let $L_{\gamma,\mu}$ denote the unique polynomial of $\mathcal{Q}_{2R-1} := \operatorname{span}\{x^a y^b : a, b \in \{0, 1, \dots, 2R-1\}\}$ that satisfies

$$L_{\gamma,\mu}(\lambda) = \delta_{\mu,\lambda} \qquad \forall \lambda \in S_\gamma$$

(where δ stands for the Kronecker symbol), and is constant with respect to x or y in the case where S_γ is given by (5.1) or (5.2), respectively. Finally, we set

$$\pi(\gamma,\mu) := L_{\gamma,\mu}(\gamma).$$

Remark 5.2. By using the shift invariance and self-similarity of the dyadic grids, we can check that the value of $\pi(\gamma,\mu)$ only depends on the relative positions of γ and μ. More precisely, for any R there exists a sequence $(h_{n,n'})_{n,n' \in \mathbb{Z}}$ such that

$$\pi(\gamma,\mu) = h_{k-2i,k'-2i'} \quad \text{for } \gamma = (2^{-(j+1)}k, 2^{-(j+1)}k'),\ \mu = (2^{-j}i, 2^{-j}i').$$

Moreover, we note that $h_{n,n'} = 0$ for $\max(|n|,|n'|) \geq 2R$, and it follows that the prediction coefficients $\pi(\gamma,\mu)$ are bounded by a constant that only depends on R.

Now, as previously announced, we see that by restricting and further reconstructing samples at a given level one looses information, i.e., $P^j_{j+1}P^{j+1}_j \mathbf{g}^{[j+1]} = P^j_{j+1}\mathbf{g}^{[j]}$ generally differs from $\mathbf{g}^{[j+1]}$ on the finest nodes $\gamma \in \nabla_{j+1}$. The prediction error is then stored in a sequence

$$\mathbf{d}^{[j+1]} := \mathbf{g}^{[j+1]} - P^j_{j+1}\mathbf{g}^{[j]} \equiv \{d_\gamma(g) := (\mathbf{g}^{[j+1]} - P^j_{j+1}\mathbf{g}^{[j]})(\gamma) : \gamma \in \nabla_{j+1}\},$$

and introducing the (Radon) measures $\tilde{\varphi}_\gamma := \delta_\gamma - \sum_{\mu\in S_\gamma} \pi(\gamma,\mu)\delta_\mu$, every $d_\gamma(g)$ rewrites as

$$d_\gamma(g) = g(\gamma) - \sum_{\mu\in S_\gamma} \pi(\gamma,\mu)g(\mu) = \langle\tilde{\varphi}_\gamma, g\rangle. \tag{5.4}$$

In the wavelet terminology these coefficients are called *details*, as they are seen as the additional information needed to recover the exact values of g from a coarse sampling. Intuitively, one would expect these coefficients to be small in the regions were g is smooth, and indeed it is easy to write a rigorous estimate: By observing that the prediction is exact for polynomials of order $2R$, i.e., we have $\langle\tilde{\varphi}_\gamma, p\rangle = 0$ for any $p \in \mathcal{Q}_{2R-1}$, we can exploit the bound on the coefficients $\pi(\gamma,\mu)$ (see Remark 5.2) together with the fact that every $\tilde{\varphi}_\gamma$ vanishes outside

$$\Sigma_\gamma := \overline{B}_{\ell^\infty}(\gamma, 2^{-|\gamma|}(2R-1)). \tag{5.5}$$

This yields (with constants that only depend on R)

$$|d_\gamma(g)| \leq \inf_{p\in\mathcal{Q}_{2R-1}} |\langle\tilde{\varphi}_\gamma, g-p\rangle| \lesssim \inf_{p\in\mathcal{Q}_{2R-1}} \|g-p\|_{L^\infty(\Sigma_\gamma)} \lesssim 2^{-\nu|\gamma|}|g|_{W^{\nu,\infty}(\Sigma_\gamma)} \tag{5.6}$$

for any integer $\nu \leq 2R$, where the third inequality follows from the Deny–Lions theorem, see (3.20). Note that it is also possible to estimate the details by using the modulus of smoothness of g already introduced in Section 3.1. According to a local variant of a theorem by Hassler Whitney (see, e.g., [5] or [11]), we have indeed

$$|d_\gamma(g)| \lesssim \inf_{p\in\mathcal{Q}_{2R-1}} \|g-p\|_{L^\infty(\Sigma_\gamma)} \lesssim \omega_\nu(g, 2^{-|\gamma|}, \Sigma_\gamma)_\infty \tag{5.7}$$

with again $\nu \leq 2R$, see (3.9).

From the above iterative interpolation scheme, it is possible to define a hierarchical wavelet basis for the full space $\mathcal{C}(\mathbb{R}^2)$ in which the details $d_\gamma(g)$ will play the role of the coefficients of g. In order to make this statement more precise, we introduce for any j and $\gamma \in \Gamma_j$ a sequence (of sequences) $\phi^{[j']}_{j,\gamma} \in \ell^\infty(\Gamma_{j'})$, $j' \geq j$, defined by

$$\phi^{[j]}_{j,\gamma}(\lambda) := \delta_{\gamma,\lambda} \quad \forall\lambda \in \Gamma_j \qquad \text{and} \qquad \phi^{[j'+1]}_{j,\gamma} := P^{j'}_{j'+1}\phi^{[j']}_{j,\gamma} \quad \forall j' \geq j. \tag{5.8}$$

Note that by definition of the prediction operator, we have

$$\phi^{[j'']}_{j,\gamma}(\lambda) = \phi^{[j']}_{j,\gamma}(\lambda) \qquad \forall\lambda \in \Gamma_{j'} \text{ and } j'' \geq j' \geq j,$$

therefore the above process essentially consists in refining growing sets of values. Now, as we will see in the next section (and as is well known), for any j and $\gamma \in \Gamma_j$ this process converges towards a continuous function $\varphi_{j,\gamma} : \mathbb{R}^2 \to \mathbb{R}$ in the sense that

$$\phi_{j,\gamma}^{[j']}(\lambda) = \varphi_{j,\gamma}(\lambda) \qquad \forall j' \geq j,\ \lambda \in \Gamma_{j'}. \tag{5.9}$$

Moreover the limit functions span nested spaces

$$V_j := \operatorname{span}\{\varphi_{j,\gamma} : \gamma \in \Gamma_j\} \subset V_{j+1} \subset \cdots \tag{5.10}$$

which have a dense union in $\mathcal{C}(\mathbb{R}^2)$, and by keeping only the functions of the type

$$\varphi_\gamma := \varphi_{|\gamma|,\gamma}, \qquad \gamma \in \Gamma_\infty,$$

one obtains as announced a hierarchical basis of $\mathcal{C}(\mathbb{R}^2)$, i.e., every continuous function g reads as

$$g = \sum_{\gamma \in \Gamma_{j_0}} g(\gamma)\varphi_\gamma + \sum_{j \geq j_0+1} \sum_{\gamma \in \Gamma_j} d_\gamma(g)\varphi_\gamma, \tag{5.11}$$

for any $j_0 \in \mathbb{N}$, the convergence of the infinite sum holding in a pointwise sense.

5.2 Convergence of the iterative interpolation scheme

We shall now give a proof of the above claims (5.9), (5.10) (and defer (5.11) to the next section). As (5.8) is a particular instance of what is referred to as *stationary subdivision schemes*, its properties can be analyzed by using general tools such as those presented in the review articles [8] and [18]. In [15] and [11], the connections between wavelets and subdivision schemes are investigated in more details. Here we shall adapt the arguments given in [11] for our particular case of interest.

To begin with, we observe that it suffices to consider the case where $j = 0$ and $\gamma = 0$. Indeed, by using the shift invariance and the self-similarity of the dyadic grids, we can check that for all j, j' with $0 \leq j \leq j'$ and all γ, λ in Γ_j, $\Gamma_{j'}$, respectively, we have

$$\phi_{j,\gamma}^{[j']}(\lambda) = \phi_{j,0}^{[j']}(\lambda - \gamma) = \phi_{j-1,0}^{[j'-1]}(2(\lambda - \gamma)) = \cdots = \phi_{0,0}^{[j'-j]}(2^j(\lambda - \gamma)).$$

Hence the convergence of $(\phi_{j,\gamma}^{[j']})_{j' \geq j}$ will follow from that of $(\phi_{0,0}^{[j']})_{j' \geq 0}$. Moreover,

$$\varphi_{j,\gamma}(x) = \varphi_{j,0}(x - \gamma) = \varphi_{j-1,0}(2(x - \gamma)) = \cdots = \varphi_{0,0}(2^j(x - \gamma)) \tag{5.12}$$

will be established for all $x \in \mathbb{R}^2$. In the sequel, we shall drop the subscripts $_{0,0}$.

In a nutshell, it is possible to establish the convergence of (5.8) with two arguments: the first one consists in saying that it is equivalent to the existence of a continuous function satisfying an appropriate two-scale equation, and the second one in proving that such a *scaling function* indeed exists (and in this case it is the limit of the scheme). For the sake of simplicity, we shall give a detailed proof in the one-dimensional case (gathering arguments from [11]) and leave the two-dimensional case for the reader. Then the prediction operator is defined by univariate Lagrange interpolations, and the stencil corresponding to a node of level $j+1$, say $\gamma = 2^{-(j+1)}(2k+1)$, reads

$$S_\gamma = \{2^{-j}(k+l) : -R+1 \le l \le R\}.$$

In particular, for any integer k, we have (writing $\phi^{[j]} = \phi^{[j]}_{0,0}$)

$$\phi^{[j+1]}(2^{-(j+1)}(2k+1)) = \sum_{i=k-R+1}^{k+R} \pi(2^{-(j+1)}(2k+1), 2^{-j}i)\phi^{[j]}(2^{-j}i) \tag{5.13}$$

whereas the values remain unchanged on nodes of level j, i.e.,

$$\phi^{[j+1]}(2^{-j}k) := \phi^{[j]}(2^{-j}k). \tag{5.14}$$

As in Remark 5.2, we then observe that there exist coefficients h_n, $n \in \mathbb{Z}$, such that $\pi(2^{-(j+1)}k, 2^{-j}i) = h_{k-2i}$ for any odd integer k, so that by setting the remaining (even) values to $h_{2n} := \delta_{n,0}$, our iterative scheme reads

$$\phi^{[j+1]}(2^{-(j+1)}k) = \sum_{i\in\mathbb{Z}} h_{k-2i}\phi^{[j]}(2^{-j}i) \qquad \forall k \in \mathbb{Z},\ j \in \mathbb{N}. \tag{5.15}$$

Remark 5.3. Before going further, let us list a few important properties of the sequence $(h_n)_{n\in\mathbb{Z}}$ that will be of great importance in the sequel. From the fact that the prediction stencils are symmetric and local, one easily infers that h shares the same properties: its only non-zero terms are $h_0 = 1$ and $h_{-1} = h_1, \dots, h_{-(2R-1)} = h_{2R-1}$. Moreover, it is possible to express the polynomial reproduction properties of the iterative scheme in terms of discrete moments of h. As already observed, defining $\mathbf{g}^{[0]}$ as samples of $p_l(x) := (2x+1)^l$ indeed leads to samples of the same polynomial for $P_1^0\mathbf{g}^{[0]}$, as long as $l \le 2R-1$. By linearity of P_1^0, this gives

$$\sum_{n\in\mathbb{Z}} h_{2n+1}(2n+1)^l = \sum_{n\in\mathbb{Z}} h_{-1-2n}\mathbf{g}^{[0]}(n) = P_1^0\mathbf{g}^{[0]}\Big(-\frac{1}{2}\Big) = p_l\Big(-\frac{1}{2}\Big) = \delta_{l,0} \tag{5.16}$$

for $l = 0, \dots, 2R-1$. As $h_{2n} = \delta_{n,0}$, the left hand side is exactly $\sum_{n\in\mathbb{Z}} h_n n^l$.

The connection with the scaling function $\varphi = \varphi_{0,0}$ can then be stated as follows.

Lemma 5.4. *Let $(h_n)_{n\in\mathbb{Z}}$ be a sequence of real coefficients. If there exists a continuous function φ satisfying $\varphi(k) = \delta_{k,0}$, $k \in \mathbb{Z}$, and the two-scale equation*

$$\varphi(x) = \sum_{n\in\mathbb{Z}} h_n\varphi(2x-n) \qquad \forall x \in \mathbb{R}, \tag{5.17}$$

then the iterative scheme defined by the refinement rule (5.15) *and the initial condition $\phi^{[0]}(k) = \delta_{k,0}$, $k \in \mathbb{Z}$, satisfies* (5.14) *and converges towards φ. Conversely, if the above iterative scheme satisfies* (5.14) *and if it converges towards a continuous function φ, then the latter satisfies $\varphi(k) = \delta_{k,0}$, $k \in \mathbb{Z}$, as well as* (5.17).

Proof. If φ satisfies the two-scale equation (5.17), then by induction we see that it belongs to $\operatorname{span}\{x \mapsto \varphi(2^j x - n) : n \in \mathbb{Z}\}$, for any $j \in \mathbb{N}$. In particular, there exist coefficients $c_{j,n}$ such that $\varphi(x) = \sum_{n\in\mathbb{Z}} c_{j,n}\varphi(2^j x - n)$ for all $x \in \mathbb{R}$, and by using that $\varphi(k) = \delta_{k,0}$ for $k \in \mathbb{Z}$, we find $c_{j,k} = \sum_{n\in\mathbb{Z}} c_{j,n}\varphi(k-n) = \varphi(2^{-j}k)$. Therefore, we have

$$\varphi(x) = \sum_{n\in\mathbb{Z}} \varphi(2^{-j}n)\varphi(2^j x - n) \qquad \forall x \in \mathbb{R},\ j \in \mathbb{N}. \tag{5.18}$$

Now, assume that $\varphi(2^{-j}k) = \phi^{[j]}(2^{-j}k)$, $k \in \mathbb{Z}$, holds for a given $j \in \mathbb{N}$, as it is the case for $j = 0$. By inserting the two-scale relation (5.17) into (5.18), we find

$$\begin{aligned}
\varphi(x) &= \sum_{n\in\mathbb{Z}} \phi^{[j]}(2^{-j}n)\varphi(2^j x - n) \\
&= \sum_{n\in\mathbb{Z}} \phi^{[j]}(2^{-j}n) \sum_{n'\in\mathbb{Z}} h_{n'}\varphi(2^{j+1}x - 2n - n') \\
&= \sum_{k\in\mathbb{Z}} \Big(\sum_{n\in\mathbb{Z}} h_{k-2n}\phi^{[j]}(2^{-j}n)\Big)\varphi(2^{j+1}x - k) \\
&= \sum_{k\in\mathbb{Z}} \phi^{[j+1]}(2^{-(j+1)}k)\varphi(2^{j+1}x - k).
\end{aligned}$$

According to (5.18), this yields $\varphi(2^{-(j+1)}k) = \phi^{[j+1]}(2^{-(j+1)}k)$ for all $k \in \mathbb{Z}$, hence the convergence follows. In particular, note that $\phi^{[j+1]}(2^{-j}k) = \varphi(2^{-j}k) = \phi^{[j]}(2^{-j}k)$ for all $k \in \mathbb{Z}$, which is (5.14).

In the other direction, we first observe that the convergence of $\phi^{[j]}$ towards φ reads $\varphi(2^{-j}k) = \phi^{[j]}(2^{-j}k)$ for all $k \in \mathbb{Z}$ and $j \in \mathbb{N}$. In particular, we have $\varphi(k) = \delta_{k,0}$ for all $k \in \mathbb{Z}$, and

$$\varphi(2^{-1}k) = \phi^{[1]}(2^{-1}k) = \sum_{i\in\mathbb{Z}} h_{k-2i}\phi^{[0]}(i) = h_k = \sum_{n\in\mathbb{Z}} h_n\varphi(k-n),$$

which shows that the two-scale relation (5.17) holds for any $x \in \Gamma_1$. Now, by assuming that it holds for any $x \in \Gamma_j$, i.e.,

$$\phi^{[j]}(2^{-j}i) = \sum_{n\in\mathbb{Z}} h_n\phi^{[j-1]}(2^{-(j-1)}i - n) \qquad \forall i \in \mathbb{Z},$$

and by applying the latter to (5.15), we find

$$\begin{aligned}
\phi^{[j+1]}(2^{-(j+1)}k) &= \sum_{i\in\mathbb{Z}} h_{k-2i} \sum_{n\in\mathbb{Z}} h_n\phi^{[j-1]}(2^{-(j-1)}i - n) \\
&= \sum_{n\in\mathbb{Z}} h_n \sum_{m\in\mathbb{Z}} h_{(k-2^j n)-2m}\phi^{[j-1]}(2^{-(j-1)}m) \\
&= \sum_{n\in\mathbb{Z}} h_n\phi^{[j]}(2^{-j}(k - 2^j n)) = \sum_{n\in\mathbb{Z}} h_n\phi^{[j]}(2^{-j}k - n),
\end{aligned}$$

where we have used (5.15) again in the third equality. This shows that (5.17) also holds for all $x \in \Gamma_{j+1}$ ($j \in \mathbb{N}$), and thus, by density, for all $x \in \mathbb{R}$ since φ is continuous. □

Remark 5.5. By using $\phi^{[0]}(k) = \delta_{k,0}$, one easily checks that (5.14) is equivalent with

$$h_{2n} = \delta_{n,0} \qquad \forall n \in \mathbb{Z}, \tag{5.19}$$

which can also be inferred from the two-scale relation (5.17) and $\varphi(k) = \delta_{k,0}$.

It thus remains to prove that such a scaling function indeed exists. For that purpose, we first observe that φ is a continuous solution of (5.17) if and only if its Fourier transform $\hat{\varphi} : \omega \mapsto \int_{\mathbb{R}} \varphi(x) e^{-ix\omega}\,\mathrm{d}x$ is a $L^1(\mathbb{R})$-function satisfying

$$\hat{\varphi}(\omega) = m\Big(\frac{\omega}{2}\Big)\hat{\varphi}\Big(\frac{\omega}{2}\Big), \tag{5.20}$$

where the trigonometric polynomial $m(\omega) := \frac{1}{2}\sum_{n\in\mathbb{Z}} h_n e^{-in\omega}$ is sometimes called the *symbol* of φ (remember that the sequence $(h_n)_{n\in\mathbb{Z}}$ is finite), and is non-negative on $\mathbb{R}$, as we shall see soon.

Formally, (5.20) gives rise to consider $\hat{\varphi}(\omega) := \prod_{j=1}^{\infty} m(2^{-j}\omega)$. In order to make this rigorous, let $\hat{\varphi}_J(\omega) := \big[\prod_{j=1}^{J} m(2^{-j}\omega)\big]\chi_{[-\pi,\pi]}(2^{-J}\omega)$ and observe that (5.19) yields

$$m(\omega) + m(\omega+\pi) = \frac{1}{2}\sum_{n\in\mathbb{Z}} h_n e^{-in\omega}(1+(-1)^n) = \sum_{n\in\mathbb{Z}} h_{2n} = 1. \tag{5.21}$$

By using this identity together with the periodicity of m, following [11] we calculate

$$\begin{aligned}
\int_{\mathbb{R}} \hat{\varphi}_J &= \int_{-2^J\pi}^{2^J\pi} \Big[\prod_{j=1}^{J} m(2^{-j}\omega)\Big]\,\mathrm{d}\omega = 2^J \int_{-\pi}^{\pi} \Big[\prod_{j'=0}^{J-1} m(2^{j'}\xi)\Big]\,\mathrm{d}\xi \\
&= 2^J \int_0^{\pi} (m(\xi)+m(\xi+\pi))\Big[\prod_{j'=1}^{J-1} m(2^{j'}\xi)\Big]\,\mathrm{d}\xi = 2^J \int_0^{\pi} \Big[\prod_{j'=1}^{J-1} m(2^{j'}\xi)\Big]\,\mathrm{d}\xi \\
&= 2^J \int_{-\pi/2}^{\pi/2} \Big[\prod_{j'=1}^{J-1} m(2^{j'}\xi)\Big]\,\mathrm{d}\xi = \int_{-2^{J-1}\pi}^{2^{J-1}\pi} \Big[\prod_{j=1}^{J-1} m(2^{-j}\omega)\Big]\,\mathrm{d}\omega = \int_{\mathbb{R}} \hat{\varphi}_{J-1} \\
&= \cdots = \int_{\mathbb{R}} \hat{\varphi}_1 = 2\int_{-\pi}^{-\pi} m(\omega)\,\mathrm{d}\omega = 2\pi.
\end{aligned}$$

Now, if we have $m(\omega) \geq 0$ as claimed above, the latter equality shows that $\hat{\varphi}_J$ is uniformly bounded in $L^1(\mathbb{R})$, and it is an easy matter to check that $\hat{\varphi}_J$ converges towards $\hat{\varphi}$ in a pointwise sense. It thus follows by using Fatou's lemma that $\hat{\varphi} \in L^1(\mathbb{R})$, which establishes the continuity of its Fourier transform φ and the two-scale equation (5.17) follows. According to Lemma 5.4 this proves the convergence of the iterative interpolation scheme.

Therefore, it only remains to verify the claim $m(\omega) \geq 0$, and for this purpose let us show that there exists a polynomial P_R of degree not larger than $R-1$ such that

$$m(\omega) = \Big(\cos^2\frac{\omega}{2}\Big)^R P_R\Big(\sin^2\frac{\omega}{2}\Big). \tag{5.22}$$

Indeed, according to Remark 5.3 we can write

$$m(\omega) = \frac{1}{2}\Big(h_0 + 2\sum_{n\geq 1}^{2R-1} h_n \cos(n\omega)\Big) = Q\Big(\cos^2\frac{\omega}{2}\Big)$$

with a polynomial Q of degree not larger than $2R-1$, as we know that

$$\cos(n\omega) = \text{Re}\left[(e^{i\frac{\omega}{2}})^{2n}\right] = \text{Re}\left[\sum_{k=0}^{2n}\binom{2n}{k}\big(i\sin\frac{\omega}{2}\big)^k\big(\cos\frac{\omega}{2}\big)^{2n-k}\right]$$
$$= \sum_{k=0}^{n}\binom{2n}{2k}\big(\cos^2\frac{\omega}{2}-1\big)^k\big(\cos^2\frac{\omega}{2}\big)^{n-k}.$$

Now, from Remark 5.3 we can also infer that

$$m^{(l)}(0) = \frac{1}{2}\sum_{n\in\mathbb{Z}}(-in)^l h_n = \frac{(-i)^l}{2}\Big(\delta_{l,0} + \sum_{n\in\mathbb{Z}}(2n+1)^l h_{2n+1}\Big) = \delta_{l,0}$$

for $l = 0,\dots,2R-1$. In particular, it follows by differentiating (5.21) that $m = m(\omega)$ vanishes with order $2R-1$ at $\omega = \pi$, which in turn implies that $Q(X^2)$ can be factorized by X^{2R}, and this establishes (5.22). In order to identify P_R, we next observe that equation (5.21) leads to see it as a polynomial solution to

$$X^R P(1-X) + (1-X)^R P(X) = 1. \tag{5.23}$$

For such an equation, Bézout's theorem ensures the existence of a minimal degree solution, but we shall give a direct argument: Indeed, we have

$$1 = (X+(1-X))^{2R-1} = \sum_{k=0}^{2R-1}\binom{2R-1}{k}X^k(1-X)^{R-1-k}$$
$$= (1-X)^R\sum_{k=0}^{R-1}\left[\binom{2R-1}{k}X^k(1-X)^{R-1-k}\right]$$
$$+ X^R\sum_{k=0}^{R-1}\left[\binom{2R-1}{k}X^{R-1-k}(1-X)^k\right],$$

so that $P(X) := \sum_{k=0}^{R-1}\binom{2R-1}{k}X^k(1-X)^{R-1-k}$ is a solution to (5.23) of degree $R-1$ or lower. Since such a solution is clearly unique, it coincides with P_R and the non-negativity of $m(\omega)$ follows.

In conclusion, we have proved (5.9) and (5.10) as well, since it follows from (5.12) and (5.17) that any $\varphi_{j,\gamma}$ can be written as a linear combination of $\varphi_{j+1,\lambda}$, $\lambda \in \Gamma_{j+1}$.

5.3 The hierarchical wavelet basis

In order to establish the validity of the hierarchical decomposition (5.11), we now study some properties of the scaling functions $\varphi_{j,\gamma}$. According to the previous section, it is easily seen that they are interpolatory in the sense that

$$\varphi_{j,\gamma}(\lambda) = \delta_{\gamma,\lambda}, \quad \forall\gamma,\lambda \in \Gamma_j. \tag{5.24}$$

In particular they form a nodal basis for the space $V_j := \operatorname{span}\{\varphi_{j,\gamma} : \gamma \in \Gamma_j\}$, in the sense that every $g \in V_j$ reads as $g = \sum_{\gamma\in\Gamma_j} g(\gamma)\varphi_{j,\gamma}$.

Remark 5.6 (Polynomial exactness). By using the polynomial reproduction properties of the linear prediction operator (as, for instance, in Remark 5.3) one easily sees that polynomials of coordinate degree less than $2R$ belong to the space V_0 (and hence to any V_j, $j \geq 0$).

Let us now estimate the support of $\varphi_{j,\gamma}$ from the locality of the prediction stencils. To do so, we set

$$\Sigma^{[j]}_{j,\gamma} := \{\gamma\} \quad \text{and} \quad \Sigma^{[\ell+1]}_{j,\gamma} := \{\lambda \in \Gamma_{\ell+1} : (\{\lambda\} \cup S_\lambda) \cap \Sigma^{[\ell]}_{j,\gamma} \neq \emptyset\} \quad \text{for } \ell \geq j.$$

Clearly, the sequence $(\Sigma^{[\ell]}_{j,\gamma})_{\ell\geq j}$ is increasing in the sense that $\Sigma^{[\ell]}_{j,\gamma} \subset \Sigma^{[\ell+1]}_{j,\gamma}$ for all $\ell \geq j$, and, by using the size of the stencils, we can check that $\Sigma^{[j+i]}_{j,\gamma} \subset B_{\ell^\infty}(\gamma, r_i)$ with $r_i = 2^{-j}(R - 1/2)(1 + 2^{-1} + \cdots + 2^{-i})$ for $i \geq 0$, hence every $\Sigma^{[\ell]}_{j,\gamma}$ is a subset of $B_{\ell^\infty}(\gamma, 2^{-j}(2R-1))$. More precisely, we observe that

$$\Sigma^{[\ell]}_{j,\gamma} = \Big(\{\gamma\} \cup \bigcup_{i=j+1}^{\ell} \nabla_i\Big) \cap B_{\ell^\infty}(\gamma, 2^{-j}(2R-1)). \tag{5.25}$$

Therefore, the density of the dyadic points yields

$$\Sigma_{j,\gamma} := \overline{\bigcup_{\ell\geq j} \Sigma^{[\ell]}_{j,\gamma}} = \overline{B}_{\ell^\infty}(\gamma, 2^{-j}(2R-1)). \tag{5.26}$$

In particular, we have $\Sigma_{|\gamma|,\gamma} = \Sigma_\gamma$, see (5.5). Since $\operatorname{supp}(\phi^{[\ell]}_{j,\gamma}) \subset \Sigma^{[\ell]}_{j,\gamma}$ by construction of the sets $\Sigma^{[\ell]}_{j,\gamma}$, it follows that $\operatorname{supp}(\varphi_{j,\gamma}) \subset \Sigma_{j,\gamma}$. Hence the functions $\varphi_{j,\gamma}$, $\gamma \in \Gamma_j$, satisfy a bounded overlapping property, namely

$$\#(\{\lambda \in \Gamma_j : \operatorname{supp}(\varphi_{j,\gamma}) \cap \operatorname{supp}(\varphi_{j,\lambda}) \neq \emptyset\}) \leq C_s \quad \forall\gamma \in \Gamma_j \tag{5.27}$$

with a constant $C_s = C_s(R)$ independent of j. One major consequence of this fact is that the basis $\{\varphi_{j,\gamma} : \gamma \in \Gamma_j\}$ is *stable* in the sense that

$$c\|g\|_{L^\infty(\mathbb{R}^2)} \leq \sup_{\gamma\in\Gamma_j} |g(\gamma)| \leq \|g\|_{L^\infty(\mathbb{R}^2)} \qquad \forall g = \sum_{\gamma\in\Gamma_j} g(\gamma)\varphi_{j,\gamma} \in V_j \tag{5.28}$$

holds with $c = (C_s \|\varphi\|_{L^\infty(\mathbb{R}^2)})^{-1}$, see (5.12). Note that it is also possible to write a so-called Bernstein (i.e., inverse) estimate for the functions in V_j, according to

$$\Big| \sum_{\gamma \in \Gamma_j} c_\gamma \varphi_{j,\gamma} \Big|_{W^{\nu,\infty}(\mathbb{R}^2)} \le C_s \sup_{\gamma \in \Gamma_j} |c_\gamma \varphi_{j,\gamma}|_{W^{\nu,\infty}(\mathbb{R}^2)} \le 2^{\nu j} C \sup_{\gamma \in \Gamma_j} |c_\gamma| \tag{5.29}$$

with $C = C_s |\varphi|_{W^{\nu,\infty}(\mathbb{R}^2)}$. On every space V_j, we next define a linear projector P_j by $P_j g := \sum_{\gamma \in \Gamma_j} g(\gamma)\varphi_{j,\gamma}$ which, due to (5.24), is an interpolation. Moreover, P_j is uniformly stable in the sense that

$$\|P_j g\|_{L^\infty(\mathbb{R}^2)} \lesssim \sup_{\gamma \in \Gamma_j} |g(\gamma)| \le \|g\|_{L^\infty(\mathbb{R}^2)} \quad \forall g \in \mathcal{C}(\mathbb{R}^2) \tag{5.30}$$

holds with a constant independent of j, by using (5.28). Note that the stability is local, i.e., for any cell $\Omega_{j,\gamma} := B_{\ell^\infty}(\gamma, 2^{-j})$ we have

$$\|P_j g\|_{L^\infty(\Omega_{j,\gamma})} \le \sum_{\mu \in \widetilde{\Omega}_{j,\gamma}} \|g(\mu)\varphi_{j,\mu}\|_{L^\infty(\mathbb{R}^2)} \lesssim \sup_{\mu \in \widetilde{\Omega}_{j,\gamma}} |g(\mu)| \lesssim \|g\|_{L^\infty(\widetilde{\Omega}_{j,\gamma})} \tag{5.31}$$

with $\widetilde{\Omega}_{j,\gamma} := \{\mu :\in \Gamma_j, \Sigma_{j,\mu} \cap \Omega_{j,\gamma} \ne \emptyset\}$. By using arguments similar to those in Section 3.2, we can establish an error estimate for this uniform interpolation.

Lemma 5.7. *For any $g \in W^{\nu,\infty}(\mathbb{R}^2)$ and any integer $\nu \le 2R$, we have*

$$\|g - P_j g\|_{L^\infty(\mathbb{R}^2)} \lesssim 2^{-\nu j} |g|_{W^{\nu,\infty}(\mathbb{R}^2)} \tag{5.32}$$

with a constant independent of j.

Proof. For any $\gamma \in \Gamma_j$, let $p_{j,\gamma}$ be a polynomial in $\mathcal{Q}_{2R-1}$ such that

$$\|g - p_{j,\gamma}\|_{L^\infty(\widetilde{\Omega}_{j,\gamma})} \le 2 \inf_{p \in \mathcal{Q}_{2R-1}} \|g - p\|_{L^\infty(\widetilde{\Omega}_{j,\gamma})} \lesssim 2^{-\nu j} |g|_{W^{\nu,\infty}(\widetilde{\Omega}_{j,\gamma})},$$

where the second inequality follows from the Deny–Lions theorem, see (3.20). By using the fact that polynomials of degree less than $2R$ are contained in every V_j (see Remark 5.6), the local error is then estimated by

$$\begin{aligned} \|g - P_j g\|_{L^\infty(\Omega_{j,\gamma})} &\le \|g - p_{j,\gamma}\|_{L^\infty(\Omega_{j,\gamma})} + \|P_j(p_{j,\gamma} - g)\|_{L^\infty(\Omega_{j,\gamma})} \\ &\lesssim \|g - p_{j,\gamma}\|_{L^\infty(\Omega_{j,\gamma})} \lesssim 2^{-\nu j} |g|_{W^{\nu,\infty}(\widetilde{\Omega}_{j,\gamma})}, \end{aligned}$$

and the global estimate (5.32) follows by taking the supremum over $\gamma \in \Gamma_j$ and observing that the sets $\widetilde{\Omega}_{j,\gamma}$ also satisfy a bounded overlapping property, see (5.27). □

Remark 5.8. Since $W^{1,\infty}(\mathbb{R}^2)$ is dense in $\mathcal{C}(\mathbb{R}^2)$, the previous estimate also shows that $P_j g \mapsto g$ uniformly as $j \to \infty$. Indeed, if $g_\varepsilon \in W^{1,\infty}(\mathbb{R}^2)$ is such that $\|g - g_\varepsilon\|_{L^\infty} \le \varepsilon$, then for any $j \ge \ln(|g_\varepsilon|_{W^{1,\infty}}/\varepsilon)$, we have

$$\|g - P_j g\|_{L^\infty(\mathbb{R}^2)} \le \|g - g_\varepsilon\|_{L^\infty(\mathbb{R}^2)} + \|g_\varepsilon - P_j g_\varepsilon\|_{L^\infty(\mathbb{R}^2)} + \|P_j(g_\varepsilon - g)\|_{L^\infty(\mathbb{R}^2)} \lesssim \varepsilon,$$

where we have used (5.30) and Lemma 5.7 in the second inequality.

Now, the detail coefficients defined in (5.4) also describe the fluctuations between successive continuous levels.

Lemma 5.9. *Remember that we have set* $\varphi_\gamma := \varphi_{|\gamma|,\gamma}$. *For any* $J \geq j_0$ *and any* $g \in \mathcal{C}(\mathbb{R}^2)$, *we have*

$$P_J g = P_{j_0} g + \sum_{j=j_0+1}^{J} \sum_{\gamma \in \nabla_j} d_\gamma(g) \varphi_\gamma.$$

Proof. Clearly, it suffices to show that for any $j \geq 1$, we have

$$P_j g = P_{j-1} g + \sum_{\gamma \in \nabla_j} d_\gamma(g) \varphi_\gamma. \tag{5.33}$$

To see this, let $\tilde{\mathbf{g}}^{[j]} := P_j^{j-1} \mathbf{g}^{[j-1]} + \sum_{\gamma \in \nabla_j} d_\gamma(g) \phi_{j,\gamma}^{[j]}$ and observe that for $\lambda \in \Gamma_{j-1}$ the definition (5.8) of $\phi_{j,\gamma}^{[j]}$ yields

$$\tilde{\mathbf{g}}^{[j]}(\lambda) = P_j^{j-1} \mathbf{g}^{[j-1]}(\lambda) = \mathbf{g}^{[j-1]}(\lambda) = g(\lambda) = \mathbf{g}^{[j]}(\lambda).$$

By using the definition of $d_\lambda(g)$ and again (5.8), for any $\lambda \in \nabla_j$, we next see that

$$\tilde{\mathbf{g}}^{[j]}(\lambda) = (P_j^{j-1} \mathbf{g}^{[j-1]})(\lambda) + d_\lambda(g) = \mathbf{g}^{[j]}(\lambda),$$

hence $\mathbf{g}^{[j]} = \tilde{\mathbf{g}}^{[j]}$. Applying then $P_J^j := P_J^{J-1} \cdots P_{j+1}^j$ to the latter equality gives

$$\sum_{\mu \in \Gamma_j} g(\mu) \phi_{j,\mu}^J = P_J^j \tilde{\mathbf{g}}^{[j]} = \sum_{\mu \in \Gamma_{j-1}} g(\mu) \phi_{j-1,\mu}^J + \sum_{\gamma \in \nabla_j} d_\gamma(g) \phi_{j,\gamma}^J.$$

So, letting $J \to \infty$ finally yields (5.33). □

Remark 5.10. From Remark 5.8 and Lemma 5.9, we easily infer the validity of the hierarchical decomposition (5.11).

5.4 Adaptive interpolations

Summing up, we now have a representation of any continuous function g in terms of the multilevel nodal functions φ_γ, $\gamma \in \Gamma_\infty$, with small coefficients in the regions where g is smooth, according to (5.6) or (5.7). Adaptivity will be achieved by discarding small coefficients in this expansion. In order to study such approximation schemes, we introduce the following definition.

Definition 5.11. Let $j_0 \in \mathbb{N}$. A grid $\Lambda \subset \Gamma_\infty$ is said to be *admissible* if it contains the coarse grid Γ_{j_0} and if it satisfies

$$\gamma \in \Lambda \quad \Longrightarrow \quad S_\gamma \subset \Lambda.$$

Next we define a mapping $P_\Lambda : \mathcal{C}(\mathbb{R}^2) \to V_\Lambda := \mathrm{span}\{\varphi_\lambda : \lambda \in \Lambda\}$ by

$$P_\Lambda g := \sum_{\lambda\in\Lambda} d_\lambda(g)\varphi_\lambda. \tag{5.34}$$

Clearly this makes sense for any grid $\Lambda \subset \Gamma_\infty$, but if in addition Λ is admissible, then P_λ is an interpolation, which might be of interest for practical implementations.

Lemma 5.12. *If the grid Λ is admissible, then $P_\Lambda g(\gamma) = g(\gamma)$ for all $\gamma \in \Lambda$.*

Proof. First, since (5.34) is a wavelet expansion of $P_\Lambda g$, we clearly have

$$d_\gamma(P_\Lambda g) = d_\gamma(g) \quad \forall \gamma \in \Lambda. \tag{5.35}$$

Observe next that any φ_μ, $|\mu| \geq j_0 + 1$, vanishes on any $\gamma \in \Gamma_{j_0}$, and calculate

$$P_\Lambda g(\gamma) = \sum_{|\lambda|\leq j_0} d_\lambda(g)\varphi_\lambda(\gamma) = \sum_{|\lambda|\geq 0} d_\lambda(g)\varphi_\lambda(\gamma) = g(\gamma), \quad \gamma \in \Gamma_{j_0}.$$

Now assume that $P_\Lambda g$ and g coincide on $\Lambda \cap \Gamma_{j-1}$, and consider $\gamma \in \Lambda \cap \nabla_j$: since Λ is admissible, we have by definition of the details d_γ

$$P_\Lambda g(\gamma) = \sum_{\mu\in S_\gamma} \pi(\gamma,\mu)P_\Lambda g(\mu) + d_\gamma(P_\Lambda g) = \sum_{\mu\in S_\gamma} \pi(\gamma,\mu)g(\mu) + d_\gamma(g) = g(\gamma).$$

□

As we said before, the error resulting from "discarding the small details" should be controlled by the amplitude of these details. Here is the precise statement that we shall use for an approximation in the supremum norm.

Lemma 5.13. *The approximation of g associated with the grid Λ is bounded by*

$$\|g - P_\Lambda g\|_{L^\infty(\mathbb{R}^2)} \leq C \sum_{j\geq 0} \sup_{\gamma\in\nabla_j\setminus\Lambda} |d_\gamma(g)|$$

with $C = C_s\|\varphi\|_{L^\infty(\mathbb{R}^2)}$, see (5.27).

Proof. In view of (5.11), g can be written as an infinite wavelet expansion. The approximation error thus satisfies

$$\begin{aligned}
\|g - P_\Lambda g\|_{L^\infty(\mathbb{R}^2)} &= \Big\| \sum_{j\geq 0} \sum_{\gamma\in\nabla_j\setminus\Lambda} d_\gamma(g)\varphi_\gamma \Big\|_{L^\infty(\mathbb{R}^2)} \\
&\leq \sum_{j\geq 0} \Big\| \sum_{\gamma\in\nabla_j\setminus\Lambda} d_\gamma(g)\varphi_\gamma \Big\|_{L^\infty(\mathbb{R}^2)} \\
&\leq C_s\|\varphi\|_{L^\infty(\mathbb{R}^2)} \sum_{j\geq 0} \sup_{\gamma\in\nabla_j\setminus\Lambda} |d_\gamma(g)|,
\end{aligned}$$

where we have employed (5.28) in the last inequality. □

Remark 5.14. The foregoing estimate is sharp. Indeed, consider the one-dimensional case (for the sake of simplicity), where for $R = 1$, the reference scaling function is given by $\varphi(x) = \max(1 - |x|, 0)$. Now let $\gamma_i := 2^{-2i}(1 + 4 + \cdots + 4^{(i-1)}) \in \nabla_{2i}$ and check that

$$\varphi_{\gamma_i} = \varphi(2^{2i}(x - \gamma_i)) \geq \frac{1}{2} \quad \text{on} \quad \left[\frac{4^i - 1}{3 \cdot 4^i}, \frac{4^i - 1}{3 \cdot 4^i} + \frac{1}{2 \cdot 4^i}\right].$$

In particular, $\varphi_{\gamma_i}(1/3) \geq 1/2$ for all i, hence $\| \sum_{i \leq J} \varphi_{\gamma_i} \|_{L^\infty(\mathbb{R}^2)} \geq J/2$ for all J.

5.5 Connection with trees and meshes

In order to transport the numerical solutions along the flow, we will need to associate every (admissible) adaptive grid with a partition of the phase space. Moreover, our scheme will be based on tree algorithms. Hence we need to equip the dyadic grids with a tree structure. Here we describe how we shall do this.

First, we introduce the set of *-nodes of level $j \geq 1$ that correspond to a refinement of Γ_j in both directions,

$$\nabla_j^* := \{(2^{-j}(2k + 1), 2^{-j}(2k' + 1)) : k, k' \in \{0, 2^{j-1} - 1\}\} \subset \nabla_j,$$

and associate an (open) square cell to every *-node by setting

$$\Omega_\gamma := B_{\ell^\infty}(\gamma, 2^{-|\gamma|}) \qquad \forall \gamma \in \Gamma_\infty^* := \bigcup_{j \geq 1} \nabla_j^*.$$

Next, we equip the set of dyadic nodes with a tree structure by defining for every $\gamma \in \nabla_j$ one set of *children* in ∇_{j+1} as follows: If γ is a *-node of level j, we define its children as

$$\mathcal{C}(\gamma) := \{\gamma + 2^{-(j+1)}(l, l') : (l, l') \in \{-1, 0, 1\}^2 \setminus (0, 0)\} \subset \nabla_{j+1}.$$

If $\gamma = (2^{-j}k, 2^{-(j-1)}k')$ (hence with k odd), we define its children as

$$\mathcal{C}(\gamma) := \{\gamma + 2^{-(j+1)}(l, 0) : l \in \{-1, 1\}\} \subset \nabla_{j+1}.$$

If $\gamma = (2^{-(j-1)}k, 2^{-j}k')$ (hence with k' odd), we define its children as

$$\mathcal{C}(\gamma) := \{\gamma + 2^{-(j+1)}(0, l') : l' \in \{-1, 1\}\} \subset \nabla_{j+1}.$$

Finally, we say that λ is a *parent* of γ if $\gamma \in \mathcal{C}(\lambda)$. Note that this process partitions the levels in the sense that every dyadic node γ of positive level has one (and only one) parent $\mathcal{P}(\gamma)$, moreover $|\mathcal{P}(\gamma)| = |\gamma| - 1$. Now, as it can be checked, every parent of a *-node is also a *-node but the converse is not true, i.e., not every children of a *-node is a *-node itself. Hence we introduce the notion of **-children* and set

$$\mathcal{C}^*(\gamma) := \{\gamma + 2^{-(j+1)}(m, m') : (m, m') \in \{-1, 1\}^2\} = \mathcal{C}(\gamma) \cap \nabla_{j+1}^*$$

for every $\gamma \in \nabla_j^*$. Let us adopt the following definition in the wavelet framework.

Definition 5.15. A grid $\Lambda \subset \Gamma_\infty$ is said to be a *W-tree* if it contains the coarse grid Γ_{j_0} and if it satisfies

$$\gamma \in \Lambda \quad \Longrightarrow \quad \mathcal{P}(\gamma) \subset \Lambda.$$

Clearly, the simplest way to build a tree Λ consists in starting from the coarsest grid Γ_{j_0} and adding recursively children, according to some criterion. Observe that by doing so, one also builds a non-uniform partition of $\mathbb{R}^2$, given by

$$M(\Lambda) := \{\Omega_\gamma : \gamma \in \mathcal{L}_{\text{out}}(\Lambda) \cap \Gamma_\infty^*\}, \tag{5.36}$$

where

$$\mathcal{L}_{\text{out}}(\Lambda) := \{\gamma \notin \Lambda : \mathcal{P}(\lambda) \in \Lambda\}$$

denotes the set of *outer leaves* of the tree Λ. For later purposes, we will also need that the trees satisfy a stronger property.

Definition 5.16. A W-tree Λ is said to be *graded* if

$$\gamma \in \Lambda \cap \Gamma_\infty^* \quad \Longrightarrow \quad \{\mu \in \Gamma_{|\gamma|-1} : \Sigma_\mu \cap \Omega_\gamma \neq \emptyset\} \subset \Lambda,$$

see (5.5).

As we shall see, this definition is mostly motivated by the accuracy of the transported meshes. However, it turns out that the use of graded trees is already imposed by the admissibility of the dyadic wavelet grids.

Lemma 5.17. *Every admissible wavelet grid is a graded W-tree.*

Proof. Clearly, every admissible grid is a W-tree (simply observe that the parent of γ is always contained in the stencil S_γ). Let us then show that for any $\gamma \in \Gamma_\infty^*$, $|\gamma| = j$, and $\mu \in \Gamma_{j-1}$, we have

$$\Omega_\gamma \cap \Sigma_\mu \neq \emptyset \iff \Omega_\gamma \subset \Sigma_\mu \iff \gamma \in (\Sigma_\mu)^\circ = B_{\ell^\infty}(\mu, 2^{-|\mu|}(2R-1)), \tag{5.37}$$

where $(\Sigma_\mu)^\circ$ denotes the interior of Σ_μ. Indeed, since

$$\Omega_\gamma := B_{\ell^\infty}(\gamma, 2^{-j}) = 2^{-(j-1)}((k, k+1) \times (k', k'+1))$$

with k, $k' \in \mathbb{N}$, and $\Sigma_\mu = \overline{B}_{\ell^\infty}(\mu, 2^{-|\mu|}(2R-1)) = 2^{-|\mu|}([m_1, m_2] \times [m_3, m_4])$ with $m_1, \ldots, m_4 \in \mathbb{N}$, the equivalences in (5.37) easily follow from $|\mu| \leq j-1$.

Now assume in additon that γ belongs to an admissible grid Λ. Since $\lambda \in \nabla_j$ with $j \geq |\mu|+1$, according to (5.25) we see that (5.37) implies the existence of one minimal $\ell \geq |\mu|+1$ such that $\gamma \in \Sigma_{|\mu|,\mu}^{[\ell]}$, and this in turn implies that there exists $\gamma' \in S_\gamma \cap \Sigma_{|\mu|,\mu}^{[\ell-1]}$. By using that Λ is admissible, we see that γ' is also in Λ, and by repeating the argument, we show that $\Lambda \cap \Sigma_{|\mu|,\mu}^{[|\mu|]} \neq \emptyset$, i.e., $\mu \in \Lambda$. Hence Λ is graded. □

In order to accurately transport the grids with a low-cost algorithm, we now introduce an important property which involve the *neighbors* of a *-node, i.e.,

$$\mathcal{N}(\gamma) := \{\mu \in \Gamma_\infty : \Sigma_\mu \cap \Omega_\gamma \neq \emptyset\}, \qquad \gamma \in \Gamma^*_\infty.$$

For the remainder of this lecture, we fix one positive constant $\kappa < 1$, and we remind that $\varepsilon > 0$ is an arbitrary tolerance.

Definition 5.18. A W-tree Λ is said to be *weakly ε-adapted* to $g \in \mathcal{C}(\mathbb{R}^2)$ if

$$|d_\mu(g)| \leq 2^{\kappa(|\gamma|-|\mu|)}\varepsilon \tag{5.38}$$

for all $\gamma \in \mathcal{L}_{\text{out}}(\Lambda) \cap \Gamma^*_\infty$ and all $\mu \in \mathcal{N}(\gamma) \setminus \Lambda$. If (5.38) holds for all $\mu \in \mathcal{N}(\gamma)$, Λ is said to be *strongly ε-adapted* to g.

We note that the foregoing criterion (5.38) is somehow related to the prediction strategy suggested by Albert Cohen, Sidi Mahmoud Kaber, Siegfried Müller and Marie Postel [12] in the context of wavelet-based finite volume schemes with guaranteed error estimates. A first property associated with these definitions is the following.

Lemma 5.19. *If Λ is admissible and weakly ε-adapted to g, then the associated interpolation error satisfies*

$$\|g - P_\Lambda g\|_{L^\infty(\mathbb{R}^2)} \lesssim \varepsilon$$

with a constant independent of g.

Proof. Let us begin with the following observation: If Λ is admissible (hence graded), then for any $\gamma \in \mathcal{L}_{\text{out}}(\Lambda) \cap \Gamma^*_\infty$ and $\mu \in \Gamma_{|\gamma|-2} = \Gamma_{|\mathcal{P}(\gamma)|-1}$ with $\mu \notin \Lambda$, we have $\Sigma_\mu \cap \Omega_\gamma \subset \Sigma_\mu \cap \Omega_{\mathcal{P}(\gamma)} = \emptyset$ since $\mathcal{P}(\gamma) \in \Lambda$. In other words, we see that

$$|\mu| \geq |\gamma| - 1, \qquad \forall \mu \in \mathcal{N}(\gamma) \setminus \Lambda. \tag{5.39}$$

Now using that $\operatorname{supp}(\varphi_\mu) \subset \Sigma_\mu$, we find

$$\begin{aligned}
\|g - P_\Lambda g\|_{L^\infty(\Omega_\gamma)} &= \Big\| \sum_{j \geq j_0+1} \sum_{\substack{\mu \in \nabla_j \setminus \Lambda \\ \Sigma_\mu \cap \Omega_\gamma \neq \emptyset}} d_\mu(g)\varphi_\mu \Big\|_{L^\infty(\Omega_\gamma)} \\
&\lesssim \sum_{j \geq j_0+1} \sup_{\substack{\mu \in \nabla_j \setminus \Lambda \\ \Sigma_\mu \cap \Omega_\gamma \neq \emptyset}} |d_\mu(g)| \lesssim \sum_{j \geq |\gamma|-1} \sup_{\substack{\mu \in \nabla_j \setminus \Lambda \\ \Sigma_\mu \cap \Omega_\gamma \neq \emptyset}} |d_\mu(g)| \\
&\lesssim \sum_{j \geq |\gamma|-1} 2^{\kappa(|\gamma|-j)}\varepsilon \lesssim \varepsilon,
\end{aligned}$$

where the first inequality follows from Lemma 5.13, the second one from the above observation (which uses the gradedness of Λ), and the third one is the weak ε-adaptivity. The assertion follows since $\{\Omega_\gamma : \gamma \in \mathcal{L}_{\text{out}}(\Lambda) \cap \Gamma^*_\infty\}$ is a partition of $\mathbb{R}^2$. □

6 Dynamic adaptivity

In Sections 4 and 5, we have defined tree-structured adaptive discretizations of $\mathcal{C}(\Omega)$ using multilevel meshes and wavelets, respectively. Note that wavelets have been constructed on the entire $\mathbb{R}^2$, therefore, in order to represent a function $g \in \mathcal{C}(\Omega)$ in the wavelet basis, we first need to extend it outside Ω. Due to the boundary conditions that we have considered in Theorem 2.2, we extend it periodically in the x-direction and by zero in the v-direction. As the numerical flow is assumed to map Ω into itself, we note that this does not spoil the continuity of the numerical solutions inside Ω (but it might deteriorate the sparsity of smooth functions in the wavelet basis, at least in the vicinity of the boundary). Next, for both discretizations we have introduced a notion of ε-adaptivity to a given function and we have seen that interpolating on the corresponding adaptive mesh and wavelet grid, respectively, is $C\varepsilon$ accurate in the supremum norm. In this section, we describe the algorithms that allow to build an ε-adapted mesh or grid to a given function and that transport these meshes along a smooth flow, while conserving the property of being adapted to the transported solution. We also describe algorithms that build graded refinements of given trees.

6.1 Adapting the trees

Remember that in the multilevel mesh case, the FE-trees Λ consist of indices corresponding to dyadic quadrangles and that the associated meshes $M(\Lambda)$ are defined as the inner leaves of Λ, see (4.10). In the wavelet case, the W-trees Λ consist of dyadic points, and the associated quad-meshes $M(\Lambda)$, see (5.36), are defined as the outer leaves of the subtree consisting of the *-nodes of Λ. Note that this latter subtree has an FE-tree structure.

For constructing adapted multilevel meshes, we will use the following algorithm.

Algorithm 6.1 ($\mathbb{A}^{\mathrm{FE}}_\varepsilon(g)$: ε-adaption of FE-trees). Starting from $\Lambda_0 := \mathcal{I}_0$, set

$$\Lambda_{\ell+1} := \Lambda_\ell \cup \big\{\beta \in \mathcal{C}^*(\alpha) : \alpha \in M(\Lambda_\ell) \ \text{ such that } \ |g|_{W^*(\alpha)} > \varepsilon\big\}$$

for $\ell = 0, 1, \ldots$ until $\Lambda_{L+1} = \Lambda_L$, and finally set $\mathbb{A}^{\mathrm{FE}}_\varepsilon(g) := \Lambda_L$.

The following algorithm builds a W-tree (only composed of *-nodes) that is strongly ε-adapted to g.

Algorithm 6.2 ($\mathbb{A}^{\mathrm{W}}_\varepsilon(g)$: strong ε-adaption of W-trees). Starting from $\Lambda^*_0 := \Gamma^*_{j_0}$, set

$$\Lambda^*_{\ell+1} := \Lambda^*_\ell \cup \big\{\gamma \in M(\Lambda^*_\ell) : \max\{2^{\kappa|\mu|}|d_\mu(g)| : \mu \in \mathcal{N}(\gamma)\} > 2^{\kappa|\gamma|}\varepsilon\big\}$$

for $\ell = 0, 1, \ldots$ until $\Lambda^*_{L+1} = \Lambda^*_L$, and finally set $\mathbb{A}^{\mathrm{W}}_\varepsilon(g) := \Lambda^*_L$.

As the resulting adapted trees have no reason to be graded, we also need algorithms that build graded refinements of any given tree.

Algorithm 6.3 ($\mathbb{G}^{\mathrm{FE}}(\Lambda)$: graded refinement of FE-trees). Starting from $\Lambda_0 = \Lambda$, build

$$\Lambda_{\ell+1} := \Lambda_\ell \cup \{\lambda \in \mathcal{C}(\gamma) : \gamma \in \Lambda_\ell \cap \mathcal{Q}_\ell,\ \exists \mu \in \Lambda_\ell \cap \mathcal{Q}_{\ell+2},\ \overline{\Omega}_\gamma \cap \overline{\Omega}_\mu \neq \emptyset\}$$

for $\ell = 0, 1, \dots$ until $\Lambda_{L-1} = \Lambda_L$, and set $\mathbb{G}^{\mathrm{FE}}(\Lambda) := \Lambda_{L-1}$.

As graded W-trees have a structure which involves stencils of possibly high order R, we use another approach for building them.

Algorithm 6.4 ($\mathbb{G}^{\mathrm{W}}(\Lambda)$: graded refinement of W-trees). Given any tree Λ, set

$$\mathbb{G}^{\mathrm{W}}(\Lambda) := \bigcup_{\gamma \in \Lambda} \{\gamma + 2^{-|\gamma|}(m_1, m_2) : m_1, m_2 \in \{-(2R-1), \dots, (2R-1)\}\}.$$

Remark 6.5. We could also give a variant of Algorithm 6.2 that builds *weakly* ε-adapted W-trees to some given g. According to Lemma 5.19, we know that interpolating g on the resulting grid (once graded) would be $C\varepsilon$ accurate, but this is not enough to ensure the accuracy of the adaptive semi-Lagrangian scheme (indeed, think of the case where g consists of one isolated basis function φ_γ).

Now, it is clear that the trees constructed by Algorithms 6.1 and 6.2 are ε-adapted in the sense of Definitions 4.3 and 5.18, respectively. It is also clear that Algorithm 6.3 yields a graded FE-tree in the sense of Definition 4.2, but the gradedness of W-trees resulting from Algorithm 6.4 is by no means obvious.

Lemma 6.6. *For any W-tree Λ, the resulting $\mathbb{G}^{\mathrm{W}}(\Lambda)$ is an admissible grid and hence a graded W-tree (according to Lemma 5.17).*

Proof. Let $\gamma \in \mathbb{G}^{\mathrm{W}}(\Lambda)$ and $\lambda \in \Lambda$ such that $\gamma_i = \lambda_i + 2^{-|\lambda|}m_i$ with $|m_i| \le 2R-1$, $i \in \{1,2\}$. In particular, we have $\gamma \in \Gamma_{|\lambda|}$, i.e., $j := |\gamma| \le |\lambda|$, and we can as well assume that $j = |\lambda|$. Indeed, if $j < |\lambda|$ then $\gamma \in \Gamma_{|\mathcal{P}(\lambda)|}$, i.e., there exist integers m'_i, $i \in \{1,2\}$, such that $\gamma_i = \mathcal{P}(\lambda)_i + 2^{-|\mathcal{P}(\lambda)|}m'_i$. Moreover, we have

$$|m'_i| = 2^{j-1}|\gamma_i - \mathcal{P}(\lambda)_i| \le 2^{j-1}\big(|\gamma_i - \lambda_i| + |\lambda_i - \mathcal{P}(\lambda)_i|\big) \le \frac{|m_i|+1}{2} \le R \le 2R-1.$$

Therefore, we can replace λ by its parent (which also belongs to Λ), hence assume $j = \lambda$. Now observe that any $\mu \in S_\gamma$ satisfies

$$\mu \in \Gamma_{j-1} \quad \text{and} \quad |\mu_i - \gamma_i| \le \begin{cases} 2^{-j}(2R-1) & \text{if } |\gamma_i| = j, \\ 0 & \text{if } |\gamma_i| < j. \end{cases}$$

We can thus write $\mu = \mathcal{P}(\lambda) + 2^{-(j-1)}(k_1, k_2)$ and bound $|k_i|$ by

$$|k_i| = 2^{j-1}|\mu_i - \mathcal{P}(\lambda)_i| \le 2^{j-1}\big(|\mu_i - \gamma_i| + |\gamma_i - \lambda_i| + |\lambda_i - \mathcal{P}(\lambda)_i|\big).$$

We claim that $|k_i| \le 2R-1$, which, because of $\mathcal{P}(\lambda) \in \Lambda$, implies $\mu \in \mathbb{G}^{\mathrm{W}}(\Lambda)$. Indeed, we always have

$$|\mu_i-\gamma_i| \le 2^{-j}(2R-1), \quad |\gamma_i-\lambda_i| \le 2^{-j}|m_i| \le 2^{-j}(2R-1) \quad \text{and} \quad +|\lambda_i-\mathcal{P}(\lambda)_i| \le 2^{-j}$$

and observe that three cases may occur, according to $|\gamma| = |\lambda| = j$: either $|\gamma_i| < j$, in which case $|\mu_i - \gamma_i| = 0$, or $|\lambda_i| < j$, in which case $|\lambda_i - \mathcal{P}(\lambda)_i| = 0$, or $|\lambda_i| = |\gamma_i| = j$, in which case m_i must be even, hence $|\gamma_i - \lambda_i| \leq 2^{-(j-1)}(R-1)$. By summing up, we find that

$$|\mu_i - \gamma_i| + |\gamma_i - \lambda_i| + |\lambda_i - \mathcal{P}(\lambda)_i| \leq 2^{-(j-1)}(2R-1)$$

holds in every case, which ends the proof. □

6.2 Predicting the trees

For both discretizations, the algorithm that we shall use for predicting the meshes is based on partition trees (i.e., the leaves form a partition of Ω) and essentially consists in transporting the local space resolution along the (approximate) flow $\mathcal{B}$. We write it in terms of FE-trees, but already observe that it can also be applied to any W-tree Λ via its subtree $\Lambda \cap \Gamma^*_\infty$.

Algorithm 6.7 ($\mathbb{T}^{\mathrm{FE}}_{\mathcal{B}}(\Lambda)$: prediction of FE-trees). Starting from $\Lambda_0 := \mathcal{I}_{j_0}$, set

$$\Lambda_{\ell+1} := \Lambda_\ell \cup \{\mu \in \mathcal{C}^*(\gamma) : \gamma \in \mathcal{L}_{\text{in}}(\Lambda_\ell), \min\{|\lambda| : \lambda \in \mathcal{L}_{\text{in}}(\Lambda), \mathcal{B}(x_\gamma) \in \overline{\Omega_\lambda}\} > |\gamma|\},$$

where x_γ denotes the center of Ω_γ, until $\Lambda_{L+1} = \Lambda_L$, and set $\mathbb{T}^{\mathrm{FE}}_{\mathcal{B}}(\Lambda) := \Lambda_L$.

As was previously said, the following variant of Algorithm 6.7 for W-trees is almost the same algorithm. However, in order to establish the accuracy of the predicted grids, we now introduce a fixed parameter $\delta \in \mathbb{N}$ (the value of which will be chosen in the proof of Theorem 6.9) that corresponds to a constant number of additional refinement levels. Remember that for any W-tree Λ, the set $\mathcal{L}_{\text{out}}(\Lambda) \cap \Gamma^*_\infty$ consists of its outer $*$-leaves and that $\{\Omega_\lambda : \lambda \in \mathcal{L}_{\text{out}}(\Lambda) \cap \Gamma^*_\infty\}$ forms a partition of the phase space.

Algorithm 6.8 ($\mathbb{T}^{\mathrm{W}}_{\mathcal{B}}(\Lambda)$: prediction of W-trees). Starting from $\Lambda^*_0 := \Gamma^*_{j_0}$, build

$$\Lambda^*_{\ell+1} := \Lambda^*_\ell \cup \{\gamma \in \mathcal{L}_{\text{out}}(\Lambda^*_\ell) \cap \Gamma^*_\infty : \min\{|\lambda| : \lambda \in \mathcal{L}_{\text{out}}(\Lambda) \cap \Gamma^*_\infty, \mathcal{B}(\gamma) \in \overline{\Omega_\lambda}\} > |\gamma| - \delta\}$$

until $\Lambda^*_{L+1} = \Lambda^*_L$, and set $\mathbb{T}^{\mathrm{W}}_{\mathcal{B}}(\Lambda) := \Lambda^*_L$.

The main properties of these algorithms are summarized in the following theorem.

Theorem 6.9. *Let $\mathcal{B}$ be a diffeomorphism of Ω into itself, i.e., an invertible Lipschitz continuous mapping with Lipschitz continuous inverse. Then Algorithms 6.7 and 6.8 guarantee the accuracy of the predicted trees in the following sense:*

- *If Λ is a FE-tree ε-adapted to g and if the flow $\mathcal{B}$ is stable in the sense of* (4.13), *then the FE-tree $\mathbb{T}^{\mathrm{FE}}_{\mathcal{B}}(\Lambda)$ is $C\varepsilon$-adapted to the $\mathcal{T}_{\mathcal{B}}g = g \circ \mathcal{B}$ with a constant C that only depends on $\mathcal{B}$.*
- *If Λ is a W-tree strongly ε-adapted to g, then the W-tree $\mathbb{T}^{\mathrm{W}}_{\mathcal{B}}(\Lambda)$ is weakly $C\varepsilon$-adapted to $\mathcal{T}_{\mathcal{B}}g = g \circ \mathcal{B}$ with a constant C that only depends on $\mathcal{B}$.*

In addition, the cardinalities of the predicted trees are stable in the sense that

$$\#(\mathbb{T}^{\mathrm{FE}}_{\mathcal{B}}(\Lambda)) \lesssim \#(\Lambda) \quad \textit{and} \quad \#(\mathbb{T}^{\mathrm{W}}_{\mathcal{B}}(\Lambda)) \lesssim \#(\Lambda). \tag{6.1}$$

Proof. We shall only sketch the proof for the $C\varepsilon$-adaptivity of the predicted FE-trees (for details and for a proof of the complexity estimates (6.1), we refer to [7]). First, we introduce the set

$$\mathcal{I}(\Lambda, \mathcal{B}, \gamma) := \{\lambda \in \mathcal{L}_{\mathrm{in}}(\Lambda) : \overline{\Omega}_\lambda \cap \mathcal{B}(\Omega_\gamma) \neq \emptyset\}$$

corresponding to the cells of the quad-mesh $M(\Lambda)$ that are even partly advected into Ω_γ. By using the gradedness of Λ and the smoothness of $\mathcal{B}$, one can show that there exists a constant C such that

$$\#(\mathcal{I}(\Lambda, \mathcal{B}, \gamma)) \leq C \qquad \forall \gamma \in \mathcal{L}_{\mathrm{in}}(\mathbb{T}^{\mathrm{FE}}_{\mathcal{B}}(\Lambda)). \tag{6.2}$$

According to the stability (4.13), we next estimate for any $\gamma \in \mathcal{L}_{\mathrm{in}}(\mathbb{T}^{\mathrm{FE}}_{\mathcal{B}}(\Lambda))$,

$$|g \circ \mathcal{B}|_{W^*(\Omega_\gamma)} \lesssim |g|_{W^*(\mathcal{B}(\Omega_\gamma))} \lesssim \sum_{\lambda \in \mathcal{I}(\Lambda, \mathcal{B}, \gamma)} |g|_{W^*(\overline{\Omega}_\lambda)} \lesssim \varepsilon,$$

where the second inequality follows from the fact that the cells $\overline{\Omega}_\lambda, \lambda \in \mathcal{I}(\Lambda, \mathcal{B}, \gamma)$, cover $\mathcal{B}(\Omega_\gamma)$, and the third inequality follows from the ε-adaptivity of Λ together with relation (6.2).

We now give a detailed proof of the weak $C\varepsilon$-adaptivity of the W-tree $\mathbb{T}^{\mathrm{W}}_{\mathcal{B}}(\Lambda)$. Let us recall that this amounts in proving that for any $\tilde{\lambda} \in \mathcal{L}_{\mathrm{out}}(\mathbb{T}^{\mathrm{W}}_{\mathcal{B}}(\Lambda)) \cap \Gamma^*_\infty$ and any $\tilde{\mu} \in \mathcal{N}(\tilde{\lambda}) \setminus \mathbb{T}^{\mathrm{W}}_{\mathcal{B}}(\Lambda)$, we have

$$|d_{\tilde{\mu}}(g \circ \mathcal{B})| \leq 2^{\kappa(\ell - j)} C\varepsilon, \quad \text{where} \quad \ell := |\tilde{\lambda}|, \ j := |\tilde{\mu}|, \tag{6.3}$$

for a constant C that only depends on $\mathcal{B}$. By using (5.7) and Lemma 3.4, we obtain

$$|d_{\tilde{\mu}}(g \circ \mathcal{B})| \lesssim \omega_1(g \circ \mathcal{B}, 2^{-j}, \Sigma_{\tilde{\mu}})_\infty \lesssim \omega_1(g, 2^{-j}, \Sigma^{\mathcal{B}}_{\tilde{\mu}})_\infty \leq \sum_{i \in \mathbb{N}} \omega_1(g_i, 2^{-j}, \Sigma^{\mathcal{B}}_{\tilde{\mu}})_\infty,$$

where $\Sigma^{\mathcal{B}}_{\tilde{\mu}}$ denotes the set $(\Sigma_{\tilde{\mu}})^{\mathcal{B}, \tau}$ with $\tau = 2^{-j}$, see Lemma 3.4, and where the "single layer" functions defined by

$$g_i := \sum_{\mu \in \mathcal{N}^{\mathcal{B}}_i(\tilde{\mu})} d_\mu(g) \varphi_\mu \quad \text{with} \quad \mathcal{N}^{\mathcal{B}}_i(\tilde{\mu}) := \{\mu \in \nabla_i : \Sigma_\mu \cap \Sigma^{\mathcal{B}}_{\tilde{\mu}} \neq \emptyset\}$$

clearly satisfy $\sum_{i \in \mathbb{N}} g_i = g$ on $\Sigma^{\mathcal{B}}_{\tilde{\mu}}$. Note that since $\mathcal{B}$ is assumed to be Lipschitz continuous, $\Sigma^{\mathcal{B}}_{\tilde{\mu}}$ is included in a ball of center $\tilde{\mu}$ and radius $C2^{-j}$ with a constant C depending on R only. Next we estimate the modulus of smoothness following rather classical techniques (see, e.g., [11, p. 183]). First, we observe that

$$\omega_1(g_i, 2^{-j}, \Sigma^{\mathcal{B}}_{\tilde{\mu}})_\infty \lesssim \|g_i\|_{L^\infty(\Omega)} \lesssim \sup_{\mu \in \mathcal{N}^{\mathcal{B}}_i(\tilde{\mu})} |d_\mu(f)|$$

by using the definition of ω_1 and (5.28), and that

$$\omega_1(g_i, 2^{-j}, \Sigma^{\mathcal{B}}_{\tilde{\mu}})_\infty \le \inf_{p\in\mathcal{Q}_0} \omega_1(g_i - p, 2^{-j}, \Sigma^{\mathcal{B}}_{\tilde{\mu}})_\infty \lesssim \inf_{p\in\mathcal{Q}_0} \|g_i - p\|_{L^\infty(\Sigma^{\mathcal{B}}_{\tilde{\mu}})}$$
$$\lesssim 2^{-j}|g_i|_{W^{1,\infty}} \lesssim 2^{i-j} \sup_{\mu\in\mathcal{N}^{\mathcal{B}}_i(\tilde{\mu})} |d_\mu(g)|$$

by using the definition of ω_1, the Deny–Lions theorem and the Bernstein inequality (5.29). Note that the above estimates yield

$$|d_{\tilde{\mu}}(g\circ\mathcal{B})| \lesssim \sum_{i\in\mathbb{N}} \min\{2^{i-j}, 1\} \sup_{\mu\in\mathcal{N}^{\mathcal{B}}_i(\tilde{\mu})} |d_\mu(g)|. \tag{6.4}$$

Next, we observe that there exists some $\lambda \in \mathcal{L}_{\text{out}}(\Lambda)\cap\Gamma^*_\infty$ such that $\mathcal{B}(\tilde{\lambda}) \in \overline{\Omega}_\lambda$ and

$$|\lambda| \le \ell - \delta, \tag{6.5}$$

otherwise $\tilde{\lambda}$ would have been added to $\mathbb{T}^{\mathrm{W}}_{\mathcal{B}}(\Lambda)$. We claim that any $\mu \in \mathcal{N}^{\mathcal{B}}_i(\tilde{\mu})$ is a neighbor node of some leaf node adjacent to λ; in other words, we claim that there exists $\gamma \in \mathcal{L}_{\text{out}}(\Lambda)\cap\Gamma^*_\infty$ that satisfies both

$$\overline{\Omega}_\gamma \cap \overline{\Omega}_\lambda \ne \emptyset \quad \text{and} \quad \mu\in\mathcal{N}(\gamma), \ \text{ i.e., } \ \Sigma_\mu \cap \overline{\Omega}_\gamma \ne \emptyset. \tag{6.6}$$

Note that this would permit establishing the desired estimate (6.3): Indeed, if (6.6) holds then we have in particular $\lambda \in \mathcal{N}(\gamma)\setminus\Lambda$, and thus $|\gamma| \le |\lambda| + 1$ according to (5.39). By using the strong ε-adaptivity of Λ to g, we have

$$|d_\mu(g)| \le 2^{\kappa(|\gamma|-|\mu|)}\varepsilon \lesssim 2^{\kappa(\ell-i)}\varepsilon \qquad \forall \mu\in\mathcal{N}^{\mathcal{B}}_i(\tilde{\mu}),$$

where we employed (6.5) in the second inequality. Inserting this estimate into (6.4), we finally find

$$|d_{\tilde{\mu}}(g\circ\mathcal{B})| \lesssim 2^{\kappa\ell-j}\varepsilon \sum_{i\le j} 2^{(1-\kappa)i} + 2^{\kappa\ell}\varepsilon \sum_{i\ge j+1} 2^{-\kappa i} \lesssim 2^{\kappa(\ell-j)}\varepsilon,$$

which is (6.3). It thus only remains to establish the claim (6.6). For this purpose, we let $\text{dist}(A,B) := \inf_{a\in A, b\in B}\|a-b\|_{\ell^\infty}$ denote the minimal distance between two sets A, $B \subset \Omega$ (although this does not define a distance in the classical sense), for which we easily check that $\text{dist}(A,B) = 0$ if and only if $\overline{A}\cap\overline{B} \ne \emptyset$, and $\text{dist}(A,B) \le \text{dist}(A,C) + \text{dist}(B,C) + \text{diam}(C)$. By observing that none of the intersections

$$\Sigma_\mu \cap \Sigma^{\mathcal{B}}_{\tilde{\mu}}, \qquad \Sigma^{\mathcal{B}}_{\tilde{\mu}} \cap \mathcal{B}(\Sigma_{\tilde{\mu}}), \qquad \mathcal{B}(\Sigma_{\tilde{\mu}}) \cap \mathcal{B}(\Omega_{\tilde{\lambda}}) \quad \text{or} \quad \mathcal{B}(\Omega_{\tilde{\lambda}}) \cap \Omega_\lambda$$

is empty, we find

$$\text{dist}(\Sigma_\mu, \Omega_\lambda) \le \text{diam}(\Sigma^{\mathcal{B}}_{\tilde{\mu}}) + \text{diam}(\mathcal{B}(\Sigma_{\tilde{\mu}})) + \text{diam}(\mathcal{B}(\Omega_{\tilde{\lambda}})) \lesssim 2^{-j} + 2^{-\ell} \le 2^{-\ell} C_{\mathcal{B}}$$

with a constant $C_{\mathcal{B}}$ that only depends on the smoothness of $\mathcal{B}$. Therefore, choosing $\delta \ge \ln_2(C_{\mathcal{B}})$ yields $\text{dist}(\Sigma_\mu, \Omega_\lambda) \le 2^{-\ell+\delta} \le 2^{-|\lambda|}$. In order to conclude, we infer from (5.39) that, because $\lambda \in \mathcal{L}_{\text{out}}(\Lambda)\cap\Gamma^*_\infty$, every mesh cell $\overline{\Omega}_{\lambda'}$, $\lambda' \in \mathcal{L}_{\text{out}}(\Lambda)\cap\Gamma^*_\infty$, touching $\overline{\Omega}_\lambda$ is an ℓ^∞-ball of diameter $2^{1-|\lambda'|} \ge 2^{-|\lambda|}$. In particular, Σ_μ is in contact with one such cell $\overline{\Omega}_{\lambda'}$. Call it $\overline{\Omega}_\gamma$, we have just proved (6.6). □

6.3 The "predict and readapt" semi-Lagrangian scheme

Let us summarize what we have seen so far: In Section 3.1, we have described a time splitting scheme for computing an approximated backward flow $\mathcal{B}[f_n]$ from a given approximation f_n to the exact solution $f(t_n)$. This allows to compute every point value of $\mathcal{T} f_n := f_n \circ \mathcal{B}[f_n]$. Next, we have defined in Sections 4 and 5 two tree-structured discretizations of finite element and wavelet type, respectively, both suitable for adaptive interpolations. Finally, we have introduced in Section 6 algorithms for (i) building graded FE- and admissible W-trees which are ε-adapted to a given function g, and (ii) given a computable flow $\mathcal{B}$, predicting trees that stay well-adapted to the transported $g \circ \mathcal{B}$. Note that in view of Theorem 6.9, the quality of being well-adapted to the transported solution is only preserved by the predicted trees up to a multiplicative constant C that might be larger than 1. Hence it is necessary to readapt the trees once in a while in order to guarantee that *all* the interpolation errors stay within a bound of the order ε.

The resulting semi-Lagrangian scheme, which involves the algorithms $\mathbb{A}_\varepsilon = \mathbb{A}_\varepsilon^{\mathrm{FE}}$ or $\mathbb{A}_\varepsilon^{\mathrm{W}}$, $\mathbb{G} = \mathbb{G}^{\mathrm{FE}}$ or $\mathbb{G}^{\mathrm{W}}$ and $\mathbb{T}_{\mathcal{B}} = \mathbb{T}_{\mathcal{B}}^{\mathrm{FE}}$ or $\mathbb{T}_{\mathcal{B}}^{\mathrm{W}}$, depending on whether one desires to implement an adaptive multilevel mesh scheme or a wavelet scheme, is as follows. Given an initial datum f^0, compute first

$$f_0 := P_{\Lambda^0} f^0, \tag{6.7}$$

where P_{Λ^0} denotes the adaptive (finite element or wavelet) interpolation associated with

$$\Lambda^0 := \mathbb{G}(\mathbb{A}_\varepsilon(f^0)), \tag{6.8}$$

then, for $n = 1, \dots, N = T/\Delta t$,

$$f_n := P_{\Lambda^n} P_{\Lambda_p^n} \mathcal{T} f_{n-1}, \tag{6.9}$$

where the predicted and readapted trees are given by

$$\Lambda_p^n := \mathbb{G}(\mathbb{T}_{\mathcal{B}[f_{n-1}]}(\Lambda^{n-1})) \tag{6.10}$$

and

$$\Lambda^n := \mathbb{G}(\mathbb{A}_\varepsilon(f_{p,n}) \quad \text{with} \quad f_{p,n} := P_{\Lambda_p^n} \mathcal{T} f_{n-1}. \tag{6.11}$$

Remark 6.10. If the flow $\mathcal{B}[f_n]$ is defined by the splitting method described in Section 3.1, we need to compute an auxiliary electric field from the intermediate solution $\mathcal{T}_x^{1/2} f_n$. The adaptive semi-Lagrangian scheme, therefore, should decompose into substeps corresponding to every one-dimensional transport operator. Such a decomposition and the corresponding error analysis (for first order interpolations) is carried out in [7].

6.4 Error and complexity estimates (results and conjectures)

Our main result is that the above adaptive semi-Lagrangian schemes – of either multilevel mesh or wavelet type with arbitrary interpolation order – satisfy the following error estimate.

Theorem 6.11. *Under Assumptions 3.2 and 3.3, the numerical solution given by the adaptive semi-Lagrangian scheme* (6.7)–(6.11) *satisfies the estimate*

$$\|f(t_n) - f_n\|_{L^\infty(\Omega)} \lesssim (\Delta t)^{r-1} + \varepsilon/\Delta t \tag{6.12}$$

for $n = 0, \dots, N = T/\Delta t$ *if the initial datum* f^0 *is in* $W^{1,\infty}(\Omega)$.

Proof. The arguments are similar to those in Section 3.2 and, as we shall see, they also apply to high-order interpolations. Let us describe them in detail. Again we decompose the error $e_{n+1} := \|f(t_{n+1}) - f_{n+1}\|_{L^\infty(\Omega)}$ into three parts, which are now as follows. A first term is $e_{n+1,1} := \|f(t_{n+1}) - \mathcal{T} f(t_n)\|_{L^\infty(\Omega)}$ as in Section 3.2, bounded for the same reasons by

$$e_{n+1,1} \lesssim |f(t_n)|_{W^{1,\infty}(\Omega)}(\Delta t)^r \lesssim (\Delta t)^r.$$

A second term is $e_{n+1,2} := \|(I - P_{\Lambda^{n+1}} P_{\Lambda_p^{n+1}})\mathcal{T} f_n\|_{L^\infty(\Omega)}$, which again corresponds to the interpolation error, and a third term $e_{n+1,3} := \|\mathcal{T} f(t_n) - \mathcal{T} f_n\|_{L^\infty(\Omega)}$, which again represents the (nonlinear) transport of the numerical error at time step n. Note that this decomposition slightly differs from that of Section 3.2 in that the interpolation error now involves the numerical solution instead of the exact one. This is in order to exploit the main properties of the predicted grids, i.e., the $C\varepsilon$-adaptivity to $\mathcal{T} f_n$. The good news is that it allows the third term, which propagates the error from the previous time steps, *not* to involve the interpolation operator. Hence for any interpolation order, we have

$$\begin{aligned} e_{n+1,3} &= \big\|f(t_n) \circ \mathcal{B}[f(t_n)] - f(t_n) \circ \mathcal{B}[f_n]\big\|_{L^\infty(\Omega)} + \big\|(f(t_n) - f_n) \circ \mathcal{B}[f_n]\big\|_{L^\infty(\Omega)} \\ &\le |f(t_n)|_{W^{1,\infty}(\Omega)}\big\|\mathcal{B}[f(t_n)] - \mathcal{B}[f(t_n)]\big\|_{L^\infty(\Omega)} + e_n \le (1 + C\Delta t)e_n \end{aligned}$$

by only using the stability (3.3) of the mapping $\mathcal{B}[\cdot]$. For the second term, we find

$$e_{n+1,2} \lesssim \|(I - P_{\Lambda_p^{n+1}})\mathcal{T} f_n\|_{L^\infty(\Omega)} + \|(I - P_{\Lambda^{n+1}})P_{\Lambda_p^{n+1}}\mathcal{T} f_n\|_{L^\infty(\Omega)} \lesssim \varepsilon.$$

Here we have used the fact that according to Theorem 6.9 and by construction, respectively, the trees Λ_p^{n+1} and Λ^{n+1} are $C\varepsilon$ and ε-adapted to $\mathcal{T} f_n$ and $P_{\Lambda_p^{n+1}}\mathcal{T} f_n$, respectively. The above inequality follows then from (4.15) (in the multilevel mesh case) or Lemma 5.13 (in the wavelet case). Note that in the wavelet case, the above properties of the predicted and readapted trees are weak and strong, respectively, which suffices for our purposes. The error estimate follows then by gathering the above bounds and applying a discrete Gronwall lemma. □

Of course, it remains to precisely determine which schemes and which initial data yield approximate flows satisfying Assumption 3.3. We shall leave this issue to further studies.

Remark 6.12. In contrast to what happens with uniform schemes, an a priori L^∞-bound is available for high-order adaptive semi-Lagrangian schemes where the interpolations may, in general, increase the supremum norm. Indeed, with $P_n := P_{\Lambda^n} P_{\Lambda_p^n}$, we always find for $\tilde{e}_n := \|\mathcal{T}^n f_0 - f_n\|_{L^\infty(\Omega)}$ the estimate

$$\tilde{e}_n \le \|\mathcal{T}^n f_0 - \mathcal{T} f_{n-1}\|_{L^\infty(\Omega)} + \|(I - P_n)\mathcal{T} f_{n-1}\|_{L^\infty(\Omega)} \le (1 + C\Delta t)\tilde{e}_{n-1} + C'\varepsilon$$

by using arguments from the above proof (and, in particular, with $C = 0$ in the case of a linear transport). This yields, again by employing a discrete Gronwall lemma,

$$\|f_n\|_{L^\infty(\Omega)} \leq \|\mathcal{T}^n f_0\|_{L^\infty(\Omega)} + \tilde{e}_n \leq \|f_0\|_{L^\infty(\Omega)} + C\varepsilon/\Delta t$$

with a constant independent of ε and Δt.

In addition to the above error estimate, we can bound the cardinalities of the predicted and readapted trees as follows:

$$\#(\Lambda^{n+1}) \lesssim \#(\Lambda_p^{n+1}) \lesssim \#(\Lambda^n),$$

but at the present stage we do not know how to estimate the growth of these cardinalities on the overall time period. Our conjecture is that, in the case of first order interpolations, they stay bounded by $C\varepsilon^{-1}$ as in (4.5). Hence balancing $\varepsilon \sim (\Delta t)^r$ in estimate (6.12) would yield

$$\sup_{n \leq N} \|f(t_n) - f_n\|_{L^\infty(\Omega)} \lesssim \varepsilon^{1-\frac{1}{r}} \lesssim \sup_{n \leq N} \left(\#(\Lambda^n)\right)^{\frac{1}{r}-1}.$$

This "adaptive" convergence rate should require less regularity than its "uniform" version (3.22) – involving the $W^{2,\infty}(\Omega)$-seminorm of f^0 – and this would express an advantage of the adaptive method over its uniform counterpart. As for high-order interpolations, we expect them to achieve high convergence rates. In any case, a rigorous complexity analysis of our scheme remains still to be done.

Acknowledgments. It is a pleasure to thank my colleagues from Strasbourg, as well as Albert Cohen, for very fruitful discussions. I would also like to thank Petra Wittbold and Etienne Emmrich for their kind invitation at the Technische Universität Berlin, and for the careful proof-reading which significantly improved the presentation of this work.

References

[1] R. A. Adams and J. J. F. Fournier, *Sobolev spaces*, second ed, Elsevier/Academic Press, Amsterdam, 2003.

[2] N. Besse, Convergence of a semi-Lagrangian scheme for the one-dimensional Vlasov–Poisson system, *SIAM J. Numer. Anal.* **42** (2004), pp. 350–382.

[3] N. Besse, F. Filbet, M. Gutnic, I. Paun and E. Sonnendrücker, *An adaptive numerical method for the Vlasov equation based on a multiresolution analysis*, Numerical Mathematics and Advanced Applications ENUMATH 2001 (F. Brezzi, A. Buffa, S. Escorsaro and A. Murli, eds.), Springer, 2001, pp. 437–446.

[4] N. Besse and M. Mehrenberger, Convergence of classes of high-order semi-Lagrangian schemes for the Vlasov–Poisson system, *Math. Comp.* **77** (2008), pp. 93–123.

[5] Yu. A. Brudnyj, Approximation of functions of n variables by quasi-polynomials, *Izv. Akad. Nauk SSSR Ser. Mat.* **34** (1970), pp. 564–583.

[6] M. Campos Pinto and M. Mehrenberger, *Adaptive numerical resolution of the Vlasov equation*, Numerical methods for hyperbolic and kinetic problems, CEMRACS 2003/IRMA Lectures in Mathematics and Theoretical Physics (S. Cordier, T. Goudon, M. Gutnic and E. Sonnendrücker, eds.), European Mathematical Society, 2005.

[7] ———, Convergence of an adaptive semi-Lagrangian scheme for the Vlasov–Poisson system, *Numer. Math.* **108** (2008), pp. 407–444.

[8] A. S. Cavaretta, W. Dahmen and C. A. Micchelli, Stationary subdivision, *Mem. Amer. Math. Soc.* **93** (1991).

[9] C. Z. Cheng and G. Knorr, The integration of the Vlasov equation in configuration space, *J. Comput. Phys.* **22** (1976), pp. 330–351.

[10] P. G. Ciarlet, *Basic error estimates for elliptic problems*, Handbook of numerical analysis, Vol. II, North-Holland, Amsterdam, 1991, pp. 17–351.

[11] A. Cohen, *Numerical analysis of wavelet methods*, North-Holland, Amsterdam, 2003.

[12] A. Cohen, S. M. Kaber, S. Müller and M. Postel, Fully adaptive multiresolution finite volume schemes for conservation laws, *Math. Comp.* **72** (2003), pp. 183–225.

[13] J. Cooper and A. Klimas, Boundary value problems for the Vlasov–Maxwell equation in one dimension, *J. Math. Anal. Appl.* **75** (1980), pp. 306–329.

[14] G.-H. Cottet and P.-A. Raviart, Particle methods for the one-dimensional Vlasov–Poisson equations, *SIAM J. Numer. Anal.* **21** (1984), pp. 52–76.

[15] I. Daubechies, *Ten lectures on wavelets*, CBMS-NSF Regional Conference Series in Applied Mathematics 61, SIAM, Philadelphia, 1992.

[16] G. Deslauriers and S. Dubuc, Symmetric iterative interpolation processes, *Constr. Approx.* **5** (1989), pp. 49–68.

[17] R. DeVore, Nonlinear approximation, *Acta Numerica* **7** (1998), pp. 51–150.

[18] N. Dyn and D. Levin, Subdivision schemes in geometric modelling, *Acta Numerica* **11** (2002), pp. 73–144.

[19] R. T. Glassey, *The Cauchy problem in kinetic theory*, SIAM, Philadelphia, 1996.

[20] M. Gutnic, M. Haefele, I. Paun and E. Sonnendrücker, Vlasov simulations on an adaptive phase-space grid, *Comput. Phys. Comm* **164** (2004), pp. 214–219.

[21] S. V. Iordanskij, The Cauchy problem for the kinetic equation of plasma, *Amer. Math. Soc. Transl. Ser. 2,* **35** (1964), pp. 351–363.

[22] G. Rein, *Collisionless kinetic equations from astrophysics – The Vlasov–Poisson system*, Handbook of Differential Equations, Evolutionary Equations, Vol. 3 (C. M. Dafermos and E. Feireisl, eds.), Elsevier, Oxford, 2005.

[23] E. Sonnendrücker, J. Roche, P. Bertrand and A. Ghizzo, The semi-Lagrangian method for the numerical resolution of the Vlasov equation, *J. Comput. Phys.* **149** (1999), pp. 201–220.

[24] A. A. Vlasov, A new formulation of the many particle problem (Russian), *Akad. Nauk SSSR. Zhurnal Eksper. Teoret. Fiz.* **18** (1948), pp. 840–856.

Author information

Martin Campos Pinto, IRMA institute (Université de Strasbourg & CNRS), 7 rue René Descartes, 67084 Strasbourg, France.
E-mail: campos@math.u-strasbg.fr

Coupling of a scalar conservation law with a parabolic problem

Julien Jimenez

Abstract. We deal with the mathematical analysis of the coupling between a purely quasilinear hyperbolic problem and a parabolic one. First, we use the Schauder–Tikhonov fixed point theorem and the Holmgren-type duality method to show existence and uniqueness for a class of nonlinear parabolic problems. In a second part, we recall the method of doubling variables and the vanishing viscosity method to obtain uniqueness and existence for a class of hyperbolic problems. Then we combine and adapt the methods stated in the two previous parts to obtain an existence and uniqueness result for the coupled problem we consider.

Keywords. Parabolic equation, hyperbolic equation, entropy solution, entropy process.

AMS classification. 35A05, 35K65, 35L65, 35R05.

1 Introduction

These lecture notes are devoted to the mathematical analysis of the coupling of scalar conservation laws with parabolic problems. The main motivation for considering such problems stems from the study of infiltration processes in a heterogeneous porous medium. For instance, in a stratified subsoil made up of layers with different geological characteristics, the effects of diffusivity may be negligible in some layers. We will focus on the case of two layers.

Indeed, let Ω be a bounded domain of $\mathbb{R}^n$, $n \geq 1$, with a smooth boundary Γ. We assume that $\overline{\Omega} = \overline{\Omega}_h \cup \overline{\Omega}_p$, Ω_h and Ω_p being two disjoint bounded domains of $\mathbb{R}^n$, with Lipschitz boundary denoted by $\Gamma_l = \partial\Omega_l$, $l \in \{h, p\}$. We suppose that the interface $\Gamma_{hp} = \Gamma_h \cap \Gamma_p$ is such that the Lebesgue measure of the set $(\overline{\Gamma}_{hp} \cap (\overline{\Gamma_l \setminus \Gamma_{hp}}))$ is zero. The time under consideration is denoted by $T > 0$. For $l \in \{h, p\}$, $\boldsymbol{\nu}_l$ is the outward unit normal vector defined a.e. on Γ_l. We set $Q_l = (0, T) \times \Omega_l$, $\Sigma_l = (0, T) \times \Gamma_l$. Finally, the interface including the time variable is denoted by $\Sigma_{hp} = (0, T) \times \Gamma_{hp}$.

The mathematical model under consideration originates from a combination of conservation laws and Darcy's law and can be described as follows: Find a measurable and essentially bounded function u on $Q := (0, T) \times \Omega$ such that in the sense of distributions

$$\begin{cases} \partial_t u + \nabla \cdot (\boldsymbol{b}(x) f(u)) &= 0 & \text{in } Q_h, \\ \partial_t u + \nabla \cdot (\boldsymbol{b}(x) f(u)) &= \Delta\phi(u) & \text{in } Q_p, \\ u &= 0 & \text{on } (0, T) \times \Gamma, \\ u(0, \cdot) &= u_0 & \text{on } \Omega. \end{cases} \tag{1.1}$$

Of course, suitable conditions on u across the interface Σ_{hp} must be added. These transmission conditions include the continuity of the flux through the interface, formally written here as

$$-f(u)\boldsymbol{b}\cdot\boldsymbol{\nu}_h = (\nabla\phi(u) + f(u)\boldsymbol{b})\cdot\boldsymbol{\nu}_p \text{ on } \Sigma_{hp}.$$

Regarding the data of the problem, we assume the following: The initial datum u_0 belongs to $L^\infty(\Omega)$ and there exist $m, M \in \mathbb{R}$ such that for almost all $x \in \Omega$

$$m \leq u_0(x) \leq M.$$

The vector field $\boldsymbol{b} = (b_1, \ldots, b_n)$ is an element of $W^{2,\infty}(\Omega)^n$. We denote by M_{b_i} the Lipschitz constant of b_i, $i = 1, \ldots, n$, and set $M_b = \sum_{i=1}^n M_{b_i}$. To simplify the analysis, we assume

$$\nabla\cdot\boldsymbol{b}(x) = 0 \text{ for almost all } x \in \Omega. \tag{1.2}$$

The continuous function $f : \mathbb{R} \to \mathbb{R}$ is assumed to be Lipschitz continuous on $[m, M]$. We denote by M_f its Lipschitz constant. The right-hand side ϕ is an increasing continuously differentiable function on $\mathbb{R}$. By normalisation, we suppose $\phi(0) = 0$. We also suppose that there exists $\alpha > 0$ such that for all $\tau \in \mathbb{R}$

$$\phi'(\tau) \geq \alpha. \tag{1.3}$$

Assumption (1.3) means that the partial differential operator set in Q_h is not degenerated. This hypothesis is not necessary (we may suppose that this operator is weakly degenerated) but it avoids technical difficulties.

This lecture is organized as follows. In Section 2, we introduce some classical methods used in the theory of nonlinear parabolic equations. Indeed, thanks to the Schauder–Tikhonov fixed point theorem and a Holmgren-type duality method, we prove existence and uniqueness of weak solutions for a class of parabolic problems. In Section 3, we present the method of doubling variables in order to prove a uniqueness result for a class of hyperbolic problems. Existence is obtained via the vanishing viscosity method based upon the notion of entropy process solutions. Then, in Section 4, we show an existence and uniqueness result for problem (1.1) by combining and adapting the methods stated in the two previous sections.

For the reader's convenience, we collect in the following some notations that will be used in the sequel. We also give some classical results of functional analysis. However, we suppose that the definition and the main properties of Banach, Hilbert, Lebesgue and Sobolev spaces are known.

The standard norm in $H_0^1(\Omega)$ shall be denoted by $\|\cdot\|$. The duality pairing between $H_0^1(\Omega)$ and its dual space $H^{-1}(\Omega)$ is denoted by $\langle\cdot,\cdot\rangle$. The energetic extension of the classical differential operator $(-\Delta)$, supplemented by homogeneous Dirichlet boundary conditions, is an isomorphism from $H_0^1(\Omega)$ onto $H^{-1}(\Omega)$. The $L^\infty(\Omega)$-norm is denoted by $\|\cdot\|_\infty$.

Let X be a Banach space and $T > 0$. We denote by $L^p(0, T; X)$, $1 \leq p < \infty$, the space of functions $f : [0,T] \to X$ such that

$$\begin{cases} f \text{ is Bochner measurable,} \\ \|f\|_{L^p(0,T;X)} := \left(\int_0^T \|f\|_X^p dt \right)^{\frac{1}{p}} < \infty. \end{cases}$$

We denote by $\mathcal{C}([0,T];X)$ the space of continuous functions $u : [0,T] \to X$ with norm

$$\|u\|_{\mathcal{C}([0,T];X)} := \max_{t \in [0,T]} \|u(t)\|_X.$$

The function space

$$W(0,T) := \{u \in L^2(0,T;H_0^1(\Omega));\ u' \in L^2(0,T;H^{-1}(\Omega))\},$$

endowed with the standard norm $\|u\|_{W(0,T)} = (\|u\|^2_{L^2(0,T;H_0^1(\Omega))} + \|u'\|^2_{L^2(0,T;H^{-1}(\Omega))})^{1/2}$ and the corresponding inner product, is a Hilbert space, which is continuously embedded in $\mathcal{C}([0,T];L^2(\Omega))$ and compactly embedded in $L^2(0,T;L^2(\Omega)) \cong L^2((0,T)\times\Omega)$. In particular, if $u \in W(0,T)$ then $u(0) \in L^2(\Omega)$ and so $u(0)$ is defined a.e. on Ω. We will also use the fact that for all $u, v \in W(0,T)$ and almost all $t \in [0,T]$,

$$\frac{d}{dt} \int_\Omega u(t)v(t)dx = \langle u'(t), v(t)\rangle + \langle u(t), v'(t)\rangle. \tag{1.4}$$

We refer, e.g., to [5, Chap. XVIII] (see also [6] for an English translation of [5]) for more details on the function space $W(0,T)$. If $u \in W(0,T)$ then u is identified with a function $\tilde{u}$ defined on $[0,T]\times\overline{\Omega}$ by setting, for all $t \in [0,T]$, $x \in \overline{\Omega}$, $\tilde{u}(t,x) = [u(t)](x)$. The time derivative u' can be identified with the partial derivative $\partial_t \tilde{u}$. In the sequel, we will skip the "tilde", and $\partial_t u$ will denote the derivative u' as well as the partial derivative $\partial_t \tilde{u}$.

A sequence $(\rho_j)_{j\in\mathbb{N}}$ is called a sequence of mollifiers in $\mathbb{R}^n$ if, for all $j \in \mathbb{N}$, ρ_j is nonnegative, $\rho_j \in \mathcal{C}^\infty(\mathbb{R}^n;\mathbb{R})$, supp $\rho_j \subset \overline{B}(0, \frac{1}{j})$ and $\int_{\mathbb{R}^n} \rho_j(x)dx = 1$.

Let $x \in \mathbb{R}^n$. Then x is said to be a Lebesgue point of $f \in L^1_{\mathrm{loc}}(\mathbb{R}^n)$ if

$$\lim_{r\to 0^+} \frac{1}{\mathrm{meas}(B(x,r))} \int_{B(x,r)} |f(x) - f(y)|dy = 0,$$

where $B(x,r) = \{y \in \mathbb{R}^n;\ \|x-y\| < r\}$. We refer to [7] for the properties of a Lebesgue point and only mention that if f is continuous at x then x is a Lebesgue point of f. For $f \in L^1_{\mathrm{loc}}(\mathbb{R}^n)$, almost every point is a Lebesgue point. Therefore, if $v \in L^1(\mathcal{O})$ ($\mathcal{O}$ is an open bounded subset of $\mathbb{R}^n$) then

$$\lim_{j\to\infty} \int_{\mathcal{O}\times\mathcal{O}} (v(y) - v(x))\rho_j(y-x)dxdy = 0.$$

The following lemma is due to Mignot and Bamberger (see [9, p. 31]); a generalization can be found in [18].

Lemma 1.1. *Let $\beta : \mathbb{R} \to \mathbb{R}$ be a continuous and increasing function such that*

$$\limsup_{|\lambda|\to\infty} \frac{|\beta(\lambda)|}{|\lambda|} < \infty.$$

Then, for any function $u \in L^2(0,T;L^2(\Omega))$ with $\partial_t u \in L^2(0,T;H^{-1}(\Omega))$ and $\beta(u) \in L^2(0,T;H_0^1(\Omega))$, the following integration-by-parts formula holds in $L^1(0,T)$:

$$\langle \partial_t u, \beta(u)\rangle = \frac{d}{dt}\int_\Omega \left(\int_0^{u(\cdot,x)} \beta(r)dr\right)dx.$$

For the following well-known Gronwall lemma, we refer, e.g., to [5, p. 672].

Lemma 1.2 (Gronwall). *Let $T > 0$, $\phi \in L^\infty(0,T)$, $\phi \geq 0$ a.e. on $(0,T)$, $\mu \in L^1(0,T)$, $\mu \geq 0$ a.e. on $(0,T)$. Suppose there exists $\kappa > 0$ such that, for almost all $s \in (0,T)$,*

$$\phi(s) \leq \kappa + \int_0^s \mu(t)\phi(t)dt.$$

Then, for almost all $s \in (0,T)$,

$$\phi(s) \leq \kappa \exp\left(\int_0^s \mu(t)dt\right).$$

The following auxiliary result can be found in [16].

Proposition 1.3. *Let $w \in H^1(\Omega)$ and let $g : \mathbb{R} \to \mathbb{R}$ be a Lipschitz continuous function. Then $g \circ w \in H^1(\Omega)$ and for all $i = 1,\dots,n$*

$$\partial_{x_i} g(w) = g'(w)\partial_{x_i} w \quad \text{a.e. on } \Omega.$$

We also make use of a variant of the well-known Schauder–Tikhonov fixed point theorem that can be found in [9, p. 30].

Theorem 1.4. *Let X be a reflexive and separable Banach space and $K \subset X$, $K \neq \emptyset$, be a closed, bounded and convex set. Let $\mathcal{T} : K \to K$ be a "weakly-weakly" sequentially continuous mapping, i.e., for any sequence $(x_n)_{n\in\mathbb{N}} \subset K$ that converges weakly towards some $x \in X$, the sequence $(\mathcal{T}(x_n))_{n\in\mathbb{N}}$ converges weakly towards $\mathcal{T}(x)$. Then $\mathcal{T}$ has at least one fixed point in K.*

Finally, we employ the following result that goes back to Eymard, Gallouët and Herbin, see [8].

Theorem 1.5. *Let $\mathcal{O}$ be an open bounded set of $\mathbb{R}^n$ and $(u_j)_{j\in\mathbb{N}}$ a sequence of measurable essentially bounded functions on $\mathcal{O}$. If there exists $A > 0$ such that*

$$\|u_j\|_{L^\infty(\mathcal{O})} \leq A \text{ for all } j \in \mathbb{N}$$

then there exist a subsequence $(u_{\varphi(j)})_{j\in\mathbb{N}}$ and a function $\pi \in L^\infty((0,1)\times\mathcal{O})$ such that for any continuous and bounded function $f : \mathcal{O}\times(-A,A) \to \mathbb{R}$

$$\lim_{j\to\infty}\int_{\mathcal{O}} f(x,u_{\varphi(j)}(x))\xi dx = \int_0^1\int_{\mathcal{O}} f(x,\pi(\alpha,x))\xi dx d\alpha \quad \forall\, \xi \in L^1(\mathcal{O}).$$

Let sgn : $\mathbb{R} \to \mathbb{R}$ be given by

$$\operatorname{sgn}(x) = \begin{cases} -1 & \text{if } x < 0, \\ 0 & \text{if } x = 0, \\ 1 & \text{if } x > 0, \end{cases}$$

and denote by sgn_η ($\eta > 0$) its Lipschitz continuous approximation given by

$$\operatorname{sgn}_\eta(x) = \begin{cases} -1 & \text{if } x \leq -\eta, \\ \dfrac{x}{\eta} & \text{if } -\eta < x < \eta, \\ 1 & \text{if } x \geq \eta. \end{cases}$$

Furthermore, for all $x \in \mathbb{R}$, $\eta > 0$, we set

$$\operatorname{sgn}_\eta^+(x) := \max(\operatorname{sgn}_\eta(x), 0) \text{ and } \operatorname{sgn}_\eta^-(x) := \max(-\operatorname{sgn}_\eta(x), 0).$$

2 The parabolic problem

In this part, we collect some tools that will be useful in the sequel in order to obtain existence and uniqueness results for a class of nonlinear parabolic equations. One may assume that we treat the special case $\Omega_h = \emptyset$ so that $\Omega_p = \Omega$. We consider the initial-boundary value problem

$$\begin{cases} \partial_t u + \nabla \cdot (\boldsymbol{b}(x) f(u)) &= \Delta\phi(u) & \text{in } Q, \\ u &= 0 & \text{on } (0,T) \times \Gamma, \\ u(0,\cdot) &= u_0 & \text{on } \Omega. \end{cases} \tag{2.1}$$

The assumptions on the data are the same as in Section 1. In particular, $u_0 \in L^\infty(\Omega)$ with $m \leq u_0(x) \leq M$ for almost all $x \in \Omega$. The vector field $\boldsymbol{b} \in W^{2,\infty}(\Omega)^n$ satisfies (1.2). The function f is Lipschitz continuous on $[m, M]$ and ϕ is an increasing continuously differentiable function on $\mathbb{R}$ that fulfills (1.3) with $\phi(0) = 0$.

First, we provide a definition of a weak solution to (2.1).

Definition 2.1. A function $u \in W(0,T)$ with

$$m \leq u(t,x) \leq M \text{ for almost all } (t,x) \in Q \tag{2.2}$$

is said to be a weak solution to (2.1) if for almost all $t \in (0,T)$ and all $v \in H_0^1(\Omega)$ the equation

$$\langle \partial_t u(t), v \rangle + \int_\Omega (\nabla\phi(u(t)) - \boldsymbol{b}(x) f(u(t))) \cdot \nabla v dx = 0 \tag{2.3}$$

is fulfilled and if

$$u(0) = u_0 \text{ in } L^2(\Omega). \tag{2.4}$$

Remark 2.2. As is mentioned in Section 1, we have identified in Definition 2.1 the function $u \in W(0,T)$ with the function $\tilde{u}$ defined on $Q = (0,T) \times \Omega$ such that $\tilde{u}(t,x) = [u(t)](x)$ for almost all $t \in (0,T)$, $x \in \Omega$.

2.1 Existence: the fixed point method

This part is devoted to the proof of the following theorem:

Theorem 2.3. *There exists at least one weak solution to* (2.1).

The main idea is to use the Schauder–Tikhonov fixed point theorem. To do so, we need a result about linear parabolic problems due to J. L. Lions that can be found in [5, Chap. XVIII, p. 629] and also in [19, Chap. IV, p. 397].

Theorem 2.4 (Lions). *For any $t \in [0,T]$, let $a(t;\cdot,\cdot)$ be a bilinear form on $H_0^1(\Omega) \times H_0^1(\Omega)$ such that $t \mapsto a(t;u,v)$ is measurable on $[0,T]$ for all $u, v \in H_0^1(\Omega)$. The bilinear form is assumed to be uniformly bounded and strongly positive, i.e., there are constants $\beta, \gamma > 0$ such that for all $u, v \in H_0^1(\Omega)$, $t \in [0,T]$,*

$$|a(t;u,v)| \leq \beta\|u\|\|v\| \text{ and } a(t;u,u) \geq \gamma\|u\|^2.$$

Then, for any $u_0 \in L^2(\Omega)$ and $g \in L^2(0,T;H^{-1}(\Omega))$, there exists a unique solution $u \in W(0,T)$ such that for almost all $t \in (0,T)$ and all $v \in H_0^1(\Omega)$

$$\langle u'(t), v\rangle + a(t;u(t),v) = \langle g(t), v\rangle$$

and $u(0) = u_0$ in $L^2(\Omega)$.

Furthermore there exists a constant $C > 0$ depending on β and γ such that

$$\|u\|_{W(0,T)} \leq C(\|u_0\|_{L^2(\Omega)} + \|g\|_{L^2(0,T;H^{-1}(\Omega))}). \tag{2.5}$$

Proof of Theorem 2.3. The proof can be divided into three steps. In a first step, we introduce a modified problem with bounded coefficients (see (2.6)) that is equivalent to (2.2)–(2.4). In a second step, we consider a linearized problem related to (2.6). With the help of Theorem 2.4, we prove that this linear problem has a unique solution. This allows us to define a mapping $\mathcal{T}$ on $W(0,T)$ whose possible fixed points are solutions to (2.6). Finally, in the third step, we show that Theorem 1.4 can be applied to the mapping $\mathcal{T}$.

Step 1: Truncation process. We introduce the following auxiliary problem: Find u in $W(0,T)$ such that a.e. on $(0,T)$ and for any $v \in H_0^1(\Omega)$,

$$\begin{cases} \langle \partial_t u(t), v\rangle + \displaystyle\int_\Omega (\phi'(u^*(t))\nabla u(t) - \boldsymbol{b}(x) f(u^*(t))) \cdot \nabla v dx = 0, \\ u(0) = u_0 \text{ in } L^2(\Omega), \end{cases} \tag{2.6}$$

where, for almost all $(t,x) \in Q$,

$$u^*(t,x) = \begin{cases} m & \text{if } u(t,x) < m, \\ u(t,x) & \text{if } m \leq u(t,x) \leq M, \\ M & \text{if } u(t,x) > M. \end{cases}$$

Let us prove that (2.2)–(2.4) and (2.6) are equivalent. It is clear that if u satisfies (2.2)–(2.4) then u fulfills (2.6).

Conversely, suppose that u is a solution to (2.6). Let us show that u satisfies (2.2). We consider in (2.6) the test function $v_\eta = \mathrm{sgn}_\eta^+(u - M)$ (thanks to Proposition 1.3, $v_\eta \in W(0,T)$) and we integrate over $(0,s)$ for any s of $(0,T]$. This yields

$$\int_0^s \langle \partial_t u, v_\eta \rangle dt + \int_0^s \int_\Omega (\phi'(u^*)\nabla u - \boldsymbol{b}(x) f(u^*)) \cdot \nabla v_\eta dx = 0. \tag{2.7}$$

For the evolution term, we use Lemma 1.1 and the fact that $u_0 \leq M$ a.e. on Ω to obtain

$$\int_0^s \langle \partial_t u, v_\eta \rangle dt = \int_0^s \langle \partial_t (u - M), v_\eta \rangle dt = \int_\Omega \left(\int_0^{u(s,x)-M} \mathrm{sgn}_\eta^+(r) dr \right) dx.$$

To treat the convection term, we refer to the definition of sgn_η^+ and u^*. We find

$$\begin{aligned}\int_0^s \int_\Omega \boldsymbol{b}(x) f(u^*) \nabla v_\eta dx dt &= \int_0^s \int_{\{x\in\Omega;\, 0<u(t,x)-M<\eta\}} \boldsymbol{b}(x) f(u^*) \nabla v_\eta dx dt \\ &= \int_0^s \int_\Omega \boldsymbol{b}(x) f(M) \nabla v_\eta dx dt.\end{aligned}$$

By the Gauss–Green formula and (1.2), we have

$$\int_0^s \int_\Omega \boldsymbol{b}(x) f(M) \nabla v_\eta dx dt = - \int_0^s \int_\Omega f(M) \nabla \cdot \boldsymbol{b}(x) v_\eta dx dt = 0.$$

We use Proposition 1.3 and the definition of sgn_η^+ for the diffusion term to deduce

$$\int_0^s \int_\Omega \phi'(u^*) \nabla u \cdot \nabla v_\eta dx dt = \frac{1}{\eta} \int_0^s \int_{\{x\in\Omega;\, 0<u(t,x)-M<\eta\}} \phi'(u^*) \nabla u \cdot \nabla u dx dt \geq 0.$$

Then, from (2.7), for all $s \in (0,T]$, $\eta > 0$,

$$\int_\Omega \left(\int_0^{u(s,x)-M} \mathrm{sgn}_\eta^+(r) dr \right) dx \leq 0.$$

Thus, for all $s \in (0,T]$, we find

$$\lim_{\eta \to 0^+} \int_\Omega \left(\int_0^{u(s,x)-M} \mathrm{sgn}_\eta^+(r) dr \right) dx = \int_\Omega (u(s,x) - M)^+ dx \leq 0,$$

where $(u(s,x) - M)^+ = \mathrm{sgn}^+(u(s,x) - M)((u(s,x) - M))$ for almost all $x \in \Omega$. It follows that $u \leq M$ a.e. on Q.

To prove that $u \geq m$ a.e. on Q, the reasoning is similar: We just consider the test function $\mathrm{sgn}_\eta^-(u - m)$ in (2.6). Then u satisfies (2.2), (2.4) and, consequently,

(2.3) (since $u^* = u$ a.e. on Q). Thus, the existence result for (2.1) is equivalent to an existence result to (2.6).

Step 2: Study of a linear problem. For any given $w \in W(0,T)$, we consider now the linear problem: Find $U_w \in W(0,T)$ such that, a.e. on $(0,T)$ and for all $v \in H_0^1(\Omega)$

$$\begin{cases} \langle \partial_t U_w, v\rangle + \int_\Omega (\phi'(w^*)\nabla U_w - f(w^*)\boldsymbol{b}(x)) \cdot \nabla v dx = 0, \\ U_w(0) = u_0 \text{ in } L^2(\Omega). \end{cases} \tag{2.8}$$

In order to prove the existence of a unique solution $U_w \in W(0,T)$ to problem (2.8), we set for all $t \in [0,T]$ and all $u, v \in H_0^1(\Omega)$

$$a(t; u, v) := \int_\Omega \phi'(w^*(t,x))\nabla u(x) \cdot \nabla v(x) dx$$

and

$$\langle g(t), v\rangle := \int_\Omega f(w^*(t,x))\boldsymbol{b}(x) \cdot \nabla v(x) dx.$$

We have $g \in L^2(0,T; H^{-1}(\Omega))$ (by the Cauchy–Schwarz inequality) and, for almost all $t \in (0,T)$,

$$|a(t;u,v)| \le \|\phi'(w^\star)\|_\infty \|u\|\|v\| \le \|\phi'\|_{\infty,[m,M]}\|u\|\|v\| \text{ and } a(t;u,u) \ge \alpha\|u\|^2,$$

where $\|\phi'\|_{\infty,[m,M]} := \max_{\tau\in[m,M]} |\phi'(\tau)|$.

Hence, Theorem 2.4 ensures that there exists a unique solution $U_w \in W(0,T)$ to (2.8). Moreover, it also follows from Theorem 2.4 that there exists $C > 0$ such that

$$\|U_w\|_{W(0,T)} \le C. \tag{2.9}$$

Since, for any $w \in W(0,T)$, the problem (2.8) has a unique solution $U_w \in W(0,T)$, we may define the operator

$$\mathcal{T} : W(0,T) \to W(0,T),\ w \mapsto U_w.$$

The aim is now to prove that $\mathcal{T}$ admits at least one fixed point. Indeed, if there exists $u \in W(0,T)$ such that $\mathcal{T}(u) = u$ then u will be a weak solution to (2.6).

Step 3: Use of the Schauder–Tikhonov theorem. We set

$$K := \{v \in W(0,T);\ v(0) = u_0 \text{ and } \|v\|_{W(0,T)} \le C\},$$

where C is the a priori bound given by (2.9). It is clear that K is non-empty, bounded, convex and closed. By (2.9), we also have $\mathcal{T}(K) \subset K$. Therefore, in order to apply the Schauder–Tikhonov fixed point theorem (Theorem 1.4), it remains to prove that $\mathcal{T}$ is "weakly-weakly" sequentially continuous. Let $(w_n)_{n\in\mathbb{N}} \subset K$ be a sequence that converges weakly towards $w \in K$. We have to prove that $(\mathcal{T}(w_n))_{n\in\mathbb{N}}$ converges weakly towards $\mathcal{T}(w)$.

By definition of $\mathcal{T}$, the function $U_{w_n} = \mathcal{T}(w_n)$ satisfies, a.e. in $(0,T)$ and for all $v \in H_0^1(\Omega)$,

$$\begin{cases} \langle \partial_t U_{w_n}, v\rangle + \displaystyle\int_\Omega (\phi'(w_n^*)\nabla U_{w_n} - f(w_n^*)\boldsymbol{b}(x)) \cdot \nabla v dx = 0, \\ U_{w_n}(0) = u_0 \text{ in } L^2(\Omega). \end{cases} \tag{2.10}$$

In order to take the limit with respect to n in (2.10), we need to specify some convergence results for the sequences $(w_n)_{n\in\mathbb{N}}$ and $(U_{w_n})_{n\in\mathbb{N}}$. Since $W(0,T) \overset{c}{\hookrightarrow} L^2(0,T;L^2(\Omega)) \cong L^2(Q)$, there exists a subsequence, still denoted by $(w_n)_{n\in\mathbb{N}}$, that converges strongly in $L^2(Q)$ to w.

From (2.9), we have $\|U_{w_n}\|_{W(0,T)} \le C$. Hence, there exists a subsequence, still denoted by $(U_{w_n})_{n\in\mathbb{N}}$, such that $U_{w_n} \rightharpoonup U$ in $W(0,T)$, i.e.,

$$\partial_t U_{w_n} \rightharpoonup \partial_t U \text{ in } L^2(0,T;H^{-1}(\Omega)) \text{ and } U_{w_n} \rightharpoonup U \text{ in } L^2(0,T;H_0^1(\Omega)). \tag{2.11}$$

Since $W(0,T) \overset{c}{\hookrightarrow} L^2(Q)$, the strong convergence

$$U_{w_n} \to U \text{ in } L^2(Q)$$

follows. Moreover, since $W(0,T) \hookrightarrow \mathcal{C}([0,T],L^2(\Omega))$, we have

$$U_{w_n}(0) \rightharpoonup U(0) \text{ in } L^2(\Omega).$$

As $U_{w_n}(0) = u_0$ for all $n \in \mathbb{N}$, we deduce that $U(0) = u_0$ in $L^2(\Omega)$.

Now it is possible to take the limit with respect to n in (2.10). Since f is Lipschitz continuous on $[m,M]$, $\boldsymbol{b}$ is bounded and $w_n \to w$ in $L^2(Q)$, we also have

$$\boldsymbol{b}f(w_n^*) \to \boldsymbol{b}f(w^*) \text{ in } L^2(Q)^n.$$

Since $m \le w_n^* \le M$ and ϕ' is continuous, the subsequence can be chosen such that $\phi'(w_n^*)\nabla U_{w_n} \rightharpoonup \phi'(w^*)\nabla U$ in $L^2(Q)^n$. Taking the limit in (2.10) now shows that $U \in W(0,T)$ fulfills a.e. in $(0,T)$ and for all $v \in H_0^1(\Omega)$

$$\langle \partial_t U, v\rangle dt + \int_\Omega (\phi'(w^*)\nabla U - \boldsymbol{b}(x)f(w^*))\nabla v dx = 0$$

as well as $U(0) = u_0$. Hence, $U = \mathcal{T}(w)$.

Up to now, we have shown that there exists a subsequence of $(\mathcal{T}(w_n))_{n\in\mathbb{N}}$ that converges towards $\mathcal{T}(w)$. To see that the whole sequence $(\mathcal{T}(w_n))_{n\in\mathbb{N}}$ converges towards $\mathcal{T}(w)$, we remark that the limit of any convergent subsequence $(\mathcal{T}(w_n))_{n\in\mathbb{N}}$ satisfies (2.8). Since (2.8) has a unique solution, we deduce by contradiction that the whole sequence converges towards $\mathcal{T}(w)$.

Thus the assumptions of the Schauder–Tikhonov theorem are satisfied, and we may conclude that there exists $u \in W(0,T)$ such that $u = \mathcal{T}(u)$. Eventually, this fixed point u is a weak solution to (2.6) and thus to (2.1). □

2.2 Uniqueness: the Holmgren-type duality method

We are now interested in the uniqueness of solutions to (2.1), which shall be proved by contradiction. Let u and $\hat{u}$ be two weak solutions to (2.1). The idea is to use the $H^{-1}(\Omega)$-norm in order to estimate $(u-\hat{u})$. Since $(-\Delta) : H_0^1(\Omega) \to H^{-1}(\Omega)$ is an isomorphism, the $H^{-1}(\Omega)$-norm of $u-\hat{u}$ is equivalent to $\|(-\Delta)^{-1}(u-\hat{u})\|$.

For any $t \in [0,T]$, let $z(t) \in H_0^1(\Omega)$ be the weak solution to

$$\begin{cases} -\Delta z(t) &= u(t) \quad \text{in } \Omega, \\ z(t) &= 0 \quad \text{on } \partial\Omega, \end{cases}$$

and analogously $\hat{z}(t) \in H_0^1(\Omega)$ be the weak solution of the foregoing Poisson problem with right-hand side $\hat{u}(t)$, i.e., for all $v \in H_0^1(\Omega)$,

$$\int_\Omega \nabla z(t) \cdot \nabla v dx = \int_\Omega u(t) v dx \quad (\text{resp. } \int_\Omega \nabla \hat{z}(t) \cdot \nabla v dx = \int_\Omega \hat{u}(t) v dx). \tag{2.12}$$

Remember that $u, \hat{u} \in \mathcal{C}([0,T]; L^2(\Omega))$.

Since $\partial_t u \in L^2(0,T;H^{-1}(\Omega))$, we are able to define $\partial_t z(t)$ (resp. $\partial_t \hat{z}(t)$) for almost all $t \in (0,T)$ as elements of $H_0^1(\Omega)$ characterized by

$$\int_\Omega \nabla(\partial_t z(t)) \cdot \nabla v dx = \langle \partial_t u, v\rangle \quad (\text{resp. } \int_\Omega \nabla(\partial_t \hat{z}(t)) \cdot \nabla v dx = \langle \partial_t \hat{u}, v\rangle) \tag{2.13}$$

for all $v \in H_0^1(\Omega)$. Indeed, the isomorphism $(-\Delta)^{-1} : H^{-1}(\Omega) \to H_0^1(\Omega)$ can be extended to a linear and continuous mapping of $L^2(0,T;H^{-1}(\Omega))$ onto $L^2(0,T;H_0^1(\Omega))$.

We subtract the equations (2.3) for the solutions u and $\hat{u}$ and take $v = (z-\hat{z})(t)$. We then integrate from 0 to s, $s \in (0,T]$, and obtain

$$\int_0^s \langle \partial_t(u-\hat{u}), z-\hat{z}\rangle dt + \int_0^s \int_\Omega (\nabla(\phi(u)-\phi(\hat{u})) - (f(u)-f(\hat{u}))\boldsymbol{b}(x)) \cdot \nabla(z-\hat{z}) dx dt = 0. \tag{2.14}$$

From (2.13), it follows with an integration by parts that

$$\begin{aligned} \int_0^s \langle \partial_t(u-\hat{u}), z-\hat{z}\rangle dt &= \int_0^s \int_\Omega \nabla(\partial_t(z-\hat{z})) \cdot \nabla(z-\hat{z}) dx dt \\ &= \frac{1}{2}\int_\Omega |\nabla(z-\hat{z})|^2(s) dx - \frac{1}{2}\int_\Omega |\nabla(z-\hat{z})|^2(0) dx. \end{aligned}$$

With (2.12) for $t=0$ and since $u(0) = \hat{u}(0) = u_0$, we find

$$\int_\Omega |\nabla(z-\hat{z})|^2(0) dx = \int_\Omega (u(0)-\hat{u}(0))^2 dx = 0.$$

We choose $v = \phi(u(t)) - \phi(\hat{u}(t))$ in (2.12) written for $z(t)$ and $\hat{z}(t)$, integrate over $(0,s)$ (with $Q_s = (0,s) \times \Omega$) and get, using (1.3),

$$\begin{aligned} \int_0^s \int_\Omega \nabla(\phi(u)-\phi(\hat{u})) \cdot \nabla(z-\hat{z}) dx dt &= \int_0^s \int_\Omega (u-\hat{u})(\phi(u)-\phi(\hat{u})) dx dt \\ &\geq \alpha\|u-\hat{u}\|^2_{L^2(Q_s)}. \end{aligned}$$

Then, from (2.14), we deduce that

$$\frac{1}{2}\int_\Omega |\nabla(z-\hat{z})|^2(s)dx + \alpha\|u-\hat{u}\|^2_{L^2(Q_s)} \le \int_0^s \int_\Omega \boldsymbol{b}(x)(f(u)-f(\hat{u}))\cdot\nabla(z-\hat{z})dxdt. \tag{2.15}$$

As $\boldsymbol{b}$ is bounded and f is Lipschitz continuous on $[m, M]$, we see that

$$\begin{aligned}
&\left|\int_0^s \int_\Omega \boldsymbol{b}(x)(f(u)-f(\hat{u}))\cdot\nabla(z-\hat{z})dx\right| \\
&\quad \le \|\boldsymbol{b}\|_{L^\infty(\Omega)^n} M_f \|u-\hat{u}\|_{L^2(Q_s)}\|z-\hat{z}\|_{L^2(0,s;H_0^1(\Omega))}. \\
&\quad \le \frac{\alpha}{2}\|u-\hat{u}\|^2_{L^2(Q_s)} + C_\alpha\|\nabla(z-\hat{z})\|^2_{L^2(Q_s)^n},
\end{aligned} \tag{2.16}$$

where $C_\alpha > 0$ depends on α, $\boldsymbol{b}$ and f.

Finally, from (2.15) and (2.16), we deduce

$$\frac{1}{2}\int_\Omega |\nabla(z-\hat{z})|^2(s)dx \le C_\alpha\|\nabla(z-\hat{z})\|^2_{L^2(Q_s)^n}.$$

The conclusion follows by using Gronwall's lemma (see Lemma 1.2).

3 The hyperbolic problem

In this section, we consider the special case $\Omega_p = \emptyset$ so that $\Omega = \Omega_h$. The problem we aim to solve can be formally written as follows: Find a measurable and essentially bounded function u such that in the sense of distributions

$$\left\{\begin{array}{rcll} \partial_t u + \nabla\cdot(\boldsymbol{b}(x)f(u)) &=& 0 & \text{in } Q, \\ u(0,\cdot) &=& u_0 & \text{on } \Omega, \\ u &=& 0 & \text{(on a part of) } \Gamma. \end{array}\right. \tag{3.1}$$

To begin with, we introduce some notations and definitions: The outward unit normal vector of Ω defined a.e. on Γ is denoted by $\boldsymbol{\nu}$. An element of Σ is denoted by σ, while $\overline{\sigma}$ is an element of Γ so that $\sigma = (t, \overline{\sigma})$. For all $a, b \in \mathbb{R}$, we set $F(a,b) = (f(a)-f(b))\mathrm{sgn}(a-b)$. The function F is Lipschitz continuous on $[m, M]^2$ and usually called the "Kruzhkov flux".

For all $\tau, k \in \mathbb{R}$, we define

$$\mathcal{F}_0(\tau, k) = \frac{1}{2}(F(\tau,0) - F(k,0) + F(\tau,k)).$$

Definition 3.1 (regular entropy pair). Let $\eta \in C^1(\mathbb{R})$ be a convex function and $q \in C^1(\mathbb{R})$. Then (η, q) is called a regular entropy pair to problem (3.1) if, for all $r \in \mathbb{R}$,

$$q'(r) = f'(r)\eta'(r).$$

Remark 3.2. In [15, Def. 3.22], (η, q) is called an entropy-entropy flux pair.

Following [17], we also give

Definition 3.3 (boundary entropy-entropy flux pair). The pair (H, J) with $H \in \mathcal{C}^2(\mathbb{R}^2)$, $J \in \mathcal{C}^2(\mathbb{R}^2)$ is said to be a boundary entropy-entropy flux pair if $(H(\cdot, w), J(\cdot, w))$ is a regular entropy pair and

$$H(w,w) = 0, \quad J(w,w) = 0, \quad \partial_1 H(w,w) = 0$$

for all $w \in \mathbb{R}$, where ∂_1 denotes the partial derivative with respect to the first argument.

In the sequel, we will use the particular boundary entropy-entropy flux pair below.

Example 3.4. Let $\delta > 0$. We define on $\mathbb{R}^2$ the functions H_δ and J_δ by

$$H_\delta(\tau, k) = ((\text{dist}(\tau, I[0,k]))^2 + \delta^2)^{1/2} - \delta$$

and

$$J_\delta(\tau, k) = \int_k^\tau \partial_1 H_\delta(\lambda, k) f'(\lambda) d\lambda.$$

Here $I[0,k]$ denotes the closed interval with the endpoints 0 and k. Then, for any δ, (H_δ, J_δ) is a boundary entropy-entropy flux pair. Moreover, the pair converges uniformly towards $(\text{dist}(\tau, I[0,k]), \mathcal{F}_0)$ as $\delta \to 0^+$. This example of boundary entropy-entropy flux pair will be used to obtain the boundary condition given for (3.1).

3.1 The notion of entropy solution

It is well known (see for example [10]) that a weak solution to (3.1) may not be unique. So we need an additional condition to select one solution among all the weak solutions, which has to be the physically relevant solution. To this end we introduce the notion of entropy solution. Another difficulty is to formulate a boundary condition for (3.1). Indeed, it is not possible to define a trace, in a strong sense, for a L^∞-function. This difficulty has been overcome by F. Otto (see [17]) using the so-called boundary entropy-entropy flux pair defined above.

Definition 3.5. A function $u \in L^\infty(Q)$ is said to be an entropy solution to (3.1) if

(i) for all nonnegative $\varphi \in \mathcal{C}_c^\infty(Q)$ and all $k \in \mathbb{R}$

$$\int_Q (|u-k|\partial_t \varphi + \text{sgn}(u-k)(f(u) - f(k))\boldsymbol{b}(x) \cdot \nabla\varphi) dx dt \geq 0, \tag{3.2}$$

(ii)

$$\operatorname*{ess\,lim}_{t \to 0^+} \int_\Omega |u(t,x) - u_0(x)| dx = 0, \tag{3.3}$$

(iii) for all nonnegative $\varphi \in L^1(\Sigma)$ and all $k \in \mathbb{R}$

$$\operatorname*{ess\,lim}_{s\to 0^-} \int_\Sigma \mathcal{F}_0(u(\sigma + s\boldsymbol{\nu}), k)\boldsymbol{b}(\overline{\sigma}) \cdot \boldsymbol{\nu}\varphi d\sigma \geq 0. \tag{3.4}$$

Remark 3.6. The entropy inequality (3.2) implies that an entropy solution u is a solution in the sense of distributions.

3.2 Uniqueness: the method of doubling variables

This part is inspired from the book of J. Málek *et al.* [15, Chap. 2] and from the article of F. Otto [17]. We present the method of doubling variables, due to S. N. Kruzhkov [13]. More precisely, we prove the following theorem.

Theorem 3.7. *Let u_1 and u_2 be two entropy solutions to* (3.1) *for initial data $u_{0,1}$ and $u_{0,2}$, respectively. Then, for almost all $t \in (0,T)$,*

$$\int_\Omega |u_1(t,x) - u_2(t,x)|dx \leq \int_\Omega |u_{0,1}(x) - u_{0,2}(x)|dx.$$

We need first some preliminary results. We refer to [15, Chap. 2, Lemma 7.12 and 7.34] and [17] for the quite technical proofs of these results.

Lemma 3.8. *Let $u \in L^\infty(Q)$ satisfying* (3.2) *and* (3.3)*. Then, for all nonnegative $\varphi \in \mathcal{C}_c^\infty(\mathbb{R}\times\mathbb{R}^n)$ and all $k \in \mathbb{R}$,*

$$\operatorname*{ess\,lim}_{s\to 0^-} \int_\Sigma F(u(\sigma + s\boldsymbol{\nu}), k)\boldsymbol{b}(\overline{\sigma}) \cdot \boldsymbol{\nu}\varphi d\sigma \text{ exists.} \tag{3.5}$$

Moreover,

$$\begin{aligned} &-\int_Q \{|u-k|\partial_t\varphi + \boldsymbol{b}(x)F(u,k)\cdot\nabla\varphi\}dxdt \\ &\quad\leq \int_\Omega |u_0 - k|\varphi(0,x)dx - \operatorname*{ess\,lim}_{s\to 0^-} \int_\Sigma F(u(\sigma + s\boldsymbol{\nu}), k)\boldsymbol{b}(\overline{\sigma}) \cdot \boldsymbol{\nu}\varphi d\sigma. \end{aligned} \tag{3.6}$$

For $j \in \mathbb{N}$, $(t,x,\tilde{t},\tilde{x}) \in Q\times Q$, we set

$$\psi_j(t,x,\tilde{t},\tilde{x}) = \varphi\left(\frac{t+\tilde{t}}{2}, \frac{x+\tilde{x}}{2}\right)\rho_j(t-\tilde{t})\prod_{i=1}^n \rho_j(x_i - \tilde{x}_i),$$

where $(\rho_j)_{j\in\mathbb{N}}$ is a standard sequence of mollifiers on $\mathbb{R}$ and $\varphi \in \mathcal{C}_c^\infty(Q)$, $\varphi \geq 0$.
To simplify the notation, we set $p = (t,x)$, $\tilde{p} = (\tilde{t},\tilde{x})$,

$$\rho_j(x-\tilde{x}) = \prod_{i=1}^n \rho_j(x_i - \tilde{x}_i) \text{ and } \rho_j(p-\tilde{p}) = \rho_j(t-\tilde{t})\rho_j(x-\tilde{x}),$$

so that

$$\psi_j(p,\tilde{p}) = \varphi\left(\frac{p+\tilde{p}}{2}\right)\rho_j(p-\tilde{p}).$$

Let us remark that, for almost all $\sigma \in \Sigma$,

$$0 \leq \int_Q \rho_j(\sigma-\tilde{p})d\tilde{p} \leq 1 \text{ and } \lim_{j\to\infty}\int_Q \rho_j(\sigma-\tilde{p})d\tilde{p} = \frac{1}{2}.$$

In this framework, the next statement holds.

Lemma 3.9. *Suppose $u \in L^\infty(Q)$ fulfills (3.6). Then there holds for all $\varphi \in \mathcal{C}(\overline{Q})$*

$$\begin{aligned}\lim_{j\to\infty}\int_Q\int_\Sigma F(u(p),0)\boldsymbol{b}(\bar{\tilde{\sigma}})\psi_j(p,\bar{\tilde{\sigma}})d\tilde{\sigma}dp \\ = \frac{1}{2}\operatorname{ess}\lim_{s\to 0^-}\int_\Sigma F(u(\sigma+s\boldsymbol{\nu}),0)\boldsymbol{b}(\overline{\sigma})\cdot\boldsymbol{\nu}\varphi(\sigma)d\sigma,\end{aligned}$$

as well as

$$\begin{aligned}\lim_{j\to\infty}\int_Q \operatorname{ess}\lim_{s\to 0^-}\int_\Sigma F(u(\sigma+s\nu),0)\boldsymbol{b}(\overline{\sigma})\cdot\boldsymbol{\nu}\psi_j(\sigma,\tilde{p})d\sigma d\tilde{p} \\ = \frac{1}{2}\operatorname{ess}\lim_{s\to 0^-}\int_\Sigma F(u(\sigma+s\boldsymbol{\nu}),0)\boldsymbol{b}(\overline{\sigma})\cdot\boldsymbol{\nu}\varphi(\sigma)d\sigma.\end{aligned}$$

We turn now to the proof of Theorem 3.7. First, we prove the following lemma that gives a so-called Kruzhkov inequality for the difference of two entropy solutions to (3.1).

Lemma 3.10. *Let u_1 and u_2 be two entropy solutions to* (3.1). *Then there holds*

$$\int_Q (|u_1-u_2|\partial_t\varphi + F(u_1,u_2)\boldsymbol{b}(x)\cdot\nabla\varphi)dxdt \geq 0 \tag{3.7}$$

for all nonnegative $\varphi \in \mathcal{C}_c^\infty(\mathbb{R}\times\mathbb{R}^n)$ with $\varphi(0,\cdot) = 0$.

Proof. First, we deduce from (3.5) that there exists $\theta_i \in L^\infty(\Sigma)$ $(i = 1,2)$ such that for any $\varphi \in L^1(\Sigma)$

$$\operatorname{ess}\lim_{s\to 0^-}\int_\Sigma F(u_i(\sigma+s\boldsymbol{\nu}),0)\boldsymbol{b}(\overline{\sigma})\cdot\boldsymbol{\nu}\varphi(\sigma)d\sigma = \int_\Sigma \theta_i(\sigma)\varphi(\sigma)d\sigma.$$

From the definition of $\mathcal{F}_0$, the boundary condition (3.4) and inequality (3.6), we find

$$\begin{aligned}&-\int_Q (|u_i-k|\partial_t\varphi + \boldsymbol{b}(x)F(u_i,k)\cdot\nabla\varphi)dxdt \\ &\quad\leq -\int_\Sigma F(k,0)\boldsymbol{b}(\overline{\sigma})\cdot\boldsymbol{\nu}\varphi d\sigma + \operatorname{ess}\lim_{s\to 0^-}\int_\Sigma F(u_i(\sigma+s\boldsymbol{\nu}),0)\boldsymbol{b}(\overline{\sigma})\cdot\boldsymbol{\nu}\varphi d\sigma \qquad (3.8)\end{aligned}$$

for all nonnegative $\varphi \in \mathcal{C}_c^\infty(\mathbb{R} \times \mathbb{R}^n)$ with $\varphi(0,\cdot) = 0$ and all $k \in \mathbb{R}$.

Now we use the method of doubling variables that can be divided into three steps:

Step 1. For $\tilde{p} = (\tilde{t}, \tilde{x})$ fixed, we choose $k = u_2(\tilde{p})$, $\varphi = \psi_j(\cdot, \tilde{p})$ in (3.8) written for $u_1(p)$. We then integrate with respect to $\tilde{p}$ over Q. It follows (recall that $p = (t,x)$)

$$
\begin{aligned}
&- \frac{1}{2}\int_Q\int_Q |u_1(p) - u_2(\tilde{p})|\partial_t\varphi\left(\frac{p+\tilde{p}}{2}\right)\rho_j(p-\tilde{p})dpd\tilde{p}\\
&- \frac{1}{2}\int_Q\int_Q \boldsymbol{b}(x)F(u_1(p),u_2(\tilde{p}))\cdot\nabla\varphi\left(\frac{p+\tilde{p}}{2}\right)\rho_j(p-\tilde{p})dpd\tilde{p}\\
&- \int_Q\int_Q |u_1(p) - u_2(\tilde{p})|\partial_t\rho_j(p-\tilde{p})\varphi\left(\frac{p+\tilde{p}}{2}\right)dpd\tilde{p}\\
&- \int_Q\int_Q \boldsymbol{b}(x)F(u_1(p),u_2(\tilde{p}))\cdot\nabla_x\rho_j(p-\tilde{p})\varphi\left(\frac{p+\tilde{p}}{2}\right)dpd\tilde{p}\\
&\qquad \le \int_Q\int_\Sigma \theta_1(\sigma)\psi_j(\sigma,\tilde{p})d\sigma d\tilde{p} - \int_Q\int_\Sigma F(u_2(\tilde{p}),0)\boldsymbol{b}(\overline{\sigma})\cdot\boldsymbol{\nu}\psi_j(\sigma,\tilde{p})d\sigma d\tilde{p}.
\end{aligned}
$$

Step 2. In the same way, for p fixed, we choose $k = u_1(p)$, $\varphi = \psi_j(p,\cdot)$ in (3.8) written for $u_2(\tilde{p})$. Now we integrate with respect to p over Q which yields (using Fubini's theorem and the fact that $-\partial_{\tilde{t}}\rho_j(p-\tilde{p}) = \partial_t\rho_j(p-\tilde{p})$, $-\nabla_{\tilde{x}}\rho_j(p-\tilde{p}) = \nabla_x\rho_j(p-\tilde{p})$)

$$
\begin{aligned}
&- \frac{1}{2}\int_Q\int_Q |u_2(\tilde{p}) - u_1(p)|\partial_t\varphi\left(\frac{p+\tilde{p}}{2}\right)\rho_j(p-\tilde{p})dpd\tilde{p}\\
&- \frac{1}{2}\int_Q\int_Q \boldsymbol{b}(\tilde{x})F(u_2(\tilde{p}),u_1(p))\cdot\nabla\varphi\left(\frac{p+\tilde{p}}{2}\right)\rho_j(p-\tilde{p})dpd\tilde{p}\\
&+ \int_Q\int_Q |u_2(\tilde{p}) - u_1(p)|\partial_t\rho_j(p-\tilde{p})\varphi\left(\frac{p+\tilde{p}}{2}\right)dpd\tilde{p}\\
&+ \int_Q\int_Q \boldsymbol{b}(\tilde{x})F(u_2(\tilde{p}),u_1(p))\cdot\nabla_x\rho_j(p-\tilde{p})\varphi\left(\frac{p+\tilde{p}}{2}\right)dpd\tilde{p}\\
&\qquad \le \int_Q\int_\Sigma \theta_2(\tilde{\sigma})\psi_j(p,\tilde{\sigma})d\tilde{\sigma}dp - \int_Q\int_\Sigma F(u_1(p),0)\boldsymbol{b}(\overline{\tilde{\sigma}})\cdot\boldsymbol{\nu}\psi_j(p,\tilde{\sigma})d\tilde{\sigma}dp.
\end{aligned}
$$

We sum up the two previous inequalities and remark that the terms involving $\partial_t\rho_j(p-\tilde{p})$ vanish. Therefore, we find

$$
\begin{aligned}
&- \int_Q\int_Q |u_1(p) - u_2(\tilde{p})|\partial_t\varphi\left(\frac{p+\tilde{p}}{2}\right)\rho_j(p-\tilde{p})\\
&- \frac{1}{2}\int_Q\int_Q F(u_1(p),u_2(\tilde{p}))(\boldsymbol{b}(x)+\boldsymbol{b}(\tilde{x}))\cdot\nabla\varphi\left(\frac{p+\tilde{p}}{2}\right)\rho_j(p-\tilde{p})dpd\tilde{p}
\end{aligned}
$$

$$
\begin{aligned}
&- \int_Q \int_Q F(u_1(p), u_2(\tilde{p}))(\boldsymbol{b}(x) - \boldsymbol{b}(\tilde{x}))\varphi\left(\frac{p+\tilde{p}}{2}\right) \cdot \nabla_x \rho_j(p-\tilde{p})dpd\tilde{p} \\
&\leq \int_Q \int_\Sigma \theta_1(\sigma)\psi_j(\sigma,\tilde{p})d\sigma d\tilde{p} + \int_Q \int_\Sigma \theta_2(\tilde{\sigma})\psi_j(p,\tilde{\sigma})d\tilde{\sigma}dp \\
&\quad - \int_Q \int_\Sigma F(u_2(\tilde{p}),0)\boldsymbol{b}(\overline{\sigma}) \cdot \boldsymbol{\nu}\psi_j(\sigma,\tilde{p})d\sigma d\tilde{p} - \int_Q \int_\Sigma F(u_1(p),0)\boldsymbol{b}(\overline{\tilde{\sigma}})\psi_j(p,\tilde{\sigma})d\tilde{\sigma}dp.
\end{aligned}
\tag{3.9}
$$

Step 3. We take the limit with respect to j. By Lemma 3.9, we have

$$
\begin{aligned}
&\lim_{j\to\infty} \int_Q \int_\Sigma F(u_1(p),0)\boldsymbol{b}(\overline{\tilde{\sigma}})\psi_j(p,\tilde{\sigma})d\tilde{\sigma}dp \\
&\qquad = \frac{1}{2}\int_\Sigma \theta_1(\sigma)\varphi(\sigma)d\sigma = \lim_{j\to\infty}\int_Q \int_\Sigma \theta_1(\sigma)\psi_j(\sigma,\tilde{p})d\sigma d\tilde{p}
\end{aligned}
$$

and

$$
\begin{aligned}
&\lim_{j\to\infty} \int_Q \int_\Sigma F(u_2(\tilde{p}),0)\boldsymbol{b}(\overline{\sigma})\psi_j(\sigma,\tilde{p})d\sigma d\tilde{p} \\
&\qquad = \frac{1}{2}\int_\Sigma \theta_2(\sigma)\varphi(\sigma)d\sigma = \lim_{j\to\infty}\int_Q \int_\Sigma \theta_2(\tilde{\sigma})\psi_j(p,\tilde{\sigma})d\tilde{\sigma}dp.
\end{aligned}
$$

Consequently, the right-hand side of (3.9) goes to zero as j tends to ∞.

We study now the first line of (3.9). To this end, we set

$$
\begin{aligned}
I_j &:= \int_Q \int_Q |u_1(p) - u_2(\tilde{p})|\rho_j(p-\tilde{p})\partial_t\varphi\left(\frac{p+\tilde{p}}{2}\right)dpd\tilde{p}, \\
I &:= \int_Q |u_1(p) - u_2(p)|\partial_t\varphi(p)dp.
\end{aligned}
$$

Let us show that $\lim_{j\to\infty} I_j = I$. By definition of ρ_j, we have

$$
\int_Q |u_1(p) - u_2(p)|\partial_t\varphi(p)dp = \int_Q \int_Q |u_1(p) - u_2(p)|\partial_t\varphi(p)\rho_j(p-\tilde{p})dpd\tilde{p}
$$

if j is sufficiently large. Therefore, we obtain

$$
\begin{aligned}
|I_j - I| = &\int_Q \int_Q (|u_1(p) - u_2(\tilde{p})| - |u_1(p) - u_2(p)|)\rho_j(p-\tilde{p})\partial_t\varphi\left(\frac{p+\tilde{p}}{2}\right)dpd\tilde{p} \\
&+ \int_Q \int_Q |u_1(p) - u_2(p)|\left(\partial_t\varphi\left(\frac{p+\tilde{p}}{2}\right) - \partial_t\varphi(p)\right)\rho_j(p-\tilde{p})dpd\tilde{p},
\end{aligned}
$$

and thus

$$
\begin{aligned}
|I_j - I| \leq \ &\|\partial_t\varphi\|_\infty \int_Q \int_Q |u_2(p) - u_2(\tilde{p})|\rho_j(p-\tilde{p})dpd\tilde{p} \\
&+ \|u_1 - u_2\|_\infty \int_Q \int_Q \left|\partial_t\varphi\left(\frac{p+\tilde{p}}{2}\right) - \partial_t\varphi(p)\right|\rho_j(p-\tilde{p})dpd\tilde{p}.
\end{aligned}
$$

The first term on the right-hand side of the foregoing inequality tends to zero in view of the property of Lebesgue points. The second one goes to zero since $\partial_t \varphi$ is continuous on Q. Therefore, we may conclude that $\lim_{j\to\infty} I_j = I$.

In a similar way, we prove that

$$\lim_{j\to\infty} \frac{1}{2}\int_Q\int_Q F(u_1(p),u_2(\tilde{p}))(\boldsymbol{b}(x)+\boldsymbol{b}(\tilde{x}))\cdot\nabla_x\varphi\left(\frac{p+\tilde{p}}{2}\right)\rho_j(p-\tilde{p})dpd\tilde{p}$$
$$= \int_Q F(u_1,u_2)\boldsymbol{b}(x)\cdot\nabla\varphi(p)dp.$$

We continue with the study of the term

$$L_j := \int_Q\int_Q F(u_1(p),u_2(\tilde{p}))(\boldsymbol{b}(x)-\boldsymbol{b}(\tilde{x}))\varphi\left(\frac{p+\tilde{p}}{2}\right)\cdot\nabla_x\rho_j(p-\tilde{p})dpd\tilde{p}$$

that can be decomposed as $L_j = L_{1,j} + L_{2,j}$ with

$$L_{1,j} := \int_Q\int_Q (F(u_1(p),u_2(\tilde{p})) - F(u_1(\tilde{p}),u_2(\tilde{p})))(\boldsymbol{b}(x)-\boldsymbol{b}(\tilde{x}))\cdot$$
$$\nabla_x\rho_j(p-\tilde{p})\varphi\left(\frac{p+\tilde{p}}{2}\right)dpd\tilde{p},$$
$$L_{2,j} := \int_Q\int_Q F(u_1(\tilde{p}),u_2(\tilde{p}))(\boldsymbol{b}(x)-\boldsymbol{b}(\tilde{x}))\nabla_x\rho_j(p-\tilde{p})\varphi\left(\frac{p+\tilde{p}}{2}\right)dpd\tilde{p}.$$

Thanks to an integration by parts and (1.2), we have

$$L_{2,j} = -\int_Q\int_Q F(u_1(\tilde{p}),u_2(\tilde{p}))(\boldsymbol{b}(x)-\boldsymbol{b}(\tilde{x}))\rho_j(p-\tilde{p})\nabla_x\varphi\left(\frac{p+\tilde{p}}{2}\right)dpd\tilde{p}$$
$$+\int_Q\int_\Sigma F(u_1(\tilde{p}),u_2(\tilde{p}))(\boldsymbol{b}(\sigma)-\boldsymbol{b}(\tilde{x}))\rho_j(\sigma-\tilde{p})\varphi\left(\frac{\sigma+\tilde{p}}{2}\right)d\sigma d\tilde{p}.$$

As $\boldsymbol{b}$ is continuous on $\overline{\Omega}$, we also find $\lim_{j\to\infty} L_{2,j} = 0$.

Moreover, using the Lipschitz continuity of f and $\boldsymbol{b}$, we may estimate

$$|L_{1,j}| \le \int_Q\int_Q M_f M_b |u_1(p)-u_1(\tilde{p})||x-\tilde{x}||\nabla_x\rho_j(x-\tilde{x})|\rho_j(t-\tilde{t})\varphi\left(\frac{p+\tilde{p}}{2}\right)dpd\tilde{p}.$$

In view of the definition of ρ_j, there exists a constant $C_1 > 0$ such that for all $(t,x)\in Q$,

$$|\nabla_x\rho_j(x-\tilde{x})|\rho_j(t-\tilde{t}) \le C_1 j^{n+2}.$$

Therefore, there exists a constant $C_2 > 0$ such that

$$|L_{1,j}| \le C_2 j^{n+1}\int_{\{(p,\tilde{p})\in Q^2;\,|p-\tilde{p}|\le\frac{1}{j}\}} |u_1(p)-u_1(\tilde{p})|dpd\tilde{p}.$$

Using the property of Lebesgue points, we find

$$\lim_{j\to\infty} j^{n+1} \int_{\{(p,\tilde{p})\in Q^2;\, |p-\tilde{p}|\le \frac{1}{j}\}} |u_1(p) - u_1(\tilde{p})| dp d\tilde{p} = 0.$$

Thus, $\lim_{j\to\infty} L_{1,j} = 0$ and $\lim_{j\to\infty} L_j = 0$. Finally, gathering all the limits in (3.9) proves the assertion. □

Proof of Theorem 3.7. We choose in (3.7) the test function $\varphi(t,x) = \alpha(t)\beta(x)$, where $\beta \in \mathcal{C}_c^\infty(\mathbb{R}^n)$, $\beta \equiv 1$ on Ω and $\alpha \in \mathcal{C}_c^\infty(\mathbb{R})$, $\alpha \geq 0$ with $\alpha(0) = 0$. This choice yields

$$\int_Q |u_1(t,x) - u_2(t,x)|\alpha'(t) dx dt \geq 0.$$

Then, for almost all $t \in (0,T)$, consider a sequence of test functions $(\alpha_\varepsilon)_{\varepsilon>0}$ approximating the characteristic function $\chi_{[0,t]}$. We may suppose that $\alpha'_\varepsilon = 0$ on $[\varepsilon, t-\varepsilon]$ and that $|\alpha'_\varepsilon| \leq \frac{2}{\varepsilon}$. From (3.3), we now obtain

$$\begin{aligned} 0 &\leq \lim_{\varepsilon\to 0^+} \int_Q |u_1 - u_2|\alpha'_\varepsilon(t) dx dt \\ &= -\int_\Omega |u_1(t,x) - u_2(t,x)| dx + \int_\Omega |u_{0,1}(x) - u_{0,2}(x)| dx. \end{aligned}$$

This completes the proof. □

3.3 Existence: the vanishing viscosity method

In this section, we prove the existence of an entropy solution to (3.1) via the vanishing viscosity method. Let μ be a positive real number. We introduce the following viscous problem.

Find a measurable and essentially bounded function u_μ such that in the sense of distributions

$$\left\{ \begin{array}{rcll} \partial_t u_\mu + \nabla\cdot(\boldsymbol{b}(x) f(u_\mu)) &=& \nabla\cdot(\mu\nabla u_\mu) & \text{in } Q, \\ u_\mu(0,\cdot) &=& u_0 & \text{on } \Omega, \\ u_\mu &=& 0 & \text{on } \Sigma. \end{array} \right. \tag{3.10}$$

The aim of the vanishing viscosity method is to study the limit of the sequence $(u_\mu)_{\mu>0}$ as μ goes to 0^+ and to show that this is an entropy solution to problem (3.1). Therefore, we need to collect some a priori estimates on the sequence $(u_\mu)_{\mu>0}$ and apply compactness arguments in order to take the limit over μ. Generally, one looks for $W^{1,1}$-estimates in order to be able to use a compactness argument. We refer to [4, 9, 10] for more details. Here, with the final aim to solve the coupled problem (1.1), we will present a method that only requires an L^∞-estimate on $(u_\mu)_{\mu>0}$. To this purpose, we use the concept of entropy process solution.

Definition 3.11. A function $\pi \in L^\infty((0,1)\times Q)$ is said to be an entropy process solution to (3.1) if

(i) for all nonnegative $\varphi \in \mathcal{C}_c^\infty(Q)$ and all $k \in \mathbb{R}$

$$\int_0^1 \int_Q (|\pi - k|\partial_t\varphi + \operatorname{sgn}(\pi - k)(f(\pi) - f(k))\boldsymbol{b}(x)\cdot\nabla\varphi)dxdtd\alpha \geq 0, \quad (3.11)$$

(ii)

$$\operatorname*{ess\,lim}_{t\to 0^+} \int_0^1 \int_\Omega |\pi(\alpha,t,x) - u_0(x)|dxd\alpha = 0, \quad (3.12)$$

(iii) for all nonnegative $\varphi \in L^1(\Sigma)$ and all $k \in \mathbb{R}$

$$\operatorname*{ess\,lim}_{s\to 0^-} \int_0^1 \int_\Sigma \mathcal{F}_0(\pi(\alpha,\sigma + s\boldsymbol{\nu}),k)\boldsymbol{b}(\overline{\sigma})\cdot\boldsymbol{\nu}\varphi d\sigma d\alpha \geq 0. \quad (3.13)$$

By using the same method as in the previous section, we can prove that the entropy process solution to (3.1) is unique.

Theorem 3.12. *Assume there exists an entropy process solution $\pi \in L^\infty((0,1)\times Q)$ to* (3.1). *Then π is unique and there exists $u \in L^\infty(Q)$ such that $u(\cdot) = \pi(\alpha,\cdot)$ a.e. on Q and for almost all $\alpha \in (0,1)$. In particular, u is an entropy solution to* (3.1).

Proof. We start with the uniqueness. Let $\pi_1(\alpha,\cdot,\cdot)$ and $\pi_2(\beta,\cdot,\cdot)$ be two entropy process solutions to (3.1) for the same initial data u_0. We use the same reasoning as in the proof of Theorem 3.7 to see that for almost all $t \in (0,T)$

$$\int_0^1 \int_0^1 \int_\Omega |\pi_1(\alpha,t,x) - \pi_2(\beta,t,x)|dxd\alpha d\beta \leq 0.$$

This estimate implies that there exists $u \in L^\infty(Q)$ such that a.e. on Q

$$u(\cdot) = \pi_1(\alpha,\cdot) = \pi_2(\beta,\cdot) \text{ for almost all } \alpha,\beta \text{ in } (0,1).$$

The conclusion follows. □

Our purpose is now to show that there exists an entropy process solution to (3.1). The existence of an entropy solution to (3.1) will follow from Theorem 3.12. To do so, we need some lemmas that we will use to obtain (3.12)–(3.13) (see [15, Chap. 2, Lemma 7.34 and 7.41] for the proofs in the framework of entropy solutions).

Lemma 3.13. *Suppose that $\pi \in L^\infty((0,1)\times Q)$ satisfies, for all nonnegative $\varphi \in \mathcal{C}_c^\infty([0,T)\times\Omega)$ and all $k \in \mathbb{R}$,*

$$-\int_0^1 \int_Q (|\pi - k|\partial_t\varphi + \boldsymbol{b}(x)F(\pi,k)\cdot\nabla\varphi)dxdtd\alpha \leq \int_\Omega |u_0 - k|\varphi(0,x)dx.$$

Then

$$\operatorname*{ess\,lim}_{t\to 0^+} \int_0^1 \int_\Omega |\pi(\alpha,t,x) - u_0(x)|dxd\alpha = 0.$$

Lemma 3.14. *Assume that* $\pi \in L^\infty((0,1)\times Q)$ *fulfills, for all nonnegative* $\varphi \in \mathcal{C}_c^\infty([0,T)\times\Omega)$ *and all* $k\in\mathbb{R}$,

$$-\int_0^1\int_Q (H_\delta(\pi,k)\partial_t\varphi + \boldsymbol{b}(x)J_\delta(\pi,k)\cdot\nabla\varphi)dxdtd\alpha \le 0,$$

where H_δ *and* J_δ *are defined in Example 3.4. Then, for any nonnegative* $\beta \in L^1(\Sigma)$

$$\operatorname*{ess\,lim}_{s\to 0^-}\int_0^1\int_\Sigma J_\delta(\pi(\alpha,\sigma+s\boldsymbol{\nu}),k)\beta\boldsymbol{b}(\overline{\sigma})\cdot\boldsymbol{\nu}d\sigma d\alpha \ge 0.$$

We come back to the viscous problem (3.10). We use the results gathered in Section 2 to prove

Lemma 3.15. *There exists a unique weak solution* $u_\mu \in W(0,T)$ *to* (3.10)*, i.e.,*

$$m \le u_\mu(t,x) \le M \textit{ for almost all } (t,x)\in Q, \tag{3.14}$$

for almost all $t\in(0,T)$ *and all* $v\in H_0^1(\Omega)$ *there holds*

$$\langle\partial_t u_\mu, v\rangle + \int_\Omega (\mu\nabla u_\mu - \boldsymbol{b}(x)f(u_\mu))\cdot\nabla v dx = 0, \tag{3.15}$$

and the initial condition

$$u_\mu(0) = u_0 \textit{ in } L^2(\Omega)$$

is fulfilled.

Proof. Setting $\phi(u_\mu) = \mu u_\mu$, we notice that ϕ satisfies (1.3). Therefore, the assumptions of Theorem 2.3 are satisfied and the conclusion follows. □

We also prove the following lemma that will be used to take the limit as $\mu\to 0^+$.

Lemma 3.16. *There exists a positive constant* C *independent of* μ *such that*

$$\mu\int_Q |\nabla u_\mu|^2 dxdt \le C. \tag{3.16}$$

Proof. We take $v = u_\mu$ in (3.15) and integrate over $(0,T)$. An integration by parts in the evolution term gives

$$\int_0^T \langle\partial_t u_\mu, u_\mu\rangle dt = \frac{1}{2}\|u_\mu(T)\|^2_{L^2(\Omega)} - \frac{1}{2}\|u_0\|^2_{L^2(\Omega)}.$$

We set $g(u_\mu) = \int_0^{u_\mu} f(s)ds$ to write the convection term as

$$\int_0^T\int_\Omega f(u_\mu)\boldsymbol{b}(x)\cdot\nabla u_\mu dxdt = \int_0^T\int_\Omega \nabla g(u_\mu)\cdot\boldsymbol{b}(x)dxdt.$$

Now we use Green's formula. From (1.2), we obtain

$$\int_0^T \int_\Omega \nabla g(u_\mu) \cdot \boldsymbol{b}(x) dxdt = 0.$$

It follows the estimate

$$\int_Q \mu |\nabla u_\mu|^2 dxdt \le \frac{1}{2} \|u_0\|_{L^2(\Omega)}. \qquad \square$$

In view of (3.14), the sequence $(u_\mu)_{\mu>0}$ is bounded in $L^\infty(Q)$. So we can apply Theorem 1.5 to deduce the existence of a subsequence, still denoted by $(u_\mu)_{\mu>0}$, and a function $\pi \in L^\infty((0,1) \times Q)$ such that

$$\lim_{\mu \to 0^+} \int_Q \varphi(t, x, u_\mu) \xi dxdt = \int_0^1 \int_Q \varphi(t, x, \pi) \xi dxdtd\alpha \tag{3.17}$$

for any continuous function φ on $Q \times [m, M]$ and any $\xi \in L^1(Q)$.

Now we choose in (3.15) the test function $v = \mathrm{sgn}_\eta(u_\mu - k)\varphi_1\varphi_2$, where $\varphi_1 \in \mathcal{C}_c^\infty([0, T))$, $\varphi_2 \in \mathcal{C}_c^\infty(\Omega)$, $\varphi_1, \varphi_2 \ge 0$, $k \in \mathbb{R}$, and integrate over $(0, T)$. We use a generalization of Lemma 1.1 (see [18]) in order to handle the evolution term,

$$\int_0^T \langle \partial_t u_\mu, \mathrm{sgn}_\eta(u_\mu - k)\varphi_1\varphi_2 \rangle dt = -\int_Q I_\eta(u_\mu)\varphi_2 \partial_t \varphi_1 dxdt - \int_\Omega I_\eta(u_0)\varphi_2\varphi_1(0) dx,$$

where $I_\eta(u_\mu) = \displaystyle\int_k^{u_\mu} \mathrm{sgn}_\eta(\tau - k) d\tau$.

For the diffusion term, we find

$$\begin{aligned} &\int_Q \mu \nabla u_\mu \cdot \nabla(\mathrm{sgn}_\eta(u_\mu - k)\varphi_1\varphi_2 dxdt \\ &= \int_Q \mu |\nabla u_\mu|^2 \mathrm{sgn}'_\eta(u_{\mu - k})\varphi_1\varphi_2 dxdt + \int_Q \mu \nabla u_\mu \cdot \nabla \varphi_2 \varphi_1 \mathrm{sgn}_\eta(u_\mu - k) dxdt. \end{aligned}$$

Note that the first term on the right-hand side is nonnegative owing to the definition of the function sgn_η.

Using Green's formula in the convection term, we get (due to (1.2))

$$\begin{aligned} &\int_Q \boldsymbol{b}(x) f'(u_\mu) \nabla u_\mu \mathrm{sgn}_\eta(u_\mu - k)\varphi_1\varphi_2 dxdt \\ &= \int_Q \boldsymbol{b}(x) \nabla F_\eta(u_\mu)\varphi_1\varphi_2 dxdt = -\int_Q \boldsymbol{b}(x) F_\eta(u_\mu) \nabla \varphi_2 \varphi_1 dxdt, \end{aligned}$$

where

$$F_\eta(u_\mu) = \int_k^{u_\mu} f'(\tau) \mathrm{sgn}_\eta(\tau - k) d\tau.$$

Therefore, we obtain for $\mu > 0$, $\eta > 0$

$$\int_Q I_\eta(u_\mu)\varphi_2\partial_t\varphi_1 dxdt + \int_\Omega I_\eta(u_0)\varphi_2\varphi_1(0)dx + \int_Q \boldsymbol{b}(x)F_\eta(u_\mu)\nabla\varphi_2\varphi_1 dxdt - \int_Q \mu\nabla u_\mu\cdot\nabla\varphi_2\varphi_1\operatorname{sgn}_\eta(u_\mu - k)dxdt \geq 0. \quad (3.18)$$

Now, we take first the limit with respect to μ in (3.18). From (3.17), we have

$$\lim_{\mu\to 0^+}\int_Q I_\eta(u_\mu)\varphi_2\partial_t\varphi_1 dxdt = \int_0^1\int_Q I_\eta(\pi)\varphi_2\partial_t\varphi_1 dxdtd\alpha,$$

and

$$\lim_{\mu\to 0^+}\int_Q \boldsymbol{b}(x)F_\eta(u_\mu)\nabla\varphi_2\varphi_1 dxdt = \int_0^1\int_Q \boldsymbol{b}(x)F_\eta(\pi)\nabla\varphi_2\varphi_1 dxdtd\alpha.$$

Moreover, the Cauchy–Schwarz inequality and Lemma 3.16 provide

$$\lim_{\mu\to 0^+}\int_Q \mu\nabla u_\mu\cdot\nabla\varphi_2\varphi_1\operatorname{sgn}_\eta(u_\mu - k)dxdt = 0.$$

In order to take the limit with respect to η, we remark that

$$\lim_{\eta\to 0^+} I_\eta(\pi) = |\pi - k| \text{ and } \lim_{\eta\to 0^+} F_\eta(\pi) = \operatorname{sgn}(\pi - k)(f(\pi) - f(k)).$$

Since $\mathcal{C}_c^\infty(0,T)\otimes\mathcal{C}_c^\infty(\Omega)$ is dense in $\mathcal{C}_c^\infty(Q)$, we may therefore deduce that π satisfies (3.11). From (3.18), we also deduce that

$$-\int_0^1\int_Q\{|\pi - k|\varphi_2\partial_t\varphi_1 + \boldsymbol{b}(x)F(\pi,k)\cdot\nabla\varphi_2\varphi_1\}d\alpha dxdt \leq \int_\Omega |u_0 - k|\varphi_2\varphi_1(0)dx.$$

Then Lemma 3.13 ensures that π fulfills (3.12).

To show that π satisfies (3.13), we use the family of boundary entropy-entropy flux pairs introduced in Example 3.4. We choose in (3.15) the test function

$$v_\delta = \partial_1 H_\delta(u_\mu, k)\varphi_1\varphi_2, \quad \delta > 0,$$

where $\varphi_1 \in \mathcal{C}_c^\infty(0,T)$, $\varphi_2 \in \mathcal{C}^\infty(\overline{\Omega})$, $\varphi_1, \varphi_2 \geq 0$.

Note that $\partial_1 H_\delta(u_\mu, k)\varphi_1\varphi_2$ is an element of $L^2(0,T; H_0^1(\Omega))$. For the convection term, we have

$$\int_Q \boldsymbol{b}(x)f(u_\mu)\cdot\nabla(\partial_1 H_\delta(u_\mu,k)\varphi_1\varphi_2)dxdt = \int_Q J_\delta(u_\mu,k)\boldsymbol{b}(x)\cdot\nabla(\varphi_1\varphi_2)dxdt.$$

Lemma 1.1 is used to transform the evolution term.

As the function $\tau \mapsto H_\delta(\tau, \cdot)$ is convex, we get for the diffusion term

$$\mu \int_Q \nabla u_\mu \cdot \nabla(\partial_1 H_\delta(u_\mu, k)\varphi_1\varphi_2) \leq \mu \int_Q \nabla u_\mu \partial_1 H_\delta(u_\mu, k) \nabla \varphi_2 \varphi_1 dxdt.$$

Therefore, we obtain

$$- \int_Q (H_\delta(u_\mu, k)\varphi_2 \partial_t \varphi_1 + J_\delta(u_\mu, k)\boldsymbol{b}(x) \cdot \nabla(\varphi_1\varphi_2) dxdt$$
$$\leq -\mu \int_Q \nabla u_\mu \partial_1 H_\delta(u_\mu, k) \nabla \varphi_2 \varphi_1 dxdt.$$

Now, we take the limit $\mu \to 0$. On the right-hand side of the foregoing inequality, we use Lemma 3.16 while on the left-hand side, we use Theorem 1.5. This yields

$$- \int_0^1 \int_Q (H_\delta(\pi, k)\varphi_2 \partial_t \varphi_1 + J_\delta(\pi, k)\boldsymbol{b}(x) \cdot \nabla(\varphi_1\varphi_2) dxdtd\alpha \leq 0.$$

Then Lemma 3.14 ensures that, for any nonnegative $\beta \in L^1(\Sigma)$,

$$\operatorname*{ess\,lim}_{s \to 0^-} \int_0^1 \int_\Sigma J_\delta(\pi(\alpha, \sigma + s\boldsymbol{\nu}), k)\beta \boldsymbol{b}(\overline{\sigma}) \cdot \boldsymbol{\nu} d\sigma d\alpha \geq 0.$$

Finally, we observe that the sequence $(J_\delta)_{\delta>0}$ uniformly converges towards $\mathcal{F}_0$ as δ goes to 0^+. Thus, π fulfills (3.13).

We conclude that there exists a unique entropy process solution to (3.1). By Theorem 3.12, it follows that there exists a unique entropy solution $u \in L^\infty(Q)$ to (3.1).

4 The coupled problem

In this part, we prove an existence and uniqueness result for (1.1). To this end, we need that the function f is nondecreasing. This implies $F(a, b) = |f(a) - f(b)|$ for $a, b \in \mathbb{R}$. We also have $\mathcal{F}_0(a, b) = \frac{1}{2}(|f(a) - f(0)| - |f(b) - f(0)| + |f(a) - f(b)|)$ so that $\mathcal{F}_0(a, b) \geq 0$.

Moreover, we suppose that $\boldsymbol{b} \cdot \boldsymbol{\nu}_h$ is nonnegative along the interface Γ_{hp}. More precisely, let

$$\Gamma_{hp} \subset \{\overline{\sigma} \in \Gamma_h;\ \boldsymbol{b}(\overline{\sigma}) \cdot \boldsymbol{\nu}_h \geq 0\}. \tag{4.1}$$

Remark 4.1. The aforementioned additional assumptions on f and $\boldsymbol{b}$ ensure that the interface is included in the set of outwards characteristics for the first-order differential operator posed in the hyperbolic domain. This is a key point that will allow us to prove uniqueness by considering first the behaviour of a solution on the hyperbolic area and then on the parabolic one.

As $\operatorname{meas}(\Gamma_p \setminus \Gamma_{hp}) \neq 0$, the function space

$$V = \{v \in H^1(\Omega_p);\ v = 0 \text{ a.e. on } \Gamma_p \setminus \Gamma_{hp}\}$$

is a Hilbert space when equipped with the standard H_0^1-inner product. The norm in V is, therefore, still $\|\cdot\|$. We denote by $\langle\cdot,\cdot\rangle_{V'\times V}$ the duality pairing between V and V'.

We provide a definition of a weak entropy solution to (1.1) by observing that the equation set in Q can be viewed as a quasilinear parabolic equation that strongly degenerates on a fixed subdomain. We propose a weak formulation relying on a global entropy inequality on the whole space-time cylinder Q, inspired by the one given in [1, 3].

Definition 4.2. A function $u : Q \to \mathbb{R}$ is said to be a weak entropy solution to (1.1) if

(i) $u \in L^\infty(Q)$, $\phi(u) \in L^2(0,T;V)$,

(ii) for all nonnegative $\varphi \in \mathcal{C}_c^\infty(Q)$ and all $k \in \mathbb{R}$

$$\int_Q (|u-k|\partial_t\varphi + (\boldsymbol{b}(x)F(u,k) - \chi_{\Omega_p}\nabla|\phi(u)-\phi(k)|)\cdot\nabla\varphi)dxdt \geq 0, \tag{4.2}$$

(iii)

$$\operatorname*{ess\,lim}_{t\to 0^+}\int_\Omega |u(t,x)-u_0(x)|dx = 0, \tag{4.3}$$

(iv) for all nonnegative $\varphi \in L^1(\Sigma_h \setminus \Sigma_{hp})$ and all $k \in \mathbb{R}$

$$\operatorname*{ess\,lim}_{s\to 0^-}\int_{\Sigma_h\setminus\Sigma_{hp}} \mathcal{F}_0(u(\sigma+\tau\boldsymbol{\nu}_h),k)\boldsymbol{b}(\overline{\sigma})\cdot\boldsymbol{\nu}_h d\sigma \geq 0. \tag{4.4}$$

Remark 4.3. (i) One can choose in (4.2) successively $k > \|u\|_\infty$ and $k < -\|u\|_\infty$, and compare the two resulting inequalities. We remark that all k-dependent terms vanish. Therefore, we find

$$\int_Q (u\partial_t\varphi + \boldsymbol{b}(x)f(u) - \chi_{\Omega_p}\nabla\phi(u))\cdot\nabla\varphi dxdt = 0 \tag{4.5}$$

for all $\varphi \in H_0^1(Q)$.

(ii) From (4.2), we deduce that for all nonnegative $\varphi \in \mathcal{C}_c^\infty((0,T)\times\Omega_h)$

$$\int_{Q_h} (|u-k|\partial_t\varphi + |f(u)-f(k)|\boldsymbol{b}(x)\cdot\nabla\varphi)dxdt \geq 0. \tag{4.6}$$

4.1 Uniqueness

The inequalities (4.2) and (4.4) are the starting point to establish the Lipschitz continuous dependence, measured in $L^1(\Omega_h)$, of the weak entropy solution to (1.1) on the corresponding initial data.

Lemma 4.4. *Let* $u \in L^\infty(Q)$ *satisfy* (4.2) *and* (4.4). *We then have for all nonnegative* $\varphi \in L^1(\Sigma_h)$ *and all* $k \in \mathbb{R}$

$$\operatorname*{ess\,lim}_{s\to 0^-}\int_{\Sigma_h} \mathcal{F}_0(u(\sigma+\tau\boldsymbol{\nu}_h),k)\boldsymbol{b}(\overline{\sigma})\cdot\boldsymbol{\nu}_h d\sigma \geq 0. \tag{4.7}$$

Proof. Let $\varphi \in L^1(\Sigma_h)$ with $\varphi \geq 0$. Since u satisfies (4.6), we can apply Lemma 3.8 to conclude that

$$\operatorname*{ess\,lim}_{s\to 0^-} \int_{\Sigma_h} |f(u(\sigma + s\boldsymbol{\nu}_h)) - f(k)| \boldsymbol{b}(\overline{\sigma}) \cdot \boldsymbol{\nu}_h \varphi d\sigma \quad \text{exists.} \tag{4.8}$$

This in turn implies that

$$\operatorname*{ess\,lim}_{s\to 0^-} \int_{\Sigma_h} \mathcal{F}_0(u(\sigma + s\boldsymbol{\nu}_h), k) \boldsymbol{b}(\overline{\sigma}) \cdot \boldsymbol{\nu}_h \varphi d\sigma \quad \text{exists.}$$

Moreover, for any $s < 0$, we have

$$\begin{aligned}
&\int_{\Sigma_h} \mathcal{F}_0(u(\sigma + s\boldsymbol{\nu}_h), k) \boldsymbol{b}(\overline{\sigma}) \cdot \boldsymbol{\nu}_h \varphi d\sigma \\
&= \int_{\Sigma_h \setminus \Sigma_{hp}} \mathcal{F}_0(u(\sigma + s\boldsymbol{\nu}_h), k) \boldsymbol{b}(\overline{\sigma}) \cdot \boldsymbol{\nu}_h \varphi d\sigma + \int_{\Sigma_{hp}} \mathcal{F}_0(u(\sigma + s\boldsymbol{\nu}_h), k) \boldsymbol{b}(\overline{\sigma}) \cdot \boldsymbol{\nu}_h \varphi d\sigma.
\end{aligned}$$

Since $\mathcal{F}_0(\cdot,\cdot)$ is nonnegative, we have, according to (4.1),

$$\int_{\Sigma_{hp}} \mathcal{F}_0(u(\sigma + s\boldsymbol{\nu}_h, k) \boldsymbol{b}(\overline{\sigma}) \cdot \boldsymbol{\nu}_h \varphi d\sigma \geq 0.$$

We use (4.4) to conclude that (4.7) is fulfilled. □

We are now able to state uniqueness on the hyperbolic zone.

Theorem 4.5. *Let u_1 and u_2 be two weak entropy solutions to* (1.1) *for initial data $u_{0,1}$ and $u_{0,2}$, respectively. Then, for almost all $t \in (0,T)$,*

$$\int_{\Omega_h} |u_1(t,x) - u_2(t,x)| dx \leq \int_{\Omega} |u_{0,1}(x) - u_{0,2}(x)| dx.$$

Proof. Since (4.6) and (4.7) hold, one can apply Theorem 3.7. □

We focus now on the parabolic zone. On Q_p, we characterize a solution to (1.1) through a variational equality including the contributions from the hyperbolic part that enter the parabolic zone.

Proposition 4.6. *Let u be a weak entropy solution to* (1.1). *Then*

$$\partial_t u \in L^2(0,T;V').$$

Furthermore, there holds for all $\varphi \in L^2(0,T;V)$

$$\begin{aligned}
&\int_0^T \langle \partial_t u, \varphi \rangle_{V' \times V} dt + \int_{Q_p} (\nabla \phi(u) - \boldsymbol{b}(x) f(u)) \cdot \nabla \varphi dx dt \\
&\qquad - \operatorname*{ess\,lim}_{s\to 0^-} \int_{\Sigma_{hp}} f(u(\sigma + s\boldsymbol{\nu}_h)) \boldsymbol{b}(\overline{\sigma}) \cdot \boldsymbol{\nu}_h \varphi d\sigma = 0.
\end{aligned} \tag{4.9}$$

Proof. By virtue of Remark 4.3, u satisfies (4.5). Since $\mathcal{D}(0,T;H_0^1(\Omega)) \subset H_0^1(Q)$, equation (4.5) is true for $\varphi \in \mathcal{D}(0,T;H_0^1(\Omega))$. Let $\zeta \in \mathcal{D}(0,T;V)$. We consider an extension $\hat{\zeta} \in \mathcal{D}(0,T;H_0^1(\Omega))$ of ζ. We also introduce a sequence $(\xi_\varepsilon)_{\varepsilon>0}$ such that $\xi_\varepsilon \in W^{1,\infty}(\Omega)$ with $0 \le \xi_\varepsilon \le 1$,

$$\xi_\varepsilon(x) = \begin{cases} 1 & \text{if } x \in \overline{\Omega}_p, \\ 0 & \text{if } x \in \Omega_h,\ \operatorname{dist}(x,\Gamma_{hp}) \ge \varepsilon, \end{cases}$$

and $\varepsilon\|\nabla_x \xi_\varepsilon\|_\infty \le C$ for some $C > 0$ independent of ε. Then, we take $\varphi = \hat{\zeta}\xi_\varepsilon$ in (4.5) and pass to the limit as $\varepsilon \to 0^+$. We use (4.8) to show that

$$\lim_{\varepsilon\to 0^+} \int_{Q_h} f(u)\hat{\zeta}\boldsymbol{b}(x)\cdot\nabla\xi_\varepsilon dxdt = \operatorname*{ess\,lim}_{s\to 0^-} \int_{\Sigma_{hp}} f(u(\sigma + s\boldsymbol{\nu}_h))\zeta\boldsymbol{b}(\overline{\sigma})\cdot\boldsymbol{\nu}_h d\sigma dt.$$

There is no difficulty to take the limit with respect to ε in the other terms of (4.5), and we obtain thereby for all $\zeta \in \mathcal{D}(0,T;V)$

$$\int_{Q_p} u\partial_t\zeta dxdt = \int_{Q_p} (\nabla\phi(u) - f(u)\boldsymbol{b}(x))\cdot\nabla\zeta dxdt \\ - \operatorname*{ess\,lim}_{s\to 0^-} \int_{\Sigma_{hp}} f(u(\sigma + s\boldsymbol{\nu}_h))\zeta\boldsymbol{b}(\overline{\sigma})\cdot\boldsymbol{\nu}_h d\sigma dt.$$

Since u is bounded and $\phi(u) \in L^2(0,T;V)$, there exists $C_1 > 0$ such that

$$\left| \int_{Q_p} (\nabla\phi(u) - f(u)\boldsymbol{b}(x))\cdot\nabla\zeta dxdt \right| \le C_1\|\zeta\|_{L^2(0,T;V)}.$$

The continuity of the trace operator as a mapping of $H^1(\Omega_p)$ into $L^2(\Gamma_p)$ ensures, as long as u is bounded, the existence of a constant $C_2 > 0$ such that, for almost all $s < 0$,

$$\left| \int_{\Sigma_{hp}} f(u(\sigma + s\boldsymbol{\nu}_h))\boldsymbol{b}(\overline{\sigma})\cdot\boldsymbol{\nu}_h\zeta d\sigma dt \right| \le C_2\|\zeta\|_{L^2(0,T;V)},$$

so that

$$\left| \operatorname*{ess\,lim}_{s\to 0^-} \int_{\Sigma_{hp}} f(u(\sigma + s\boldsymbol{\nu}_h))\boldsymbol{b}(\overline{\sigma})\cdot\boldsymbol{\nu}_h\zeta d\sigma dt \right| \le C_2\|\zeta\|_{L^2(0,T;V)}.$$

Therefore, there exists a constant $C > 0$ such that for all $\zeta \in \mathcal{D}(0,T;V)$

$$\left| \int_{Q_p} u\partial_t\zeta dxdt \right| \le C\|\zeta\|_{L^2(0,T;V)}.$$

Thus, we have $\partial_t u \in L^2(0,T;V')$, and for any $\zeta \in \mathcal{D}(0,T;V)$

$$-\int_{Q_p} u\partial_t\zeta dxdt = \int_0^T \langle\partial_t u, \zeta\rangle_{V'\times V} dt.$$

The assertion follows by density of $\mathcal{D}(0,T;V)$ in $L^2(0,T;V)$. □

We turn now to the uniqueness result. Let u_1 and u_2 be two weak entropy solutions to (1.1). We suppose that u_1 and u_2 have the same initial data on the hyperbolic part. In view of Theorem 4.5, we already know that $u_1 = u_2$ a.e. on Q_h. On the parabolic zone, the Holmgren-type duality method, described in Section 2.2, cannot be applied as we lack information about the traces of u_1 and u_2 along Σ_{hp}. For this reason, we use the method of doubling only the time variable. Moreover, to deal with the convection term, we suppose that $f \circ \phi^{-1}$ is Hölder continuous on $\mathbb{R}$ with an exponent greater or equal than $1/2$, i.e., we assume there exist $\theta \in [1/2, 1]$, $K > 0$ such that for all $x, y \in \mathbb{R}$

$$|(f \circ \phi^{-1})(x) - (f \circ \phi^{-1})(y)| \leq K|x-y|^{\theta}. \tag{4.10}$$

Theorem 4.7. *Under the assumption* (4.10), (1.1) *admits at most one weak entropy solution. Moreover, if u_1 and u_2 are two weak entropy solutions corresponding to initial data $u_{0,1}$ and $u_{0,2}$ with $u_{0,1} = u_{0,2}$ a.e. on Ω_h then, for almost all* $t \in (0, T)$,

$$\int_{\Omega} |u_1(t,x) - u_2(t,x)|dx \leq \int_{\Omega_p} |u_{0,1}(x) - u_{0,2}(x)|dx.$$

Proof. For any $j \in \mathbb{N}$ and $t, \tilde{t} \in (0, T)$, we set

$$\alpha_j(t, \tilde{t}) = \gamma\left(\frac{t+\tilde{t}}{2}\right)\rho_j\left(\frac{t-\tilde{t}}{2}\right),$$

where $\gamma \in \mathcal{C}_c^{\infty}(0,T)$, $\gamma \geq 0$, and $(\rho_j)_j$ is a sequence of mollifiers. We notice that $\alpha_j \geq 0$ and, for j large enough, $\alpha_j \in \mathcal{C}_c^{\infty}((0,T) \times (0,T))$.

In (4.9), written for the weak entropy solution u_1 and in variables (t, x), we choose the test function

$$\varphi(t, \tilde{t}, x) = \operatorname{sgn}_{\eta}(\phi(u_1(t,x)) - \phi(u_2(\tilde{t},x)))\alpha_j(t,\tilde{t})$$

and integrate with respect to $\tilde{t}$. In (4.9), written for the weak entropy solution u_2 and in variables $(\tilde{t}, x)$, we take the same test function and integrate with respect to t. Taking then the difference yields

$$\begin{aligned}
&\int_0^T \int_0^T \langle \partial_t u_1 - \partial_{\tilde{t}} \tilde{u}_2, \operatorname{sgn}_{\eta}(\phi(u_1) - \phi(\tilde{u}_2))\rangle_{V' \times V} \alpha_j dt d\tilde{t} \\
&+ \int_{]0,T[\times Q_p} \nabla(\phi(u_1) - \phi(\tilde{u}_2)) \cdot \nabla \operatorname{sgn}_{\eta}(\phi(u_1) - \phi(\tilde{u}_2))\alpha_j dx dt d\tilde{t} \\
&- \int_{]0,T[\times Q_p} (f(u_1) - f(\tilde{u}_2))\boldsymbol{b}(x) \cdot \nabla \operatorname{sgn}_{\eta}(\phi(u_1) - \phi(\tilde{u}_2))\alpha_j dx dt d\tilde{t} \\
&= \int_0^T \operatorname*{ess\,lim}_{s \to 0^-} \int_{\Sigma_{hp}} f(u_1(\sigma + s\boldsymbol{\nu}_h))\boldsymbol{b}(\overline{\sigma}) \cdot \boldsymbol{\nu}_h \operatorname{sgn}_{\eta}(\phi(u_1) - \phi(\tilde{u}_2))\alpha_j d\sigma d\tilde{t} \\
&\quad - \int_0^T \operatorname*{ess\,lim}_{s \to 0^-} \int_{\Sigma_{hp}} f(u_2(\tilde{\sigma} + s\boldsymbol{\nu}_h))\boldsymbol{b}(\overline{\sigma}) \cdot \boldsymbol{\nu}_h \operatorname{sgn}_{\eta}(\phi(u_1) - \phi(\tilde{u}_2))\alpha_j d\tilde{\sigma} dt.
\end{aligned} \tag{4.11}$$

To simplify the notation, we add a tilde to any function in $\tilde{t}$. We want to pass to the limit in (4.11), first as η goes to 0^+ and then as j tends to ∞.

In the first integral on the left-hand side, we use an integration-by-parts formula based on a generalization of Lemma 1.1 (see [18]) to obtain

$$\begin{aligned}&\int_0^T\int_0^T \langle \partial_t u_1 - \partial_{\tilde t}\tilde u_2, \mathrm{sgn}_\eta(\phi(u_1)-\phi(\tilde u_2))\rangle_{V'\times V}\alpha_j dt d\tilde t\\ &= -\int_0^T\int_{Q_p}\left(\left(\int_{\tilde u_2}^{u_1}\mathrm{sgn}_\eta(\phi(r)-\phi(\tilde u_2))dr\right)\partial_t\alpha_j\right)dxdtd\tilde t\\ &\quad -\int_0^T\int_{Q_p}\left(\left(\int_{\tilde u_2}^{u_1}\mathrm{sgn}_\eta(\phi(u_1)-\phi(r))dr\right)\partial_{\tilde t}\alpha_j\right)dxdtd\tilde t.\end{aligned}$$

In the third integral of (4.11), we write $f(u_i) = f\circ\phi^{-1}(\phi(u_i))$, $i = 1,2$. Then, due to (4.10) and Young's inequality, we derive

$$\begin{aligned}&-\int_0^T\int_{Q_p}(f(u_1)-f(\tilde u_2))\boldsymbol{b}(x)\cdot\nabla\mathrm{sgn}_\eta(\phi(u_1)-\phi(\tilde u_2))\alpha_j dxdtd\tilde t\\ &\le \frac{K^2\|\boldsymbol{b}\|_\infty}{2}\int_{]0,T[\times Q_p}|\phi(u_1)-\phi(\tilde u_2)|^{2\theta}\mathrm{sgn}'_\eta(\phi(u_1)-\phi(\tilde u_2))\alpha_j dxdtd\tilde t\\ &\quad+\frac{1}{2}\int_0^T\int_{Q_p}\mathrm{sgn}'_\eta(\phi(u_1)-\phi(\tilde u_2))|\nabla(\phi(u_1)-\phi(\tilde u_2))|^2\alpha_j dxdtd\tilde t.\end{aligned}$$

By definition of sgn_η, the first term on the right-hand side of the previous inequality is bounded by

$$C\int_0^T\int_{Q_p}|\phi(u_1)-\phi(\tilde u_2)|^{2\theta-1}\alpha_j\chi_{\{-\eta<\phi(u_1)-\phi(\tilde u_2)<\eta\}}dxdtd\tilde t,$$

where $C>0$. As $\theta\ge\frac{1}{2}$, we obtain

$$\lim_{\eta\to 0^+}\int_0^T\int_{Q_p}|\phi(u_1)-\phi(\tilde u_2)|^{2\theta-1}\alpha_j\chi_{\{-\eta<\phi(u_1)-\phi(\tilde u_2)<\eta\}}dxdtd\tilde t = 0.$$

Let us now study the right-hand side of (4.11). Since $u_1 = u_2$ a.e. on Q_h, we have $u_1(\sigma+s\boldsymbol{\nu}_h) = u_2(\sigma+s\boldsymbol{\nu}_h)$ for almost all negative s. Thus we may refer to (4.8) to show (as in the proof of Lemma 3.10) the existence of $\theta\in L^\infty(\Sigma_{hp})$ such that for any $\beta\in L^1(\Sigma_{hp})$

$$\operatorname*{ess\,lim}_{s\to 0^-}\int_{\Sigma_{hp}} f(u_1(\sigma+s\boldsymbol{\nu}_h))\boldsymbol{b}(\overline{\sigma})\cdot\boldsymbol{\nu}_h\beta(\sigma)dtd\sigma = \int_{\Sigma_{hp}}\theta(\sigma)\beta(\sigma)dtd\sigma$$

and

$$\operatorname*{ess\,lim}_{s\to 0^-}\int_{\Sigma_{hp}} f(u_2(\tilde\sigma+s\boldsymbol{\nu}_h))\boldsymbol{b}(\overline{\sigma})\cdot\boldsymbol{\nu}_h\beta(\tilde\sigma)dtd\sigma = \int_{\Sigma_{hp}}\theta(\tilde\sigma)\beta(\tilde\sigma)dtd\tilde\sigma.$$

Therefore, the right-hand side of (4.11) is equal to

$$\int_0^T \int_{\Sigma_{hp}} (\theta(t,\overline{\sigma}) - \theta(\tilde{t},\overline{\sigma}))\mathrm{sgn}_\eta(\phi(u_1)(\sigma) - \phi(u_2)(\tilde{t},\overline{\sigma}))\alpha_j(\tilde{t},t)dtd\overline{\sigma}d\tilde{t}.$$

Finally, when η goes to 0^+ in (4.11), by Lebesgue's theorem on dominated convergence, we obtain

$$-\int_{(0,T)\times Q_p} |u_1 - \tilde{u}_2|(\partial_t\alpha_j + \partial_{\tilde{t}}\alpha_j)dxdtd\tilde{t} \leq \int_0^T \int_{\Sigma_{hp}} |\theta(t,\overline{\sigma}) - \theta(\tilde{t},\overline{\sigma})|\alpha_j dtd\overline{\sigma}d\tilde{t}.$$

With the definition of α_j, we see that $\partial_t\alpha_j + \partial_{\tilde{t}}\alpha_j = \gamma'(\frac{t+\tilde{t}}{2})\rho_j(t-\tilde{t})$. Now we are able to take the limit with respect to j and find

$$-\int_{Q_p} |u_1 - u_2|\gamma'(t)dxdt \leq 0.$$

The conclusion follows by choosing a suitable test function γ (see the proof of Theorem 3.7). □

4.2 Existence

We approximate a weak entropy solution to (1.1) through a sequence of solutions to viscous problems linked to (1.1) by adding a diffusion term in accordance with the proposed physical modelling of two layers in the subsoil with different geological characteristics. Thus, for any positive real number μ, we are first interested in the uniqueness and existence of a measurable and essentially bounded function u_μ satisfying in the sense of distributions

$$\left\{\begin{array}{rcll} \partial_t u_\mu + \nabla\cdot(\boldsymbol{b}(x)f(u_\mu)) & = & \nabla\cdot(\lambda_\mu(x)\nabla\phi(u_\mu)) & \text{in } Q, \\ u_\mu(0,\cdot) & = & u_0 & \text{on } \Omega, \\ u_\mu & = & 0 & \text{on } \Sigma, \end{array}\right. \tag{4.12}$$

with

$$\lambda_\mu(x) = \chi_{\Omega_p}(x) + \mu\chi_{\Omega_h}(x).$$

We state first

Proposition 4.8. *There exists a unique weak solution $u_\mu \in W(0,T)$ to (4.12), i.e.,*

$$m \leq u_\mu(t,x) \leq M \textit{ for almost all } (t,x) \in Q, \tag{4.13}$$

for almost all $t \in (0,T)$ and all $v \in H_0^1(\Omega)$

$$\langle\partial_t u_\mu, v\rangle + \int_\Omega (\lambda_\mu(x)\nabla\phi(u_\mu) - \boldsymbol{b}(x)f(u_\mu))\cdot\nabla v dx = 0, \tag{4.14}$$

and

$$u_\mu(0) = u_0 \textit{ in } L^2(\Omega).$$

Proof. We remark that, for any fixed μ the partial differential operator is not degenerated. So one may apply the Schauder–Tikhonov fixed point theorem and the Holmgren-type duality method presented in Section 2. □

We look now for a priori estimates fulfilled by the sequence $(u_\mu)_{\mu>0}$ in order to study its limit as μ tends to 0^+.

Proposition 4.9. *There exists a constant $C > 0$, independent of μ, such that*

$$\|\lambda_\mu^{1/2}\nabla\widehat{\phi}(u_\mu)\|^2_{L^2(Q)^n} \le C, \tag{4.15}$$

$$\|\partial_t u_\mu\|_{L^2(0,T;H^{-1}(\Omega))} \le C, \tag{4.16}$$

where $\widehat{\phi}(u_\mu) = \int_0^{u_\mu} \sqrt{\phi'(\tau)}d\tau.$

Proof. We choose $v = u_\mu$ in (4.14) and integrate over $(0,T)$. We observe that

$$\int_0^T \langle\partial_t u_\mu, u_\mu\rangle dt = \frac{1}{2}\|u_\mu(T)\|^2_{L^2(\Omega)} - \frac{1}{2}\|u_0\|^2_{L^2(\Omega)}$$

and

$$\int_0^T\int_\Omega f(u_\mu)\boldsymbol{b}(x)\cdot\nabla u_\mu dxdt = 0.$$

The diffusion term can be written as

$$\int_Q(\lambda_\mu(x)^{1/2}\sqrt{\phi'(u_\mu)}\nabla u_\mu)^2 dxdt = \|\lambda_\mu^{1/2}\nabla\widehat{\phi}(u_\mu)\|^2_{L^2(Q)^n},$$

and (4.15) follows. The second assertion then follows from the first one together with the equation. □

Our aim is now to describe the behaviour of the sequence $(u_\mu)_{\mu>0}$ when μ goes to 0^+. On the hyperbolic domain we take advantage of (4.13) and Theorem 1.5. On the parabolic area, estimates (4.13), (4.15) and (4.16) are not sufficient to study the behaviour of $(u_\mu)_{\mu>0}$. For this reason, we use an additional assumption on ϕ:

$$\phi^{-1} \text{ is Hölder continuous with an exponent } \tau \in (0,1). \tag{4.17}$$

Proposition 4.10. *Under* (4.17)*, there exists a function $u \in L^\infty(Q)$ with $\phi(u)$ in $L^2(0,T;V)$, and a subsequence, still denoted by $(u_\mu)_{\mu>0}$, such that*

$$u_\mu \overset{*}{\rightharpoonup} u \text{ in } L^\infty(Q), \quad u_\mu \to u \text{ in } L^q(Q_p) \text{ for any } 1\le q<\infty \text{ and a.e. on } Q_p.$$

Moreover, we also have

$$\nabla\phi(u_\mu) \rightharpoonup \nabla\phi(u) \text{ in } L^2(Q_p)^n, \quad \mu\nabla\phi(u_\mu) \to 0 \text{ in } L^2(Q_h)^n.$$

In order to prove this proposition, we need the following lemma:

Lemma 4.11. *Let $G : \mathbb{R} \to \mathbb{R}$ be Hölder continuous with exponent $\alpha \in (0, 1)$. Then $G(v) \in W^{\alpha s, \frac{p}{\alpha}}(\Omega)$ for any $v \in W^{s,p}(\Omega)$, $0 < s < 1$, $1 < p < \infty$, and the mapping $G : W^{s,p}(\Omega) \to W^{\alpha s, \frac{p}{\alpha}}(\Omega)$ is bounded.*

Proof of Proposition 4.10. Note first that (4.13) implies that there exists a subsequence, still denoted by $(u_\mu)_{\mu>0}$, such that $u_\mu \overset{*}{\rightharpoonup} u$ in $L^\infty(Q)$. The two convergence results about $\lambda_\mu \nabla\phi(u_\mu)$ are an immediate consequence of (4.15). So we just have to prove the strong convergence of $(u_\mu)_{\mu>0}$ in $L^q(Q_p)$. To do so, we refer to the arguments used in [9, Chapt. 2].

From (4.16), $(\partial_t u_\mu)_{\mu>0}$ is bounded in $L^2(0, T; H^{-1}(\Omega_p))$. Moreover, due to (4.13) and (4.15), the sequence $(\phi(u_\mu))_{\mu>0}$ is bounded in $L^2(0, T; V)$. For any $s \in (0, 1)$, we also have

$$L^2(0, T; V) \hookrightarrow L^2(0, T; H^1(\Omega_p)) \hookrightarrow L^2(0, T; W^{s,2}(\Omega_p)).$$

Since $u_\mu = \phi^{-1}(\phi(u_\mu))$, by (4.17) and Lemma 4.11, we assert that u_μ is bounded in $L^{2/\tau}(0, T; W^{\tau s, 2/\tau}(\Omega_p))$. The compactness of the embedding of $W^{\tau s,2/\tau}(\Omega_p)$ in $L^{2/\tau}(\Omega_p)$ and the Lions–Aubin compactness theorem (see [14, p. 57]) ensure that

$$\mathcal{W} := \{v \in L^{2/\tau}(0, T; W^{\tau s,2/\tau}(\Omega_p));\ \partial_t v \in L^2(0, T; H^{-1}(\Omega_p))\}$$

is compactly embedded in $L^{2/\tau}(0, T; L^{2/\tau}(\Omega_p))$. Hence, the conclusion follows. □

We are now able to state the main theorem of this section.

Theorem 4.12. *Problem* (1.1) *has a weak entropy solution that is the limit of the sequence $(u_\mu)_{\mu>0}$ of solutions to* (4.12) *in $L^q(Q)$, $1 \le q < \infty$, as $\mu \to 0^+$.*

Proof. We consider the function u given in Proposition 4.10. Because of (4.13), we can apply Theorem 1.5 to assert that there exists a function $\pi \in L^\infty((0, 1) \times Q_h)$ such that for any continuous and bounded function ψ on $Q_h \times [m, M]$ and $\xi \in L^1(Q_h)$

$$\lim_{\mu\to 0^+} \int_{Q_h} \psi(t, x, u_\mu(t, x))\xi dxdt = \int_0^1 \int_{Q_h} \psi(t, x, \pi(\alpha, t, x))\xi dxdtd\alpha. \tag{4.18}$$

Our aim is first to establish that, as in Section 3, the process π is reduced to $u_{|Q_h}$ and, second, to prove that u is a weak entropy solution to (1.1).

In order to do so, we come back to (4.14) and, for any real k, we choose the test function $v^\mu_\eta = \operatorname{sgn}_\eta(\phi(u_\mu) - \phi(k))\varphi_1\varphi_2$, where $\varphi_1 \in \mathcal{C}^\infty(-T, T)$, $\varphi_2 \in \mathcal{C}^\infty_c(\Omega)$ with $\varphi_1, \varphi_2 \ge 0$. Thereby, we obtain

$$\langle \partial_t u_\mu, v^\mu_\eta\rangle + \int_\Omega (\lambda_\mu(x)\nabla\phi(u_\mu) - \boldsymbol{b}(x) f(u_\mu)) \cdot \nabla v^\mu_\eta dx = 0.$$

We integrate with respect to the time variable and perform the following transformations in a same way as in Section 3. Thanks to a generalization of Lemma 1.1 (see

[18]), the evolutionary term may be written as

$$\int_0^T \langle \partial_t u_\mu, \mathrm{sgn}_\eta(\phi(u_\mu) - \phi(k))\varphi_2 \rangle \varphi_1 dt$$
$$= -\int_Q I_\eta(u_\mu)\varphi_2 \partial_t \varphi_1 dxdt - \int_\Omega I_\eta(u_0)\varphi_2\varphi_1(0)dx$$

with $I_\eta(u_\mu) = \int_k^{u_\mu} \mathrm{sgn}_\eta(\phi(\tau) - \phi(k))d\tau$. For the diffusion term, we obtain

$$\int_Q \lambda_\mu(x)\nabla\phi(u_\mu)\cdot\nabla v_\eta^\mu dxdt \geq \int_Q \lambda_\mu \mathrm{sgn}_\eta(\phi(u_\mu) - \phi(k))\nabla\phi(u_\mu)\cdot\nabla\varphi_2\varphi_1 dxdt.$$

Moreover, we have

$$\int_Q \boldsymbol{b}(x)f(u_\mu)\cdot\nabla v_\eta^\mu dxdt = -\int_Q \boldsymbol{b}(x)F_\eta(u_\mu)\nabla\varphi_2\varphi_1 dxdt$$

with $F_\eta(u_\mu) = \int_k^{u_\mu} f'(\tau)\mathrm{sgn}_\eta(\phi(\tau) - \phi(k))d\tau$.

We take the limit with respect to μ separately on the hyperbolic and parabolic area. On Q_h, we take advantage of (4.18), while on Q_p, we use Proposition 4.10. In this way, we obtain

$$\int_{Q_p} I_\eta(u)\varphi_2\partial_t\varphi_1 + \boldsymbol{b}(x)F_\eta(u)\cdot\nabla\varphi_2\varphi_1 dxdt$$
$$+\int_0^1\int_{Q_h} I_\eta(\pi)\varphi_2\partial_t\varphi_1 + \boldsymbol{b}(x)F_\eta(\pi)\cdot\nabla\varphi_2\varphi_1 dxdtd\alpha$$
$$-\int_{Q_p} \mathrm{sgn}_\eta(\phi(u) - \phi(k))\nabla(\phi(u) - \phi(k))\cdot\nabla\varphi_2\varphi_1 dxdt$$
$$+\int_\Omega I_\eta(u_0)\varphi_2\varphi_1(0)dx \geq 0.$$

Then we take the limit with respect to η. We remark that

$$\lim_{\eta\to 0^+} F_\eta(u) = \mathrm{sgn}(\phi(u) - \phi(k))(f(u) - f(k)) = F(u,k) \text{ a.e. on } Q_h \text{ and } Q_p.$$

Therefore, we get in the limit

$$\int_{Q_p} |u-k|\varphi_2\partial_t\varphi_1 + \boldsymbol{b}(x)F(u,k)\cdot\nabla\varphi_2\varphi_1 dxdt$$
$$+\int_0^1\int_{Q_h} |\pi-k|\varphi_2\partial_t\varphi_1 + \boldsymbol{b}(x)F(\pi,k)\cdot\nabla\varphi_2\varphi_1 dxdtd\alpha$$
$$+\int_\Omega |u_0-k|\varphi_2\varphi_1(0)dx - \int_{Q_p} \nabla|\phi(u) - \phi(k)|\cdot\nabla\varphi_2\varphi_1 dxdt \geq 0. \tag{4.19}$$

We justify now that π satisfies the initial condition (4.3) and the boundary condition (4.4). When we consider test functions φ_2 that belong to $\mathcal{C}_c^\infty(\Omega_h)$, we deduce that

$$-\int_0^1 \int_{Q_h} |\pi - k|\varphi_2 \partial_t \varphi_1 + \boldsymbol{b}(x) F(\pi) \cdot \nabla \varphi_2 \varphi_1 dx dt d\alpha \le \int_{\Omega_h} |u_0 - k|\varphi_2 \varphi_1(0) dx.$$

Therefore, we can apply Lemma 3.13 (with $Q = Q_h$) to state that

$$\operatorname*{ess\,lim}_{t \to 0^+} \int_0^1 \int_{\Omega_h} |\pi(\alpha, t, x) - u_0(x)| dx d\alpha = 0. \tag{4.20}$$

Hence, π fulfills the initial condition (4.3) on Ω_h.

Next we choose in (4.14) the test function $v_\delta = \partial_1 H_\delta(u_\mu, k)\varphi_1\varphi_2$ (see Example 3.4 for the definition of H_δ), where $\varphi_1 \in \mathcal{C}_c^\infty(0,T)$, $\varphi_2 \in \mathcal{C}^\infty(\overline{\Omega}_h)$ with $\varphi_2 = 0$ on Γ_{hp}, $\varphi_1, \varphi_2 \ge 0$. We take the limit with respect to η and apply Lemma 3.14 to assert that, for any nonnegative $\beta \in L^1(\Sigma_h \setminus \Sigma_{hp})$

$$\operatorname*{ess\,lim}_{s \to 0^-} \int_0^1 \int_{\Sigma_h \setminus \Sigma_{hp}} J_\delta(\pi(\alpha, \sigma + s\boldsymbol{\nu}_h), k)\beta \boldsymbol{b}(\overline{\sigma}) \cdot \boldsymbol{\nu}_h d\sigma d\alpha \ge 0,$$

where J_δ is defined in Example 3.4. As $\delta \to 0^+$, we obtain that π satisfies (4.4). Therefore, π fulfills (4.2) for all $\varphi \in \mathcal{C}_c^\infty(Q_h)$ as well as (4.3) and (4.4), where integrals over Q_h, Ω_h and $\Sigma_h \setminus \Sigma_{hp}$ are replaced by integrals over $(0,1) \times Q_h$, $(0,1) \times \Omega_h$ and $(0,1) \times \Sigma_h \setminus \Sigma_{hp}$, respectively, with respect to the corresponding measure. Then, by reasoning as in Theorem 4.5, if $\pi_1(\alpha, \cdot)$ and $\pi_2(\beta, \cdot)$ are two process solutions for initial data $u_{0,1}$ and $u_{0,2}$, we obtain for almost all $t \in (0,T)$

$$\int_0^1 \int_0^1 \int_{\Omega_h} |\pi_1(\alpha, t, x) - \pi_2(\beta, t, x)| dx d\alpha d\beta \le \int_{\Omega_h} |u_{0,1} - u_{0,2}| dx.$$

Therefore, if $u_{0,1} = u_{0,2}$ on Ω_h, there exists a bounded function u_h on Q_h such that a.e. on Q_h,

$$u_h(\cdot) = \pi_1(\alpha, \cdot) = \pi_2(\beta, \cdot) \text{ for almost all } \alpha \text{ and } \beta \text{ in } (0,1).$$

Besides, $u_h = u_{|Q_h}$ a.e. on Q_h. Thus, from (4.19), it follows that u fulfills (4.2) for all $\varphi \in \mathcal{C}_c^\infty(0,T) \otimes \mathcal{C}_c^\infty(\Omega)$ and, by density, for all $\varphi \in \mathcal{C}_c^\infty(Q)$ as well as (4.4).

To complete the proof it remains to show that (4.3) holds for u. Due to (4.20), we just have to focus on Ω_p. We consider (4.19) for $\varphi_2 \in \mathcal{C}_c^\infty(\Omega_p)$. This yields

$$-\int_0^T \left(\int_{\Omega_p} |u - k|\varphi_2 dx + g(t) \right) \varphi_1'(t) dt \le \int_{\Omega_p} |u_0 - k|\varphi_2 \varphi(0) dx$$

with

$$\begin{aligned} g(t) = \int_{\Omega_p} \int_0^t & (|\phi(u(\tau, x)) - \phi(k)|\Delta\varphi_2 \\ & - \operatorname{sgn}(u(\tau, x) - k)(f(u(\tau, x) - f(k))\boldsymbol{b}(x) \cdot \nabla\varphi_2)\, d\tau dx. \end{aligned}$$

Therefore, the time-depending function

$$t \mapsto \int_{\Omega_p} |u-k|\varphi_2 dx + g(t)$$

can be identified a.e. with a nonincreasing and bounded function. Then it has an essential limit as $t \to 0^+$. Since $g(t) \to 0$ as $t \to 0^+$, it follows

$$\operatorname*{ess\,lim}_{t\to 0^+} \int_{\Omega_p} |u-k|\varphi_2 dx \le \int_{\Omega_p} |u_0-k|\varphi_2 dx,$$

for all nonnegative $\varphi_2 \in \mathcal{C}_c^\infty(\Omega_p)$, which implies (see [17]), for any $k \in L^\infty(\Omega_p)$,

$$\operatorname*{ess\,lim}_{t\to 0^+} \int_{\Omega_p} |u-k(x)| dx \le \int_{\Omega_p} |u_0-k(x)| dx.$$

We choose $k = u_0$ to ensure that u satisfies (4.3). This concludes the proof. □

5 Concluding remarks

We end these lecture notes by some remarks. We have studied a simplified coupling problem. One could, for example, also assume, as in [1, 12], that the nonlinearities are different on Q_h and Q_p. Indeed, we could consider the following problem: Find a measurable and essentially bounded function u on Q such that in the sense of distributions

$$\left\{\begin{array}{rcll} \partial_t u + \nabla\cdot(\boldsymbol{b}_h(x)f_h(u)) &=& 0 & \text{in } Q_h,\\ \partial_t u + \nabla\cdot(\boldsymbol{b}_p(x)f_p(u)) &=& \Delta\phi(u) & \text{in } Q_p,\\ u &=& 0 & \text{on } (0,T)\times\partial\Omega,\\ u(0,\cdot) &=& u_0 & \text{on } \Omega. \end{array}\right. \tag{5.1}$$

In this case, the notion of weak entropy solution in Definition 4.2 is not suitable anymore. We have to add in the entropy formulation (4.2) an interface integral that takes into account the discontinuity of the flux function along Σ_{hp}. Indeed, setting $\boldsymbol{b}(x)f(u) = \boldsymbol{b}_h(x)f_h(u)\chi_{\Omega_h} + \boldsymbol{b}_p(x)f_p(u)\chi_{\Omega_p}$, we propose the following definition of a weak entropy solution.

Definition 5.1. A function u is said to be a weak entropy solution to (5.1) if

(i) $u \in L^\infty(Q)$, $\phi(u) \in L^2(0,T;V)$,

(ii) for all nonnegative $\varphi \in \mathcal{C}_c^\infty(Q)$and all $k \in \mathbb{R}$

$$\int_Q (|u-k|\partial_t\varphi + (\boldsymbol{b}(x)F(u,k) - \nabla|\phi(u)-\phi(k)|)\cdot\nabla\varphi) dxdt$$
$$+ \int_{\Sigma_{hp}} (\boldsymbol{b}_h(\overline{\sigma})f_h(k) - \boldsymbol{b}_p(\overline{\sigma})f_p(k))\cdot\boldsymbol{\nu}_h \operatorname{sgn}(\phi(u)-\phi(k)) d\sigma \ge 0.$$

(iii) u satisfies the initial condition (4.3) and the boundary condition (4.4).

We refer to [11, Chap. 4] or [12] for the study of (5.1). When we use the vanishing viscosity method, the major difficulty relies on the study of the viscous problem related to (5.1). Roughly speaking, we are not able to prove the maximum principle without additional assumption on the flux. We also underline that the existence and uniqueness results of [12] are obtained in the framework of outwards characteristics (see Remark 4.1). The case of inwards characteristics, i.e., the case where the characteristics are entering the hyperbolic zone, is treated in [2] for (1.1). We emphasize the fact that the general case (i.e., no assumption on the characteristics) is still an open problem.

Acknowledgments. I warmly thank Petra Wittbold and Etienne Emmrich for their kind invitation at the Technische Universität Berlin. Moreover, their remarks, suggestions and comments greatly improved these lecture notes. I also would like to thank the Stiftung Luftbrückendank for supporting my stay in Berlin.

References

[1] G. Aguilar, L. Lévi and M. Madaune-Tort, Coupling of multidimensional parabolic and hyperbolic equations, *J. Hyperbolic Differ. Equ.* **3** (2006), pp. 53–80.

[2] ———, Nonlinear multidimensional parabolic-hyperbolic equations, *2006 International conference in honor of J. Fleckinger, Electron. J. Diff. Eqns.* **Conf. 16** (2007), pp. 15–28.

[3] G. Aguilar, F. Lisbona and M. Madaune-Tort, Analysis of a nonlinear parabolic-hyperbolic problem, *Adv. Math. Sci. Appl.* **7** (1997), pp. 165–181.

[4] C. Bardos, A. Y. LeRoux and J. C. Nédélec, First order quasilinear equations with boundary conditions, *Commun. Partial Differ. Equations* **4** (1979), pp. 1017–1034.

[5] R. Dautray and J. L. Lions, *Analyse mathématique et calcul numérique pour les sciences et techniques*, vol. 8, Masson, Paris, 1980.

[6] ———, *Mathematical analysis and numerical methods for science and technology*, vol. 5, Springer, Berlin, 1992.

[7] L. C. Evans and R. F. Gariepy, *Measure theory and fine properties of functions*, CRC Press, London, 1992.

[8] R. Eymard, T. Gallouët and R. Herbin, Existence and uniqueness of the nonlinear hyperbolic equation, *Chin. Ann. Math.* **Ser. B 16** (1995), pp. 1–14.

[9] G. Gagneux and M. Madaune-Tort, *Analyse mathématiques de modèles non linéaires de l'ingénierie pétrolière*, Springer, Berlin, 1996.

[10] E. Godlewski and P. A. Raviart, *Hyperbolic systems of conservation laws*, Mathématiques et Applications, S.M.A.I., Ellipses, Paris, 1991.

[11] J. Jimenez, *Modèles non linéaires de transport dans un milieu poreux hétérogène*, Ph.D. thesis, Univ. Pau, 2007.

[12] J. Jimenez and L. Lévi, Entropy formulations for a class of scalar conservation laws with space-discontinuous flux functions in a bounded domain, *J. Engrg. Math.* **60** (2008), pp. 319–335.

[13] S. N. Kružkov, First order quasilinear equations with several independent variables, *Mat. Sb.* **81:123** (1970), pp. 228–255.

[14] J. L. Lions, *Quelques méthodes de résolution des problèmes aux limites non linéaires*, Dunod, Gauthier-Villars, Paris, 1969.

[15] J. Málek, J. Nečas, M. Rokyta and M. Růžička, *Weak and measure-valued solutions to evolutionary PDEs*, Chapman & Hall, London, 1996.

[16] M. Marcus and V. J. Mizel, Every superposition mapping one Sobolev space into another is continuous, *J. Funct. Anal.* **33** (1979), pp. 217–229.

[17] F. Otto, Initial-boundary value problem for a scalar conservation law, *C. R. Acad. Sci. Paris, Sér. I* **322** (1996), pp. 729–734.

[18] ______, L^1-contraction and uniqueness for quasilinear elliptic-parabolic equations, *J. Differ. Equations* **131** (1996), pp. 20–38.

[19] J. Wloka, *Partial differential equations*, Cambridge University Press, Cambridge, 1987.

Author information

Julien Jimenez, Laboratory of applied mathematics, University of Pau, BP 1155, 64013 Pau Cedex, France.
E-mail: `julien.jimenez@univ-pau.fr`

Standing waves in nonlinear Schrödinger equations

Stefan Le Coz

Abstract. In the theory of nonlinear Schrödinger equations, it is expected that the solutions will either spread out because of the dispersive effect of the linear part of the equation or concentrate at one or several points because of nonlinear effects. In some remarkable cases, these effects counterbalance, and special solutions that neither disperse nor focus appear, the so-called *standing waves*. For the physical applications as well as for the mathematical properties of the equation, a fundamental issue is the stability of waves with respect to perturbations. Our purpose in these notes is to present various methods developed to study the existence and stability of standing waves. We prove the existence of standing waves by using a variational approach. When stability holds, it is obtained by proving a coercivity property for a linearized operator. Another approach based on variational and compactness arguments is also presented. When instability holds, we show by a method combining a virial identity and variational arguments that the standing waves are unstable by blow-up.

Keywords. Nonlinear Schrödinger equation, standing waves, orbital stability, instability, blow-up, variational methods.

AMS classification. 35Q55, 35Q51, 35B35, 35A15.

1 Introduction

In these lecture notes, we consider the nonlinear Schrödinger equation

$$i\partial_t u + \Delta u + |u|^{p-1}u = 0. \tag{1.1}$$

Equation (1.1) arises in various physical and biological contexts, for example in nonlinear optics, for Bose–Einstein condensates, in the modelling of the DNA structure, etc. We refer the reader to [14, 73] for more details on the physical background and references.

Here, the unknown u is a complex valued function of $t \in \mathbb{R}$ and $x \in \mathbb{R}^N$ for $N \geqslant 1$,

$$u : \mathbb{R} \times \mathbb{R}^N \to \mathbb{C}.$$

In most of the applications, t stands for the time and x for the space variable, but sometimes it can be the converse, for example in nonlinear optics. The number i is the imaginary unit ($i^2 = -1$), ∂_t is the first derivative with respect to the time t, Δ denotes the Laplacian with respect to the space variable x ($\Delta = \sum_{j=1}^N \frac{\partial^2}{\partial x_j^2}$). Finally, $p \in \mathbb{R}$ is such that $p > 1$, which means that the equation is superlinear.

To simplify the exposition, we have restricted ourselves to the study of nonlinear Schrödinger equations with a power-type nonlinearity, but one can also consider more general versions of these equations, see [14, 73] and the references cited therein.

The purpose of these notes is to study the properties of particular solutions of (1.1) of the form $e^{i\omega t}\varphi(x)$ with $\omega \in \mathbb{R}$ and φ satisfying

$$-\Delta\varphi + \omega\varphi - |\varphi|^{p-1}\varphi = 0. \tag{1.2}$$

Such solutions are called *standing waves*. They are part of a more general class of solutions arising for various nonlinear equations like Korteweg–de Vries, Klein–Gordon or Kadomtsev–Petviashvili equations. These equations enjoy special solutions whose *profile* remains unchanged under the evolution in time. These special solutions are called *solitary waves* or *solitons* (see [14, 18, 20, 73] for an overview of physical and mathematical questions around solitary waves).

This kind of phenomena was discovered in 1834 by John Scott Russel. The Scottish engineer was supervising work in a canal near Edinburgh when he observed that the brutal stop of a boat in the canal was creating a wave that does not seem to vanish. Indeed, he was able to follow this wave horseback on a distance of several miles. His life long, he studied this surprising phenomenon but without being able to give it a theoretical justification. In fact, most of the scientists of that time did believe that such a wave, which does not disperse, could not exist. The first theoretical justification of the existence of solitary waves was given by Korteweg and de Vries in 1895. They derived an equation for the motion of water admitting solitary wave solutions. But one had to wait until the 1950's to see the beginning of an intensive research from mathematicians and physicists about solitary waves.

Heuristically, such solutions appear because of the balance of two contradictory effects: the dispersive effect of the linear part, which tends to flatten the solution as time goes on (see Figure 1), and the focusing effect of the nonlinearity, which tends to concentrate the solution, provided the initial datum is large enough (see Figure 2). For some special initial data, each effect compensates the other and the general profile of the initial data remains unchanged (see Figure 3).

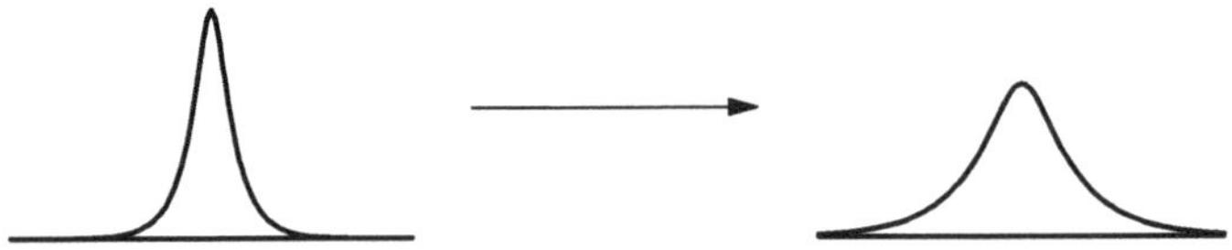

Figure 1. Dispersive effect.

In the study of standing waves, two types of questions arise naturally.

(i) Do standing waves exist, i.e., does (1.2) admit nontrivial solutions? If yes, what kind of properties for the solutions of (1.2) can be shown: Are they regular, what is their decay at infinity, can we find variational characterizations for at least some of these solutions?

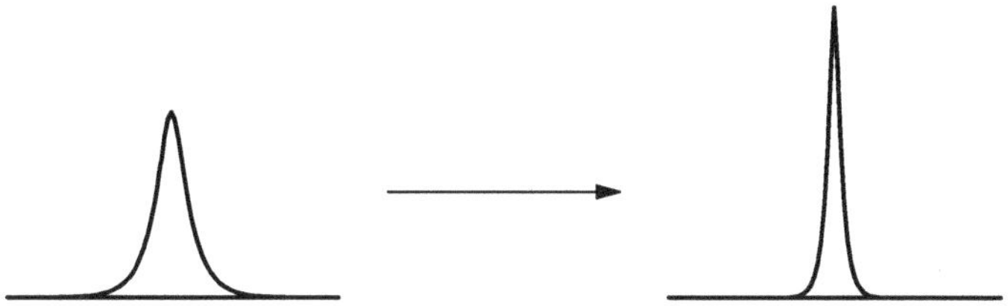

Figure 2. Focusing effect.

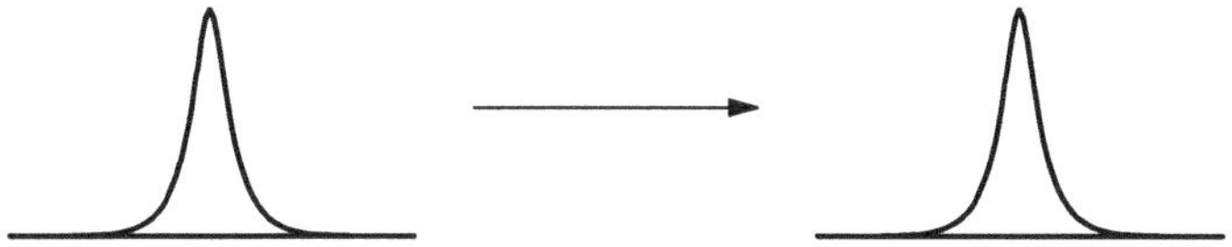

Figure 3. Balance between dispersion and focusing.

(ii) If standing waves exist, are they stable (in a sense to be made precise) as solutions of (1.1)? If they are unstable, what is the nature of instability?

The first type of questions is related to the study of existence of solutions for semilinear elliptic problems. The methods used in this context are often of variational nature and solutions are obtained by minimization under constraint or with min-max arguments. The second type of questions concerns the dynamics of the evolution equation. Nevertheless, both problems are intimately related. Indeed, informations of variational nature on the solutions of the stationary equation are essential to derive stability or instability results.

The rest of these notes is divided into four sections and an appendix. We first give basic results on the Cauchy problem for (1.1) in Section 2. In Section 3, we study the existence of standing waves. Section 4 is devoted to stability whereas Section 5 deals with instability. The proof of a stability criterion is given in the appendix.

Notation. The space of complex measurable functions whose r-th power is integrable will be denoted by $L^r(\mathbb{R}^N)$ and its standard norm by $\|\cdot\|_{L^r(\mathbb{R}^N)}$. When $r = 2$, the space $L^2(\mathbb{R}^N)$ will be endowed with the real inner product

$$(u, v)_2 = \operatorname{Re} \int_{\mathbb{R}^N} u\bar{v}dx \text{ for } u, v \in L^2(\mathbb{R}^N).$$

The space of functions from $L^r(\mathbb{R}^N)$ whose distributional derivatives of order less than or equal to m are elements of $L^r(\mathbb{R}^N)$ will be denoted by $W^{m,r}(\mathbb{R}^N)$. If $m = 1$ and

$r = 2$, we shall denote $W^{1,2}(\mathbb{R}^N)$ by $H^1(\mathbb{R}^N)$, its usual norm by $\|\cdot\|_{H^1(\mathbb{R}^N)}$ and the duality product between the dual space $H^{-1}(\mathbb{R}^N)$ and $H^1(\mathbb{R}^N)$ by $\langle\cdot,\cdot\rangle$. If $m = r = 2$, we denote $W^{2,2}(\mathbb{R}^N)$ by $H^2(\mathbb{R}^N)$.

For notational convenience, we shall sometimes identify a function and its value at some point. For example, we shall write $e^{i\omega t}\varphi(x)$ for the function $(t,x) \mapsto e^{i\omega t}\varphi(x)$. Similarly, we shall say that $|x|v \in L^2(\mathbb{R}^N)$ if the function $x \mapsto |x|v(x)$ belongs to $L^2(\mathbb{R}^N)$. For a solution u of (1.1), we shall denote by $u(t)$ the function $x \mapsto u(t,x)$. Therefore, depending on the context, u may denote either a mapping from $\mathbb{R}$ to some function space or a mapping from $\mathbb{R} \times \mathbb{R}^N$ to $\mathbb{C}$.

We make the convention that when we take a subsequence of a sequence (v_n) we denote it again by (v_n).

The letter C will denote various positive constants whose exact values may change from line to line but are not essential in the course of the analysis.

2 The Cauchy problem

For the physical properties of the model as well as for the mathematical study of the equation, it is interesting to look for quantities conserved along the time. At least formally, equation (1.1) admits three conserved quantities. The first one is the *mass* or *charge*: If u satisfies (1.1) with initial datum $u(0) = u_0$ then

$$Q(u(t)) := \frac{1}{2}\|u(t)\|^2_{L^2(\mathbb{R}^N)} \equiv Q(u_0). \tag{2.1}$$

The conservation of mass is obtained from multiplying (1.1) with $\bar{u}$, integrating over $\mathbb{R}^N$ and taking the imaginary part. The second conserved quantity is the *energy* (multiply (1.1) by $\partial_t\bar{u}$, integrate over $\mathbb{R}^N$ and take the real part),

$$E(u(t)) := \frac{1}{2}\|\nabla u(t)\|^2_{L^2(\mathbb{R}^N)} - \frac{1}{p+1}\|u(t)\|^{p+1}_{L^{p+1}(\mathbb{R}^N)} \equiv E(u_0). \tag{2.2}$$

Finally, multiplying (1.1) by $\nabla\bar{u}$, integrating over $\mathbb{R}^N$ and taking the real part, we obtain the conservation of *momentum*,

$$M(u(t)) := \operatorname{Im}\int_{\mathbb{R}^N} u(t)\nabla\bar{u}(t)dx \equiv M(u_0).$$

Among these three conserved quantities, the mass and the energy are real-valued, but the momentum may be complex-valued, which makes it less convenient to use.

To benefit from the conserved quantities, it is natural to search solutions of (1.1) in function spaces where these quantities are well-defined. Consequently, we look for solutions of (1.1) in $H^1(\mathbb{R}^N)$ and restrict the range of p to be subcritical for the Sobolev embedding of $H^1(\mathbb{R}^N)$ into $L^{p+1}(\mathbb{R}^N)$, i.e., $1 < p < 1 + \frac{4}{N-2}$ if $N \geqslant 3$ and $1 < p < +\infty$ if $N = 1, 2$. Then we have the following result concerning the well-posedness of the Cauchy problem for (1.1) (see [14] and the references cited therein).

Proposition 2.1. *Let* $1 < p < 1 + \frac{4}{N-2}$ *if* $N \geqslant 3$ *and* $1 < p < +\infty$ *if* $N = 1, 2$. *For every* $u_0 \in H^1(\mathbb{R}^N)$ *there exists a unique maximal solution* u *of* (1.1), $T_{\min} \in [-\infty, 0)$, $T^{\max} \in (0, +\infty]$ *such that* $u(0) = u_0$ *and*

$$u \in \mathcal{C}((T_{\min}, T^{\max}); H^1(\mathbb{R}^N)) \cap \mathcal{C}^1((T_{\min}, T^{\max}); H^{-1}(\mathbb{R}^N)).$$

Furthermore, we have the conservation of charge, energy and momentum*: For all* $t \in (T_{\min}, T^{\max})$,

$$Q(u(t)) = Q(u_0), \; E(u(t)) = E(u_0), \; M(u(t)) = M(u_0). \tag{2.3}$$

Finally, we have the blow-up alternative*:*

- *If* $T_{\min} > -\infty$ *then* $\lim_{t \downarrow T_{\min}} \|\nabla u(t)\|^2_{L^2(\mathbb{R}^N)} = +\infty$.
- *If* $T^{\max} < +\infty$ *then* $\lim_{t \uparrow T^{\max}} \|\nabla u(t)\|^2_{L^2(\mathbb{R}^N)} = +\infty$.

From now on, it will be understood that $1 < p < 1 + \frac{4}{N-2}$ if $N \geqslant 3$ and $1 < p < +\infty$ if $N = 1, 2$.

In view of Proposition 2.1, it is natural to ask under which circumstances global existence holds or blow-up occurs. The following proposition gives an answer to the question of global existence.

Proposition 2.2. *If* $1 < p < 1 + \frac{4}{N}$ *then* (1.1) *is globally well-posed, i.e., for any solution* u *given by Proposition 2.1,* $T_{\min} = -\infty$ *and* $T^{\max} = +\infty$.

The proof of Proposition 2.2 relies on the Gagliardo–Nirenberg inequality (see [1]): There exists a constant $C > 0$ such that for all $v \in H^1(\mathbb{R}^N)$

$$\|v\|^{p+1}_{L^{p+1}(\mathbb{R}^N)} \leqslant C \|\nabla v\|^{\frac{N(p-1)}{2}}_{L^2(\mathbb{R}^N)} \|v\|^{p+1-\frac{N(p-1)}{2}}_{L^2(\mathbb{R}^N)}. \tag{2.4}$$

Proof of Proposition 2.2. Let $1 < p < 1 + \frac{4}{N}$ and u be a solution of (1.1) as in Proposition 2.1. We prove the assertion by contradiction. Assume that $T^{\max} < +\infty$ and therefore $\lim_{t \uparrow T^{\max}} \|\nabla u(t)\|^2_{L^2(\mathbb{R}^N)} = +\infty$. By (2.4), we have

$$E(u(t)) \geqslant \|\nabla u(t)\|^2_{L^2(\mathbb{R}^N)} \left(\frac{1}{2} - C \|\nabla u(t)\|^{\frac{N(p-1)}{2} - 2}_{L^2(\mathbb{R}^N)} \|u(t)\|^{p+1-\frac{N(p-1)}{2}}_{L^2(\mathbb{R}^N)} \right).$$

In view of the conservation of charge and energy (see (2.3)), this implies

$$\|\nabla u(t)\|^2_{L^2(\mathbb{R}^N)} \left(1 - \|\nabla u(t)\|^{\frac{N(p-1)}{2} - 2}_{L^2(\mathbb{R}^N)} \right) < C \text{ for all } t \in (T_{\min}, T^{\max}). \tag{2.5}$$

Now, since $p < 1 + \frac{4}{N}$, we have $\frac{N(p-1)}{2} - 2 < 0$ and thus letting $\|\nabla u(t)\|^2_{L^2(\mathbb{R}^N)}$ go to $+\infty$ when t goes to $T^{\max}$ leads to a contradiction in (2.5). Arguing in the same way if $T_{\min} > -\infty$ leads to the same contradiction and completes the proof. □

Concerning blow-up, we can give the following sufficient condition.

Proposition 2.3. *Assume that* $p \geqslant 1 + \frac{4}{N}$ *and let* $u_0 \in H^1(\mathbb{R}^N)$ *be such that*

$$|x|u_0 \in L^2(\mathbb{R}^N) \text{ and } E(u_0) < 0.$$

Then the solution u *of* (1.1) *corresponding to* u_0 *blows up in finite time for positive and negative time, that is*

$$T_{\min} > -\infty \text{ and } T^{\max} < +\infty.$$

The proof of Proposition 2.3 relies on the *virial theorem* (the term "virial theorem" comes from the analogy to the virial theorem in classical mechanics).

Proposition 2.4 (Virial theorem). *Let* $u_0 \in H^1(\mathbb{R}^N)$ *be such that* $|x|u_0 \in L^2(\mathbb{R}^N)$ *and let* u *be the solution of* (1.1) *corresponding to* u_0. *Then* $|x|u(t) \in L^2(\mathbb{R}^N)$ *for all* $t \in (T_{\min}, T^{\max})$ *and the function* $f : t \mapsto \|xu(t)\|^2_{L^2(\mathbb{R}^N)}$ *is of class* $\mathcal{C}^2$ *and satisfies*

$$\begin{aligned} f'(t) &= 4\mathrm{Im}\int_{\mathbb{R}^N} \bar{u}(t)x \cdot \nabla u(t)dx, \\ f''(t) &= 8P(u(t)), \end{aligned}$$

where P *is given for* $v \in H^1(\mathbb{R}^N)$ *by*

$$P(v) := \|\nabla v\|^2_{L^2(\mathbb{R}^N)} - \frac{N(p-1)}{2(p+1)}\|v\|^{p+1}_{L^{p+1}(\mathbb{R}^N)}. \tag{2.6}$$

The virial theorem comes from the work of Glassey [31] in which the identities for f' and f'' were formally derived. For a rigorous proof, see [14].

Proof of Proposition 2.3. First, we remark that

$$P(u(t)) = 2E(u(t)) + \frac{4 - N(p-1)}{2(p+1)}\|u(t)\|^{p+1}_{L^{p+1}(\mathbb{R}^N)}.$$

Since $p \geqslant 1 + \frac{4}{N}$, we have in view of the conservation of energy (see (2.3))

$$P(u(t)) \leqslant 2E(u(t)) = 2E(u_0) < 0 \text{ for all } t \in (T_{\min}, T^{\max}).$$

Therefore, by Proposition 2.4, we have

$$\frac{d^2}{dt^2}\|xu(t)\|^2_{L^2(\mathbb{R}^N)} \leqslant 16E(u_0) \text{ for all } t \in (T_{\min}, T^{\max}).$$

Integrating twice in time gives

$$\|xu(t)\|^2_{L^2(\mathbb{R}^N)} \leqslant 8E(u_0)t^2 + \left(4\mathrm{Im}\int_{\mathbb{R}^N} \bar{u}_0 x \cdot \nabla u_0 dx\right)t + \|xu_0\|^2_{L^2(\mathbb{R}^N)}. \tag{2.7}$$

The right member of (2.7) is a polynomial of order two with the main coefficient negative. Hence for $|t|$ large the right-hand side of (2.7) becomes negative, which is in contradiction with $\|xu(t)\|^2_{L^2(\mathbb{R}^N)} \geqslant 0$. This implies $T_{\min} > -\infty$ and $T^{\max} < +\infty$ and finishes the proof. □

3 Existence, uniqueness and properties of solitons

We will use the following definition of standing waves for (1.1).

Definition 3.1. A *standing wave* or *soliton* of (1.1) is a solution of the form $e^{i\omega t}\varphi(x)$ with $\omega \in \mathbb{R}$ and φ satisfying

$$\begin{cases} -\Delta\varphi + \omega\varphi - |\varphi|^{p-1}\varphi = 0, \\ \varphi \in H^1(\mathbb{R}^N) \setminus \{0\}. \end{cases} \tag{3.1}$$

Many techniques have been developed to study the existence of solutions to problems of type (3.1) (see, e.g., [3]). In this section, we look for solutions of (3.1) by using variational methods (see, e.g., [71] for a general overview).

For the study of solutions of (3.1), we define a functional $S : H^1(\mathbb{R}^N) \to \mathbb{R}$ by setting for $v \in H^1(\mathbb{R}^N)$

$$S(v) := \frac{1}{2}\|\nabla v\|^2_{L^2(\mathbb{R}^N)} + \frac{\omega}{2}\|v\|^2_{L^2(\mathbb{R}^N)} - \frac{1}{p+1}\|v\|^{p+1}_{L^{p+1}(\mathbb{R}^N)}.$$

The functional S is often called *action*. It is standard that S is of class $\mathcal{C}^2$ (see, for example, [78]) and for $v \in H^1(\mathbb{R}^N)$ the Fréchet derivative of S at v is given by

$$S'(v) = -\Delta v + \omega v - |v|^{p-1}v.$$

Therefore, φ is a solution of (3.1) if and only if $\varphi \in H^1(\mathbb{R}^N) \setminus \{0\}$ and $S'(\varphi) = 0$. In other words, the nontrivial critical points of S are the solutions of (3.1). Therefore, to prove existence of solutions of (3.1) it is enough to find a nontrivial critical point of S.

This section is divided as follows. First, we prove that if solutions to (3.1) exist then they are regular, exponentially decaying at infinity and satisfy some functional identities. Next, we prove the existence of a nontrivial critical point of S. Finally, we derive various variational characterizations for some special solutions of (3.1).

3.1 Preliminaries

Before studying existence of solutions to (3.1), it is convenient to prove that, if such solutions exist, they necessarily enjoy the following properties.

Proposition 3.2. *Let $\omega > 0$. If $\varphi \in H^1(\mathbb{R}^N)$ satisfies* (3.1) *then φ is regular and exponentially decaying. More precisely,*

(i) *$\varphi \in W^{3,r}(\mathbb{R}^N)$ for all $r \in [2, +\infty)$, in particular $\varphi \in \mathcal{C}^2(\mathbb{R}^N)$;*

(ii) *there exists $\varepsilon > 0$ such that $e^{\varepsilon|x|}(|\varphi| + |\nabla\varphi|) \in L^\infty(\mathbb{R}^N)$.*

Sketch of proof. Point (i) follows from the usual elliptic regularity theory by a bootstrap argument (see [30]). We just indicate how to initiate the bootstrap and refer to [14] for a detailed proof. Let φ be a solution of (3.1). Suppose that $\varphi \in L^q(\mathbb{R}^N)$ for some $q > p$. Since $|\varphi|^{p-1}\varphi \in L^{\frac{q}{p}}(\mathbb{R}^N)$ and φ satisfies

$$-\Delta\varphi + \omega\varphi = |\varphi|^{p-1}\varphi, \tag{3.2}$$

it follows that $\varphi \in W^{2,\frac{q}{p}}(\mathbb{R}^N)$. Choosing any $q \in (p, +\infty)$ if $N = 1, 2$ and $q = \frac{2N}{N-2}$ if $N \geqslant 3$ and repeating the previous argument recursively, we obtain after a finite number of steps that $\varphi \in W^{2,r}(\mathbb{R}^N)$ for any $r \in [2, +\infty)$. This implies that $|\varphi|^{p-1}\varphi \in W^{1,r}(\mathbb{R}^N)$ for any $r \in [2, +\infty)$ and it follows from (3.2) that $\varphi \in W^{3,r}(\mathbb{R}^N)$ for all $r \in [2, +\infty)$, hence (i).

If φ is radial, i.e., $\varphi(x) = \varphi(|x|)$ then φ satisfies

$$-\varphi'' - \frac{N-1}{r}\varphi' + \omega\varphi - |\varphi|^{p-1}\varphi = 0.$$

Here, we have employed that $\Delta\varphi = \varphi'' + \frac{N-1}{r}\varphi'$ if φ is radial. In this case, the exponential decay of φ at infinity follows from the classical theory of second-order ordinary differential equations (see, for example, [37]). See [14] for a proof of (ii) for non-radial solutions of (3.1). □

Lemma 3.3. *If $\varphi \in H^1(\mathbb{R}^N)$ satisfies* (3.1) *then the following identities hold:*

$$\|\nabla\varphi\|^2_{L^2(\mathbb{R}^N)} + \omega\|\varphi\|^2_{L^2(\mathbb{R}^N)} - \|\varphi\|^{p+1}_{L^{p+1}(\mathbb{R}^N)} = 0, \tag{3.3}$$

$$\|\nabla\varphi\|^2_{L^2(\mathbb{R}^N)} - \frac{N(p-1)}{2(p+1)}\|\varphi\|^{p+1}_{L^{p+1}(\mathbb{R}^N)} = 0. \tag{3.4}$$

In the literature, (3.4) is called the *Pohozhaev identity*, since it was first derived by Pohozhaev in 1965, see [64]. Actually, one can obtain this type of identity for a large class of nonlinearities, see [9]. The set

$$\{v \in H^1(\mathbb{R}^N); v \neq 0, \|\nabla v\|^2_{L^2(\mathbb{R}^N)} + \omega\|v\|^2_{L^2(\mathbb{R}^N)} - \|v\|^{p+1}_{L^{p+1}(\mathbb{R}^N)} = 0\} \tag{3.5}$$

is called the *Nehari manifold*.

Proof of Lemma 3.3. Let $\varphi \in H^1(\mathbb{R}^N)$ be a solution of (3.1).

To obtain (3.3), we simply multiply (3.1) by $\bar{\varphi}$ and integrate over $\mathbb{R}^N$.

For the proof of (3.4), recall first that we have defined in the beginning of this section a $\mathcal{C}^2$-functional $S : H^1(\mathbb{R}^N) \to \mathbb{R}$ by setting for $v \in H^1(\mathbb{R}^N)$

$$S(v) := \frac{1}{2}\|\nabla v\|^2_{L^2(\mathbb{R}^N)} + \frac{\omega}{2}\|v\|^2_{L^2(\mathbb{R}^N)} - \frac{1}{p+1}\|v\|^{p+1}_{L^{p+1}(\mathbb{R}^N)}$$

and that if φ is a solution of (3.1) then $S'(\varphi) = 0$.

For $\lambda > 0$, let $\varphi_\lambda(\cdot) := \lambda^{N/2}\varphi(\lambda\cdot)$. It follows from straightforward calculations that

$$\|\nabla\varphi_\lambda\|^2_{L^2(\mathbb{R}^N)} = \lambda^2\|\nabla\varphi\|^2_{L^2(\mathbb{R}^N)}, \quad \|\varphi_\lambda\|^2_{L^2(\mathbb{R}^N)} = \|\varphi\|^2_{L^2(\mathbb{R}^N)},$$

$$\text{and } \|\varphi_\lambda\|^{p+1}_{L^{p+1}(\mathbb{R}^N)} = \lambda^{\frac{N(p-1)}{2}}\|\varphi\|^{p+1}_{L^{p+1}(\mathbb{R}^N)}.$$

On the one hand, we have

$$S(\varphi_\lambda) = \frac{\lambda^2}{2}\|\nabla\varphi\|^2_{L^2(\mathbb{R}^N)} + \frac{\omega}{2}\|\varphi\|^2_{L^2(\mathbb{R}^N)} - \frac{\lambda^{\frac{N(p-1)}{2}}}{p+1}\|\varphi\|^{p+1}_{L^{p+1}(\mathbb{R}^N)}$$

and

$$\frac{\partial}{\partial\lambda}S(\varphi_\lambda)_{|\lambda=1} = \|\nabla\varphi\|^2_{L^2(\mathbb{R}^N)} - \frac{N(p-1)}{2(p+1)}\|\varphi\|^{p+1}_{L^{p+1}(\mathbb{R}^N)}. \tag{3.6}$$

On the other hand, we have

$$\frac{\partial}{\partial\lambda}S(\varphi_\lambda)_{|\lambda=1} = \left\langle S'(\varphi), \frac{\partial\varphi_\lambda}{\partial\lambda}_{|\lambda=1}\right\rangle \tag{3.7}$$

and, since φ is solution of (3.1), $S'(\varphi) = 0$, which implies

$$\frac{\partial}{\partial\lambda}S(\varphi_\lambda)_{|\lambda=1} = 0. \tag{3.8}$$

Combining (3.6) and (3.8) proves (3.4).

Note that the right-hand side of (3.7) is well-defined for $\omega > 0$ since $\frac{\partial\varphi_\lambda}{\partial\lambda}_{|\lambda=1} = \frac{1}{2}\varphi + x\cdot\nabla\varphi$ is in $H^1(\mathbb{R}^N)$ because of the regularity and exponential decay of φ stated in Proposition 3.2. If $\omega \leqslant 0$, the previous calculations are only formal. However, it is possible to give a rigorous proof of (3.4) also for $\omega \leqslant 0$, see [9]. □

From Lemma 3.3 it is easy to derive a necessary condition for the existence of solutions to (3.1).

Corollary 3.4. *If* $\omega \leqslant 0$ *then* (3.1) *has no solution.*

Proof. We prove the assertion by contradiction. Let $\omega \leqslant 0$ and suppose that (3.1) has a solution $\varphi \in H^1(\mathbb{R}^N)$. By (3.3) and (3.4), φ satisfies

$$\frac{(N-2)p-(N+2)}{2(p+1)}\|\varphi\|^{p+1}_{L^{p+1}(\mathbb{R}^N)} = -\omega\|\varphi\|^2_{L^2(\mathbb{R}^N)}. \tag{3.9}$$

Since $p < 1 + \frac{4}{N-2}$ if $N \geqslant 3$ and $p < +\infty$ if $N = 1, 2$, the left-hand side of (3.9) is negative for any N, whereas the right-hand side is non negative by assumption, which is a contradiction. □

In the rest of these notes, it will be understood that $\omega > 0$.

3.2 Existence

Our main result in this section is the following.

Theorem 3.5. *There exists a nontrivial critical point* φ_ω *of* S*, that is a solution of* (3.1).

Several techniques are available to prove the existence of a nontrivial critical point of S. The functional S is clearly unbounded from above and below, and this prevents to find a critical point simply by global minimization or maximization. To overcome this difficulty, other techniques based on minimization under constraint were developed (see, e.g., [8, 9, 17, 64, 70] and the references cited therein). In what follows, we present a different approach based on the mountain pass theorem of Ambrosetti and Rabinowitz [4] and somehow inspired from [38]. This approach was suggested to us by Louis Jeanjean. We first prove that the functional S has a mountain pass geometry. Thus there exists a Palais–Smale sequence at this level. It clearly converges to a critical point of S weakly in $H^1(\mathbb{R}^N)$. Proving that this sequence is non-vanishing and taking advantage of the translation invariance of (3.1), we get the existence of a nontrivial critical point.

We define the *mountain pass level* c by setting

$$c := \inf_{\gamma\in\Gamma} \max_{s\in[0,1]} S(\gamma(s)), \tag{3.10}$$

where Γ is the set of admissible paths:

$$\Gamma := \{\gamma \in \mathcal{C}([0,1]; H^1(\mathbb{R}^N)); \gamma(0) = 0, S(\gamma(1)) < 0\}. \tag{3.11}$$

Lemma 3.6. *The functional S has a mountain pass geometry, i.e., $\Gamma \neq \emptyset$ and $c > 0$.*

Proof. Let $v \in H^1(\mathbb{R}^N) \setminus \{0\}$. Then, for any $s > 0$,

$$S(sv) = \frac{s^2}{2}(\|\nabla v\|^2_{L^2(\mathbb{R}^N)} + \omega\|v\|^2_{L^2(\mathbb{R}^N)}) - \frac{s^{p+1}}{p+1}\|v\|^{p+1}_{L^{p+1}(\mathbb{R}^N)},$$

and it is clear that if s is large enough then $S(sv) < 0$. Let $C > 0$ be such that $S(Cv) < 0$ and $\gamma : [0,1] \to H^1(\mathbb{R}^N)$ be defined by $\gamma(s) := Csv$. Then $\gamma \in \mathcal{C}([0,1]; H^1(\mathbb{R}^N))$, $\gamma(0) = 0$ and $S(\gamma(1)) < 0$; thus $\gamma \in \Gamma$ and Γ is nonempty. Now, we clearly have

$$S(v) \geqslant \frac{\min\{1,\omega\}}{2}\|v\|^2_{H^1(\mathbb{R}^N)} - \frac{1}{p+1}\|v\|^{p+1}_{L^{p+1}(\mathbb{R}^N)},$$

and by the continuous embedding of $H^1(\mathbb{R}^N)$ into $L^{p+1}(\mathbb{R}^N)$, we find

$$S(v) \geqslant \frac{\min\{1,\omega\}}{2}\|v\|^2_{H^1(\mathbb{R}^N)} - \frac{C}{p+1}\|v\|^{p+1}_{H^1(\mathbb{R}^N)}.$$

Let $\varepsilon > 0$ be small enough to have

$$\delta := \frac{\min\{1,\omega\}\varepsilon^2}{2} - \frac{C\varepsilon^{p+1}}{p+1} > 0.$$

Then for any $v \in H^1(\mathbb{R}^N)$ with $\|v\|_{H^1(\mathbb{R}^N)} < \varepsilon$, we have $S(v) > 0$. This implies that for any $\gamma \in \Gamma$, we have $\|\gamma(1)\|_{H^1(\mathbb{R}^N)} > \varepsilon$, and by continuity of γ there exists $s_\gamma \in [0,1]$ such that $\|\gamma(s_\gamma)\|_{H^1(\mathbb{R}^N)} = \varepsilon$. Therefore,

$$\max_{s\in[0,1]} S(\gamma(s)) \geqslant S(\gamma(s_\gamma)) \geqslant \delta > 0.$$

This implies for the mountain pass level c defined in (3.10) that

$$c \geqslant \delta > 0,$$

and thus S has a mountain pass geometry. □

Combined with Ekeland's variational principle (see, e.g., [78]), Lemma 3.6 immediately implies the existence of a Palais–Smale sequence at the mountain pass level. More precisely:

Corollary 3.7. *There exists a* Palais–Smale sequence $(u_n) \subset H^1(\mathbb{R}^N)$ *for* S *at the mountain pass level* c*, i.e.,* (u_n) *satisfies, as* $n \to +\infty$,

$$S(u_n) \to c \text{ and } S'(u_n) \to 0. \tag{3.12}$$

To find a critical point for S, we proceed now in two steps: First, we prove that the sequence (u_n) is, following the terminology of the concentration-compactness theory of P.-L. Lions, *non-vanishing*; second, we prove that, up to translations, (u_n) converges weakly in $H^1(\mathbb{R}^N)$ to a nontrivial critical point of S.

Lemma 3.8. *The Palais–Smale sequence* (u_n) *is non-vanishing: there exist* $\varepsilon > 0$, $R > 0$ *and a sequence* $(y_n) \in \mathbb{R}^N$ *such that for all* $n \in \mathbb{N}$*, we have*

$$\int_{B_R(y_n)} |u_n|^2 dx > \varepsilon, \tag{3.13}$$

where $B_R(y) := \{z \in \mathbb{R}^N; |y - z| < R\}$.

To prove Lemma 3.8, we will use the following lemma, which is a kind of Sobolev embedding result (see [50]).

Lemma 3.9. *Let* $R > 0$. *Then there exists* $\alpha > 0$ *and* $C > 0$ *such that for any* $v \in H^1(\mathbb{R}^N)$*, we have*

$$\|v\|_{L^{p+1}(\mathbb{R}^N)}^{p+1} \leqslant C \left(\sup_{y \in \mathbb{R}^N} \int_{B_R(y)} |v|^2 dx \right)^{\alpha} \|v\|_{H^1(\mathbb{R}^N)}^2.$$

Proof. Let $(Q_k)_{k \in \mathbb{N}}$ be a sequence of open cubes of $\mathbb{R}^N$ of same volume such that $Q_k \cap Q_l = \emptyset$ if $k \neq l$, $\overline{\cup_{k \in \mathbb{N}} Q_k} = \mathbb{R}^N$ and for each k there exists y_k with $Q_k \subset B_R(y_k)$. By Hölder's inequality and the embedding of $H^1(Q_k)$ into $L^{p+1}(Q_k)$, there exists $C > 0$ independent of k such that for any $v \in H^1(\mathbb{R}^N)$, we have

$$\int_{Q_k} |v|^{p+1} dx \leqslant C \left(\int_{Q_k} |v|^2 dx \right)^{\alpha} \int_{Q_k} \left(|\nabla v|^2 + |v|^2 \right) dx.$$

with $\alpha = \frac{N}{p+1} - \frac{N-2}{2}$ if $N \geqslant 3$ and $\alpha = 1$ if $N = 1, 2$. Summing over $k \in \mathbb{N}$, this implies that

$$\|v\|_{L^{p+1}(\mathbb{R}^N)}^{p+1} \leqslant C \left(\sup_{k \in \mathbb{N}} \int_{Q_k} |v|^2 dx \right)^{\alpha} \|v\|_{H^1(\mathbb{R}^N)}^2.$$

Now, since for any k there exists y_k with $Q_k \subset B_R(y_k)$ the conclusion follows. □

Proof of Lemma 3.8. We prove the result by contradiction. Assume that the Palais–Smale sequence (u_n) is vanishing, that is, for all $R>0$, we have

$$\lim_{n\to+\infty}\sup_{y\in\mathbb{R}^N}\int_{B_R(y)}|u_n|^2dx=0. \tag{3.14}$$

Let $\varepsilon>0$. For n large enough, we have $\left(\sup_{y\in\mathbb{R}^N}\int_{B_R(y)}|u_n|^2dx\right)<\varepsilon$ thanks to (3.14). Then Lemma 3.9 implies that, still for n large enough,

$$S(u_n)\geqslant\frac{1}{2}\|\nabla u_n\|^2_{L^2(\mathbb{R}^N)}+\frac{\omega}{2}\|u_n\|^2_{L^2(\mathbb{R}^N)}-\varepsilon\|u_n\|^2_{H^1(\mathbb{R}^N)},$$

and therefore

$$S(u_n)\geqslant\left(\frac{\min\{1,\omega\}}{2}-\varepsilon\right)\|u_n\|^2_{H^1(\mathbb{R}^N)}.$$

Consequently, taking $\varepsilon<\frac{\min\{1,\omega\}}{2}$, we infer that

$$(u_n)\text{ is bounded in }H^1(\mathbb{R}^N), \tag{3.15}$$

which allows to get

$$\langle S'(u_n),u_n\rangle\to0\text{ as }n\to+\infty. \tag{3.16}$$

Combining (3.15) with (3.14) and Lemma 3.9 implies

$$\lim_{n\to+\infty}\|u_n\|^{p+1}_{L^{p+1}(\mathbb{R}^N)}=0.$$

Consequently, we have

$$S(u_n)-\frac{1}{2}\langle S'(u_n),u_n\rangle=-\frac{p-1}{2(p+1)}\|u_n\|^{p+1}_{L^{p+1}(\mathbb{R}^N)}\to0\text{ as }n\to+\infty. \tag{3.17}$$

On the other hand, we infer from (3.12) and (3.16) that

$$S(u_n)-\frac{1}{2}\langle S'(u_n),u_n\rangle\to c\text{ as }n\to+\infty,$$

which is a contradiction with (3.17) since $c>0$. Hence, (u_n) is non-vanishing. □

Proof of Theorem 3.5. Let $v_n:=u_n(\cdot+y_n)$, where (u_n) is given by Corollary 3.7 and (y_n) by Lemma 3.8. Since the functional S is invariant under translations in space, (v_n) is still a Palais–Smale sequence for S:

$$S(v_n)\to c\text{ and }S'(v_n)\to0. \tag{3.18}$$

From (3.15) we infer that (v_n) is bounded in $H^1(\mathbb{R}^N)$. Thus there exists $v\in H^1(\mathbb{R}^N)$ such that, possibly for a subsequence only, $v_n\rightharpoonup v$ weakly in $H^1(\mathbb{R}^N)$. From (3.18), we infer that $S'(v)=0$, i.e., v is a critical point for S. To show that v is nontrivial, we remark that, since the embedding $H^1(B_R(0))\hookrightarrow L^2(B_R(0))$ is compact, (3.13) implies $v\not\equiv0$. Setting $\varphi_\omega:=v$ finishes the proof. □

The following corollary will be useful in the next subsection.

Corollary 3.10. *The critical point φ_ω is below the level c, i.e., $S(\varphi_\omega) \leqslant c$.*

Proof. First, since $S'(\varphi_\omega) = 0$, we have

$$\begin{aligned} S(\varphi_\omega) &= S(\varphi_\omega) - \frac{1}{p+1} \langle S'(\varphi_\omega), \varphi_\omega \rangle \\ &= \frac{p-1}{2(p+1)} \left(\|\nabla \varphi_\omega\|^2_{L^2(\mathbb{R}^N)} + \omega \|\varphi_\omega\|^2_{L^2(\mathbb{R}^N)} \right). \end{aligned}$$

By virtue of weak convergence of (v_n) towards φ_ω in $H^1(\mathbb{R}^N)$, this gives

$$S(\varphi_\omega) \leqslant \frac{p-1}{2(p+1)} \liminf_{n \to +\infty} \left(\|\nabla v_n\|^2_{L^2(\mathbb{R}^N)} + \omega \|v_n\|^2_{L^2(\mathbb{R}^N)} \right). \tag{3.19}$$

As in (3.16), we have

$$\langle S'(v_n), v_n \rangle \to 0 \text{ as } n \to +\infty,$$

and combining with (3.18) and (3.19), we obtain

$$S(\varphi_\omega) \leqslant \liminf_{n \to +\infty} \left(S(v_n) - \frac{1}{p+1} \langle S'(v_n), v_n \rangle \right) = c,$$

which completes the proof. □

3.3 Variational characterizations

Among the solutions of (3.1), some are of particular interest.

Definition 3.11. A solution φ of (3.1) is said to be a *ground state* or *least energy solution* if $S(\varphi) \leqslant S(v)$ for any solution v of (3.1). We define the *least energy level* m by

$$m := \inf\{S(v); v \text{ is a solution of } (3.1)\}. \tag{3.20}$$

The set of all least energy solutions is denoted by $\mathcal{G}$,

$$\mathcal{G} := \{v \in H^1(\mathbb{R}^N); v \text{ is a solution of } (3.1) \text{ and } S(v) = m\}. \tag{3.21}$$

Ground states play an important role in the theory of nonlinear Schrödinger equations. In particular, in the critical case $p = 1 + \frac{4}{N}$, they appear in an essential way in the derivation of global existence results (see [75]) and in the description of the blow-up phenomenon (see [57, 58, 59, 60] and the references cited therein).

Proposition 3.12. *The solution φ_ω of* (3.1) *found in Theorem 3.5 is a ground state and it is at the mountain pass level c defined in* (3.10)*:*

$$S(\varphi_\omega) = m = c.$$

In addition, φ_ω is a minimizer of S on the Nehari manifold (see (3.5)), i.e., it solves the following minimization problem

$$d := \min\{S(v); v \in H^1(\mathbb{R}^N) \setminus \{0\}, I(v) = 0\} \tag{3.22}$$

where $I(v) := \|\nabla v\|^2_{L^2(\mathbb{R}^N)} + \omega\|v\|^2_{L^2(\mathbb{R}^N)} - \|v\|^{p+1}_{L^{p+1}(\mathbb{R}^N)}$.

The various variational characterizations of the critical point φ_ω of S as a ground state or as a minimizer of S on the Nehari manifold will be useful when we will deal with the stability or instability issues.

Before proving Proposition 3.12, we need some preparation.

Lemma 3.13. *The following inequality holds:*

$$c \leqslant d, \tag{3.23}$$

where c is the mountain pass level defined in (3.10) *and d is given by* (3.22).

Proof. Let $v \in H^1(\mathbb{R}^N) \setminus \{0\}$ be such that $I(v) = 0$. Following [41, 42, 65], the idea is to construct a path γ in Γ (recall that Γ was defined in (3.11)) such that S reaches its maximum on γ at v. From the proof of Lemma 3.6, we know that for C large enough the path $\gamma : [0,1] \to H^1(\mathbb{R}^N)$ defined by $\gamma(s) := Csv$ belongs to Γ. It is easy to see that

$$\frac{\partial}{\partial s}S(sv) = s(\|\nabla v\|^2_{L^2(\mathbb{R}^N)} + \omega\|v\|^2_{L^2(\mathbb{R}^N)} - s^{p-1}\|v\|^{p+1}_{L^{p+1}(\mathbb{R}^N)}).$$

Therefore, we have at $s = 1$

$$\frac{\partial}{\partial s}S(sv)_{|s=1} = I(v) = 0,$$

and consequently,

$$\frac{\partial}{\partial s}S(sv) > 0 \quad \text{if} \quad s \in (0,1),$$
$$\frac{\partial}{\partial s}S(sv) < 0 \quad \text{if} \quad s \in (1,+\infty).$$

Thus S reaches its maximum on γ at v. This implies

$$c \leqslant S(v) \text{ for all } v \in H^1(\mathbb{R}^N) \setminus \{0\} \text{ with } I(v) = 0,$$

which finishes the proof. □

Proof of Proposition 3.12. We first recall that, by Corollary 3.10, we have

$$S(\varphi_\omega) \leqslant c \tag{3.24}$$

and by Lemma 3.13,

$$c \leqslant d. \tag{3.25}$$

Now, we remark that, since by (3.3) any solution v of (3.1) satisfies $I(v) = 0$, we have

$$d \leqslant m. \tag{3.26}$$

Since φ_ω is a solution of (3.1), it follows from the definition of the least energy level m (see (3.20)) that

$$m \leqslant S(\varphi_\omega). \tag{3.27}$$

Combining (3.24)–(3.27) gives

$$S(\varphi_\omega) = c = d = m$$

and completes the proof. □

Remark 3.14. For more general nonlinearities, the equality $m = c$ between the mountain pass level and the least energy level still holds (see [41, 42]).

3.4 Uniqueness

It turns out that we are able to describe explicitly the set $\mathcal{G}$ of ground states (see (3.21)).

Theorem 3.15. *There exists a real-valued, positive, spherically symmetric and decreasing function $\Psi \in H^1(\mathbb{R}^N)$ such that*

$$\mathcal{G} = \{e^{i\theta}\Psi(\cdot - y); \theta \in \mathbb{R}, y \in \mathbb{R}^N\}.$$

Therefore, the ground state is unique up to translations and phase shifts. It would exceed the scope of these notes to give a proof of Theorem 3.15 and we just indicate some references. First, a simple and general proof that any complex-valued ground state φ is of the form $e^{i\theta}\tilde{\varphi}$ with $\theta \in \mathbb{R}$ and $\tilde{\varphi}$ a *real positive* ground state was recently given in [16] (see also [36]). The fact that all real positive ground states of (3.1) are radial up to translations was first proved by Gidas, Ni and Nirenberg in 1979 (see [29]) by using the so-called moving planes method. Alternatively, the same result can be deduced by the method of Lopes [54], which relies on symmetrizations with respect to suitably chosen hyperplanes combined with a unique continuation theorem. Recently, Maris [55] developed for the symmetry of minimizers for a large class of problems a new method that furnishes a simpler proof of radial symmetry of real ground states, see [12], without even assuming a priori that they are positive. Uniqueness follows from a result of Kwong [47] in 1989.

Remark 3.16. When $N = 1$, it turns out that the set of all solutions of (3.1) (not only those of least energy) is precisely the set of ground states

$$\mathcal{G} = \{e^{i\theta}\Psi(\cdot - y); \theta \in \mathbb{R}, y \in \mathbb{R}^N\} = \{v; v \text{ is a solution of (3.1) }\}.$$

Moreover, we are able to give an explicit formula for Ψ,

$$\Psi(x) = \left[\frac{(p+1)\omega}{2}\operatorname{sech}^2\left(\frac{(p-1)\sqrt{\omega}}{2}x\right)\right]^{\frac{1}{p-1}}.$$

For $p = 3$ and $\omega = 1$, the shape of Ψ is given in Figure 4.

In higher dimensions, it is known that there exist solutions of (3.1) that are not ground states (see [8, 10, 52]), and no explicit solution is available.

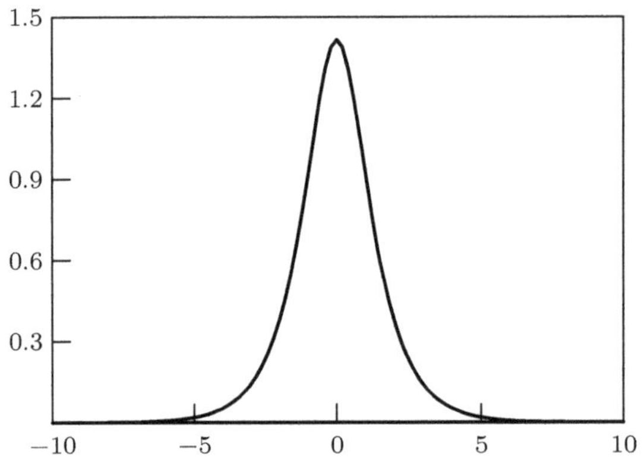

Figure 4. The function Ψ for $N = 1$, $p = 3$ and $\omega = 1$.

4 Stability

In this section, we will prove that for $\varphi \in \mathcal{G}$ the standing wave $e^{i\omega t}\varphi(x)$ is stable (in a sense to be made precise) if $1 < p < 1 + \frac{4}{N}$.

In his report [67] of 1844, Russel was already mentioning the observed remarkable stability properties of solitary waves. From the mathematical point of view, the concept of stability that comes naturally in mind for the standing waves $e^{i\omega t}\varphi(x)$ of (1.1) is *stability in the sense of Lyapunov*, which means uniformly continuous dependence on the initial data: For all $\varepsilon > 0$ there exists $\delta > 0$ such that for all $u_0 \in H^1(\mathbb{R}^N)$,

$$\|u_0 - \varphi\|_{H^1(\mathbb{R}^N)} < \delta \Longrightarrow \|u(t) - e^{i\omega t}\varphi\|_{H^1(\mathbb{R}^N)} < \varepsilon \text{ for all } t,$$

where u is the maximal solution of (1.1) with $u(0) = u_0$. However, with this definition, all standing waves would be unstable (see Remark 4.2), which is contradictory with what is observed for such phenomena. Therefore, we have to look for a different notion of stability.

The main reason for standing waves being unstable in the sense of Lyapunov is that (1.1) admits translation and phase shift invariance: If $u = u(t,x)$ is a solution of (1.1) then for all $\theta \in \mathbb{R}$ and $y \in \mathbb{R}^N$, $e^{i\theta}u(\cdot, \cdot - y)$ is still a solution of (1.1). In some sense, the equation does not prescribe the behavior of the solutions in the "direction" of translations and phase shifts. To take this fact into account, we define the concept of *orbital stability*, which is Lyapunov stability up to translations and phase shifts.

Definition 4.1. Let φ be a solution of (3.1). The standing wave $e^{i\omega t}\varphi(x)$ is said to be *orbitally stable in* $H^1(\mathbb{R}^N)$ if for all $\varepsilon > 0$ there exists $\delta > 0$ such that if $u_0 \in H^1(\mathbb{R}^N)$ satisfies $\|u_0 - \varphi\|_{H^1(\mathbb{R}^N)} < \delta$ then the maximal solution $u = u(t)$ of (1.1) with $u(0) = u_0$ exists for all $t \in \mathbb{R}$ and

$$\sup_{t\in\mathbb{R}} \inf_{\theta\in\mathbb{R}} \inf_{y\in\mathbb{R}^N} \|u(t) - e^{i\theta}\varphi(\cdot - y)\|_{H^1(\mathbb{R}^N)} < \varepsilon.$$

Otherwise, the standing wave is said to be *unstable*.

Remark 4.2. The concept of orbital stability is optimal in the following sense (see [14, 15]).

- Space translations are necessary: For a solution φ of (3.1), $\varepsilon > 0$ and $y \in \mathbb{R}^N$ with $|y| = 1$, let $\varphi_\varepsilon(x) := e^{i\varepsilon x \cdot y}\varphi(x)$. Then it is easy to check that the solution of (1.1) with initial datum φ_ε is

$$u_\varepsilon(t,x) = e^{i\varepsilon(x\cdot y - \varepsilon t)}e^{i\omega t}\varphi(x - 2\varepsilon t y).$$

We clearly have $\varphi_\varepsilon \to \varphi$ in $H^1(\mathbb{R}^N)$ as $\varepsilon \to 0$, but for any $\varepsilon > 0$

$$\sup_{t\in\mathbb{R}} \inf_{\theta\in\mathbb{R}} \|u_\varepsilon(t) - e^{i\theta}\varphi\|_{H^1(\mathbb{R}^N)} = 2\|\varphi\|_{H^1(\mathbb{R}^N)}.$$

Conversely, note that if we consider (1.1) in the subspace of $H^1(\mathbb{R}^N)$ consisting of radial functions then no space translation can occur and we can omit the translations in the definition of orbital stability.

- Phase shifts are necessary: For a solution φ of (3.1) and $\varepsilon > 0$, let $\varphi_\varepsilon(x) := (1+\varepsilon)^{\frac{1}{p-1}}\varphi((1+\varepsilon)^{\frac{1}{2}}x)$. Then it is easy to check that the solution of (1.1) with initial datum φ_ε is

$$u_\varepsilon(t,x) = e^{i\omega(1+\varepsilon)t}(1+\varepsilon)^{\frac{1}{p-1}}\varphi((1+\varepsilon)^{\frac{1}{2}}x).$$

We clearly have $\varphi_\varepsilon \to \varphi$ in $H^1(\mathbb{R}^N)$ as $\varepsilon \to 0$ but for any $\varepsilon > 0$

$$\sup_{t\in\mathbb{R}} \inf_{y\in\mathbb{R}^N} \|u_\varepsilon(t) - \varphi(\cdot - y)\|_{H^1(\mathbb{R}^N)} \geqslant \|\varphi\|_{H^1(\mathbb{R}^N)}.$$

Remark that φ_ε is a solution of (3.1) with ω being replaced by $\omega(1+\varepsilon)$.

The main result of this section is the following.

Theorem 4.3. *Let φ be a ground state of* (3.1). *If $1 < p < 1 + \frac{4}{N}$ then the standing wave $e^{i\omega t}\varphi(x)$ is orbitally stable.*

Remark 4.4. The range of p is optimal: We will see in the next section that instability occurs if $p \geqslant 1 + \frac{4}{N}$.

Remark 4.5. Except for $N = 1$, where all solutions of (3.1) are ground states, Theorem 4.3 asserts only the stability of standing waves corresponding to ground states. It is an open problem to describe the behavior of other standing waves of (1.1) under perturbations (see nevertheless [33, 74] and the references cited therein).

The first rigorous stability study is due to Benjamin [5] for solitary waves of the Korteweg–de Vries equation. Roughly speaking, two different approaches are possible in the study of stability of standing waves. The first one was introduced by Cazenave [13] and then developed by Cazenave and Lions [15]. It relies on variational and compactness arguments. The main step in this approach consists in characterizing the ground states as minimizers of the energy on a sphere of $L^2(\mathbb{R}^N)$. To use this approach, it is essential to have a uniqueness result for ground states as Theorem 3.15, otherwise a weaker result is obtained: stability of the set of ground states (see Remark 4.9). The second approach was introduced by Weinstein [76, 77] (see also [66]) and then considerably generalized by Grillakis, Shatah and Strauss [34, 35]. A criterion based on a form of coercivity for $S''(\varphi)$ is derived in this work and allows to prove stability of standing waves. Sufficient conditions for instability are also given in [34, 35].

The rest of this section is devoted to the proofs of Theorem 4.3. In Subsection 4.1, we present the proof using the Cazenave–Lions method and in Subsection 4.2 we present the proof using Grillakis–Shatah–Strauss method.

4.1 Cazenave–Lions method

The proof of Theorem 4.3 by the Cazenave–Lions method relies on the following compactness result.

Proposition 4.6. *Let $1 < p < 1 + \frac{4}{N}$. For any $\tau > 0$, define*

$$\Sigma_\tau := \{v \in H^1(\mathbb{R}^N); \|v\|^2_{L^2(\mathbb{R}^N)} = \tau\}.$$

Consider the minimization problem

$$-\nu_\tau := \inf\{E(v); v \in \Sigma_\tau\},$$

where E is the functional defined in (2.2): For $v \in H^1(\mathbb{R}^N)$,

$$E(v) = \frac{1}{2}\|\nabla v\|^2_{L^2(\mathbb{R}^N)} - \frac{1}{p+1}\|v\|^{p+1}_{L^{p+1}(\mathbb{R}^N)}.$$

Then $\nu_\tau < +\infty$, and if $(v_n) \subset H^1(\mathbb{R}^N)$ is such that

$$\|v_n\|^2_{L^2(\mathbb{R}^N)} \to \tau \text{ and } E(v_n) \to -\nu_\tau \text{ as } n \to +\infty,$$

then there exist a sequence $(y_n) \subset \mathbb{R}^N$ and a function $v \in H^1(\mathbb{R}^N)$ such that, possibly for a subsequence only,

$$v_n(\cdot - y_n) \to v \text{ strongly in } H^1(\mathbb{R}^N).$$

In particular, $v \in \Sigma_\tau$ and $E(v) = -\nu_\tau$.

The proof of Proposition 4.6 is based on the concentration-compactness principle of Lions [50, 51]. We refer to [14, 15] for a detailed proof.

Remark 4.7. For $p > 1 + \frac{4}{N}$, it is easy to see that $\nu_\tau = +\infty$. Indeed, let $v \in \Sigma_\tau$. Using the scaled functions $v_\lambda(\cdot) := \lambda^{\frac{N}{2}} v(\lambda\cdot)$, we obtain $\|v_\lambda\|^2_{L^2(\mathbb{R}^N)} = \|v\|^2_{L^2(\mathbb{R}^N)} = \tau$, but $\lim_{\lambda\to+\infty} E(v_\lambda) = -\infty$. Therefore, Proposition 4.6 cannot hold in this case.

Next we characterize the ground states of (3.1) as minimizers of the energy on a sphere of $L^2(\mathbb{R}^N)$.

Proposition 4.8. *Let* $1 < p < 1 + \frac{4}{N}$. *Then*

(i) *there exists* $\tau_{\mathcal{G}} > 0$ *such that* $\|\varphi\|^2_{L^2(\mathbb{R}^N)} = \tau_{\mathcal{G}}$ *for all* $\varphi \in \mathcal{G}$,

(ii) $\varphi \in \mathcal{G}$ *if and only if* $\varphi \in \Sigma_{\tau_{\mathcal{G}}}$ *and* $E(\varphi) = -\nu_{\tau_{\mathcal{G}}}$.

Sketch of proof. Point (i) follows immediately from Theorem 3.15. We refer to [14] for the proof of (ii). □

Proof of Theorem 4.3 by the Cazenave–Lions method. The result is proved by contradiction. Assume that there exist $\varepsilon > 0$ and two sequences $(u_{n,0}) \subset H^1(\mathbb{R}^N)$, $(t_n) \subset \mathbb{R}$ such that

$$\|u_{n,0} - \varphi\|_{H^1(\mathbb{R}^N)} \to 0 \text{ as } n \to +\infty, \tag{4.1}$$

$$\inf_{\theta \in \mathbb{R}} \inf_{y \in \mathbb{R}^N} \|u_n(t_n) - e^{i\theta}\varphi(\cdot - y)\|_{H^1(\mathbb{R}^N)} > \varepsilon \text{ for all } n \in \mathbb{N}, \tag{4.2}$$

where u_n is the maximal solution of (1.1) with initial datum $u_{n,0}$. Set $v_n(x) := u_n(t_n, x)$. By (4.1) and the conservation of charge and energy (see (2.3)), we have, as $n \to +\infty$,

$$\|v_n\|^2_{L^2(\mathbb{R}^N)} = \|u_n(t_n)\|^2_{L^2(\mathbb{R}^N)} = \|u_{n,0}\|^2_{L^2(\mathbb{R}^N)} \quad \to \quad \|\varphi\|^2_{L^2(\mathbb{R}^N)} = \tau_{\mathcal{G}}, \tag{4.3}$$

$$E(v_n) = E(u_n(t_n)) = E(u_{n,0}) \quad \to \quad E(\varphi) = -\nu_{\tau_{\mathcal{G}}}. \tag{4.4}$$

By (4.3), (4.4) and Proposition 4.6, there exist $(y_n) \subset \mathbb{R}^N$ and $v \in H^1(\mathbb{R}^N)$ such that

$$\|v_n(\cdot - y_n) - v\|_{H^1(\mathbb{R}^N)} \to 0 \text{ as } n \to +\infty, \tag{4.5}$$

$$v \in \Sigma_{\tau_{\mathcal{G}}} \text{ and } E(v) = -\nu_{\tau_{\mathcal{G}}}. \tag{4.6}$$

By (4.6) and Proposition 4.8, we have $v \in \mathcal{G}$. Therefore, by Theorem 3.15, there exist $\theta \in \mathbb{R}$ and $y \in \mathbb{R}^N$ such that $v = e^{i\theta}\varphi(\cdot - y)$. Remembering that $v_n = u_n(t_n)$ and substituting this in (4.5), we get

$$\|u_n(t_n) - e^{i\theta}\varphi(\cdot - (y - y_n))\|_{H^1(\mathbb{R}^N)} \to 0 \text{ when } n \to +\infty,$$

which is a contradiction with (4.2) and finishes the proof. □

Remark 4.9. It is essential for the proof of Theorem 4.3 by the Cazenave–Lions method that the set of ground states $\mathcal{G}$ can be explicitly described by $\{e^{i\theta}\Psi(\cdot - y); \theta \in \mathbb{R}, y \in \mathbb{R}^N\}$ as in Theorem 3.15. This uniqueness result is far from being obvious even with our simple pure power nonlinearity. For general nonlinearities, such uniqueness results are usually not available. Without this uniqueness result, one obtains a result corresponding to a weaker notion of stability: stability of the set $\mathcal{G}$. More precisely, the concept of stability would read as follows: The set of ground states $\mathcal{G}$ is said to be *stable* if for all $\varepsilon > 0$ there exists $\delta > 0$ such that if $u_0 \in H^1(\mathbb{R}^N)$ satisfies $\|u_0 - \varphi\|_{H^1(\mathbb{R}^N)} < \delta$ for some $\varphi \in \mathcal{G}$ then

$$\sup_{t\in\mathbb{R}} \inf_{\psi\in\mathcal{G}} \|u(t) - \psi\|_{H^1(\mathbb{R}^N)} < \varepsilon,$$

where u is the maximal solution of (1.1) with $u(0) = u_0$.

4.2 Grillakis–Shatah–Strauss method

The method of Grillakis–Shatah–Strauss is a powerful tool to derive stability or instability results. The results in [34, 35] state, roughly speaking, the following: If we consider a family (φ_ω) of solutions of the stationary equation, the standing wave $e^{i\omega t}\varphi_\omega(x)$ is stable if

$$\frac{\partial}{\partial\omega}\|\varphi_\omega\|^2_{L^2(\mathbb{R}^N)} > 0 \tag{4.7}$$

and unstable if

$$\frac{\partial}{\partial\omega}\|\varphi_\omega\|^2_{L^2(\mathbb{R}^N)} < 0, \tag{4.8}$$

provided some spectral assumptions on the linearized operator $S''(\varphi_\omega)$ are satisfied. Slope conditions of the type (4.7)–(4.8) can easily be checked when the stationary equation admits some scaling invariance, typically when the nonlinearity is of power type. For example, in the case of (3.1), it is easy to see that if φ_1 is a solution of (3.1) with $\omega = 1$ then the scaling $\varphi_\omega(\cdot) := \omega^{\frac{1}{p-1}}\varphi_1(\omega^{\frac{1}{2}}\cdot)$ provides a family of solutions of (3.1) for $\omega \in (0, +\infty)$ such that

$$\frac{\partial}{\partial\omega}\|\varphi_\omega\|^2_{L^2(\mathbb{R}^N)} > 0 \quad \text{if} \quad p < 1 + \frac{4}{N},$$
$$\frac{\partial}{\partial\omega}\|\varphi_\omega\|^2_{L^2(\mathbb{R}^N)} < 0 \quad \text{if} \quad p > 1 + \frac{4}{N}.$$

In such situations, the main difficulty is to handle the spectral conditions (see, e.g., [25, 26, 49]). If (3.1) has no scaling invariance, it becomes very difficult to obtain the slope conditions (4.7)–(4.8) (see nevertheless [27, 28, 56]). An alternative is to use a stability criterion derived from the work [34]. This stability criterion fits better the case of general nonlinearities, as, e.g., in [19, 23, 24, 40, 44, 46]. See also [72] for a review of the different ways to obtain stability for general nonlinearities thanks to Grillakis–Shatah–Strauss' results.

Although theses notes are restricted to nonlinear Schrödinger equations with power-type nonlinearities, our goal is to provide the reader with methods applicable in rather general situations and this is why we choose to present the proof of Theorem 4.3 using the following stability criterion.

Proposition 4.10 (Stability criterion). *Let φ be a solution of* (3.1). *Suppose that there exists $\delta > 0$ such that we have*

$$\langle S''(\varphi)v, v\rangle \geqslant \delta\|v\|^2_{H^1(\mathbb{R}^N)} \tag{4.9}$$

for all $v \in H^1(\mathbb{R}^N)$ satisfying the orthogonality conditions

$$(v, \varphi)_2 = (v, i\varphi)_2 = \left(v, \frac{\partial \varphi}{\partial x_j}\right)_2 = 0 \text{ for all } j = 1, \dots, N. \tag{4.10}$$

Then the standing wave $e^{i\omega t}\varphi(x)$ is orbitally stable in $H^1(\mathbb{R}^N)$.

Let us heuristically explain why the assumptions of Proposition 4.10 lead to stability. The idea comes from the theory of Lyapunov stability for an equilibrium in dynamical systems. A good candidate for a Lyapunov functional would be the functional S. Indeed, let us suppose for a moment that the coercivity condition (4.9) holds for any $v \in H^1(\mathbb{R}^N)$ (this is not the case, as we will see in the sequel). Let u be a solution of (1.1) with initial datum u_0 close to φ in $H^1(\mathbb{R}^N)$. A Taylor expansion gives

$$\begin{aligned} S(u(t)) - S(\varphi) &= \langle S'(\varphi), u(t) - \varphi\rangle + \frac{1}{2}\langle S''(\varphi)(u(t) - \varphi), u(t) - \varphi\rangle \\ &\quad + o(\|u(t) - \varphi\|^2_{H^1(\mathbb{R}^N)}). \end{aligned}$$

Since φ is a solution of (3.1), $S'(\varphi) = 0$. Combined with (4.9), this would give, for some constant $C > 0$ independent of t,

$$S(u(t)) - S(\varphi) \geqslant C\|u(t) - \varphi\|^2_{H^1(\mathbb{R}^N)}. \tag{4.11}$$

Since S is a conserved quantity this would give an upper bound on $\|u(t) - \varphi\|_{H^1(\mathbb{R}^N)}$, hence stability. Of course, as we already know (see Remark 4.2), this cannot be true since stability is possible only up to translations and phase shifts. In fact, translation and phase shift invariance generates, as we will see in the sequel, a kernel for $S''(\varphi)$ of the form

$$\ker\{S''(\varphi)\} = \operatorname{span}\left\{i\varphi, \frac{\partial \varphi}{\partial x_j}; j = 1, \dots, N\right\}.$$

To avoid this kernel, we require the coercivity condition (4.9) only for $v \in H^1(\mathbb{R}^N)$ satisfying

$$(v, i\varphi)_2 = \left(v, \frac{\partial \varphi}{\partial x_j}\right)_2 = 0 \text{ for all } j = 1, \dots, N,$$

which allows phase shifts and translations in the right-hand side of (4.11). The other orthogonality condition $(v, \varphi)_2 = 0$ is related to the conservation of mass. Indeed, since the mass is conserved, the evolution takes place, in some sense, in the tangent space of the sphere of $L^2(\mathbb{R}^N)$ at φ and therefore it is enough to ask for S to satisfy the coercivity condition (4.9) on this tangent space to get stability. The rigorous proof of Proposition 4.10 is involved and we have postponed it to the appendix.

In view of Proposition 4.10, it is clear that Theorem 4.3 follows immediately from the following proposition.

Proposition 4.11. *Let $1 < p < 1 + \frac{4}{N}$ and φ be a ground state of* (3.1). *Then there exists $\delta > 0$ such that for all $w \in H^1(\mathbb{R}^N)$ satisfying*

$$(w, \varphi)_2 = (w, i\varphi)_2 = \left(w, \frac{\partial \varphi}{\partial x_j}\right)_2 = 0 \text{ for all } j = 1, \dots, N, \tag{4.12}$$

we have

$$\langle S''(\varphi)w,w\rangle \geqslant \delta\|w\|^2_{H^1(\mathbb{R}^N)}.$$

Before giving the proof of Proposition 4.11, some preparation is necessary. First, from Theorem 3.15, we can assume without loss of generality that the ground state φ is real, positive, and radial. Note that uniqueness is not involved here. It is convenient to split w in the real and imaginary part. We set $w = u+iv$ for real-valued $u, v \in H^1(\mathbb{R}^N)$. Then the operator $S''(\varphi)$ can be separated into a real and an imaginary part L_1 and L_2 such that

$$\langle S''(\varphi)w,w\rangle = \langle L_1u,u\rangle + \langle L_2v,v\rangle\,.$$

Here, L_1 and L_2 are two bounded operators defined on $H^1(\mathbb{R}^N)$ restricted to real-valued functions, with values in $H^{-1}(\mathbb{R}^N)$, and given by

$$\begin{aligned} L_1u &= -\Delta u + \omega u - p\varphi^{p-1}u,\\ L_2v &= -\Delta v + \omega v - \varphi^{p-1}v. \end{aligned}$$

Until the end of this subsection, the functions considered will be real-valued. In particular, $H^1(\mathbb{R}^N)$ and $L^2(\mathbb{R}^N)$ will be restricted to real-valued functions.

Proposition 4.11 follows immediately from the two following lemmas.

Lemma 4.12. *There exists $\delta_1 > 0$ such that for all $u \in H^1(\mathbb{R}^N)$ satisfying*

$$(u,\varphi)_2 = \left(u, \frac{\partial\varphi}{\partial x_j}\right)_2 = 0 \textit{ for all } j = 1,\dots,N,$$

we have

$$\langle L_1u,u\rangle \geqslant \delta_1\|u\|^2_{H^1(\mathbb{R}^N)}.$$

Lemma 4.13. *There exists $\delta_2 > 0$ such that for all $v \in H^1(\mathbb{R}^N)$ satisfying*

$$(v,\varphi)_2 = 0,$$

we have

$$\langle L_2v,v\rangle \geqslant \delta_2\|v\|^2_{H^1(\mathbb{R}^N)}.$$

It is not hard to see that the operators L_1 and L_2 admit restrictions $\tilde{L}_1$ and $\tilde{L}_2$ with domain $H^2(\mathbb{R}^N)$ that are unbounded self-adjoint operators in $L^2(\mathbb{R}^N)$ (the so-called Friedrichs extensions, see, e.g., [43]).

The proofs of Lemmas 4.12 and 4.13 rely on the analysis of the spectra of $\tilde{L}_1$ and $\tilde{L}_2$. The following lemma gives the general structure of these spectra.

Lemma 4.14. *The spectra of $\tilde{L}_1$ and $\tilde{L}_2$ consist of essential spectrum in $[\omega,+\infty)$ and of a finite number of eigenvalues of finite multiplicity in $(-\infty,\omega']$ for all $\omega' < \omega$.*

Proof. We first remark that since $\tilde{L}_1$ and $\tilde{L}_2$ are self-adjoint operators, their spectra lie on the real line.

The spectra of $\tilde{L}_1$ and $\tilde{L}_2$ are bounded from below. Indeed, for all $u \in H^1(\mathbb{R}^N)$, we have

$$\begin{aligned}\langle L_2 u, u\rangle &= \|\nabla u\|^2_{L^2(\mathbb{R}^N)} + \omega\|u\|^2_{L^2(\mathbb{R}^N)} - \int_{\mathbb{R}^N} \varphi^{p-1}u^2 dx \\ &\geqslant \|\nabla u\|^2_{L^2(\mathbb{R}^N)} + \omega\|u\|^2_{L^2(\mathbb{R}^N)} - p\int_{\mathbb{R}^N} \varphi^{p-1}u^2 dx = \langle L_1 u, u\rangle .\end{aligned}$$

Since $\varphi \in L^\infty(\mathbb{R}^N)$, there exists $C > 0$, independent of u, such that

$$p\int_{\mathbb{R}^N} \varphi^{p-1}u^2 dx \leqslant C\|u\|^2_{L^2(\mathbb{R}^N)}$$

and therefore

$$\langle L_2 u, u\rangle \geqslant \langle L_1 u, u\rangle \geqslant (\omega - C)\|u\|^2_{L^2(\mathbb{R}^N)}.$$

Now, since φ is exponentially decaying, $\tilde{L}_1$ and $\tilde{L}_2$ are compactly perturbed versions of

$$-\Delta + \omega : H^2(\mathbb{R}^N) \subset L^2(\mathbb{R}^N) \to L^2(\mathbb{R}^N).$$

It is well known that the essential spectrum of $-\Delta + \omega$ is $\sigma_{\text{ess}}(-\Delta+\omega) = [\omega, +\infty)$, thus, by Weyl's Theorem (see, e.g., [43]),

$$\sigma_{\text{ess}}(\tilde{L}_1) = \sigma_{\text{ess}}(\tilde{L}_2) = [\omega, +\infty).$$

Since, for $j = 1, 2$, $\sigma(\tilde{L}_j) \setminus \sigma_{\text{ess}}(\tilde{L}_j)$ consists of isolated eigenvalues of finite multiplicity and $\sigma(\tilde{L}_j)$ is bounded from below, this completes the proof of the lemma. □

From now on, we consider L_1 and L_2 separately. We begin with L_2.

Lemma 4.15. *There exists $\tilde{\delta}_2 > 0$ such that for all $v \in H^1(\mathbb{R}^N)$ with $(v, \varphi)_2 = 0$, we have*

$$\langle L_2 v, v\rangle \geqslant \tilde{\delta}_2\|v\|^2_{L^2(\mathbb{R}^N)}.$$

Proof. We remark that $L_2\varphi = S'(\varphi) = 0$ since φ is a solution of (3.1). This means that 0 is an eigenvalue of L_2 with φ being an eigenfunction. But $\varphi > 0$, and it is well known (see, e.g., [11]) that this implies that 0 is the first simple eigenvalue of $\tilde{L}_2$. Let $v \in H^1(\mathbb{R}^N)$ be such that $(v, \varphi)_2 = 0$. Then, by the min-max characterization of eigenvalues (see, e.g., [11]), there exists $\tilde{\delta}_2 > 0$ independent of v (in fact, $\tilde{\delta}_2$ is the second eigenvalue of $\tilde{L}_2$) such that

$$\langle L_2 v, v\rangle \geqslant \tilde{\delta}_2\|v\|^2_{L^2(\mathbb{R}^N)}.$$

□

Proof of Lemma 4.13. The proof is carried out by contradiction. Assume the existence of a sequence $(v_n) \subset H^1(\mathbb{R}^N)$ such that

$$\|\nabla v_n\|^2_{L^2(\mathbb{R}^N)} + \omega\|v_n\|^2_{L^2(\mathbb{R}^N)} = 1,\ (v_n, \varphi)_2 = 0 \text{ and } \langle L_2 v_n, v_n\rangle \to 0 \text{ as } n \to +\infty.$$

Since $\|v_n\|^2_{H^1(\mathbb{R}^N)} \leq \max\{\omega, \omega^{-1}\}(\|\nabla v_n\|^2_{L^2(\mathbb{R}^N)} + \omega\|v_n\|^2_{L^2(\mathbb{R}^N)})$, (v_n) is bounded in $H^1(\mathbb{R}^N)$ and there exists $v \in H^1(\mathbb{R}^N)$ such that, possibly for a subsequence only,

$$v_n \rightharpoonup v \text{ weakly in } H^1(\mathbb{R}^N).$$

In particular, we have $(v, \varphi)_2 = 0$ and by Lemma 4.15

$$\langle L_2 v, v\rangle \geqslant 0. \tag{4.13}$$

By the embedding of $H^1(\mathbb{R}^N)$ into $L^{2q}(\mathbb{R}^N)$ for $q \in (1, (N-2)/N)$, we have $v_n \rightharpoonup v$ in $L^{2q}(\mathbb{R}^N)$. In view of the compact embedding of $H^1(\mathbb{R}^N)$ into $L^{2q}_{\text{loc}}(\mathbb{R}^N)$, we also get $v_n^2 \rightharpoonup v^2$ in $L^q(\mathbb{R}^N)$. Since φ is exponentially decaying, we have $\varphi^{p-1} \in L^{q'}(\mathbb{R}^N)$, where q' is the conjugate exponent of q, $\frac{1}{q} + \frac{1}{q'} = 1$. Therefore,

$$\int_{\mathbb{R}^N} \varphi^{p-1} v_n^2 dx \to \int_{\mathbb{R}^N} \varphi^{p-1} v^2 dx \text{ as } n \to +\infty. \tag{4.14}$$

From (4.14) and by the weak lower semi-continuity of the $H^1(\mathbb{R}^N)$-norm, we infer that

$$\langle L_2 v, v\rangle \leqslant \liminf_{n\to+\infty}\langle L_2 v_n, v_n\rangle = 0. \tag{4.15}$$

Combined with (4.13), (4.15) implies $\langle L_2 v, v\rangle = 0$. Since $(\varphi, v)_2 = 0$, we obtain by Lemma 4.15

$$v \equiv 0. \tag{4.16}$$

On the other hand,

$$0 = \liminf_{n\to+\infty}\langle L_2 v_n, v_n\rangle = 1 - \int_{\mathbb{R}^N} \varphi^{p-1} v^2 dx.$$

Hence $\int_{\mathbb{R}^N} \varphi^{p-1} v^2 dx = 1$, which is in contradiction with (4.16). □

We now turn our attention to L_1. The proof of Lemma 4.12 is more delicate, essentially since the spectrum of $\tilde{L}_1$ contains nonpositive eigenvalues. Furthermore, φ is no longer an eigenfunction. We first deal with the negative eigenvalues of $\tilde{L}_1$.

Lemma 4.16. *The operator $\tilde{L}_1$ has only one negative eigenvalue $-\lambda_1$ with a corresponding eigenfunction $e_1 \in H^2(\mathbb{R}^N)$ such that $\|e_1\|_{L^2(\mathbb{R}^N)} = 1$.*

Proof. Let $\tilde{S} : H^1(\mathbb{R}^N) \to \mathbb{R}$ be the restriction of S to real-valued functions. It is not hard to see that φ is also a mountain pass critical point of $\tilde{S}$. Then it is well known (see, for example, [3]) that the Morse index of $\tilde{S}$ at φ is at most 1 (recall that the Morse index is the number of negative eigenvalues of $\tilde{S}''(\varphi)$). We remark that $L_1 = \tilde{S}''(\varphi)$. Therefore, $\tilde{L}_1$ has at most one negative eigenvalue. On the other hand,

$$\begin{aligned} \langle L_1\varphi, \varphi\rangle &= \|\nabla\varphi\|^2_{L^2(\mathbb{R}^N)} + \omega\|\varphi\|^2_{L^2(\mathbb{R}^N)} - p\|\varphi\|^{p+1}_{L^{p+1}(\mathbb{R}^N)} \\ &= -(p-1)\|\varphi\|^{p+1}_{L^{p+1}(\mathbb{R}^N)} < 0, \end{aligned}$$

where the second equality follows from the Nehari identity (3.3). Therefore, $\tilde{L}_1$ has exactly one negative eigenvalue $-\lambda_1$, and we can pick up an eigenfunction $e_1 \in H^2(\mathbb{R}^N)$ such that $\|e_1\|_{L^2(\mathbb{R}^N)} = 1$. □

The following result originates from the work of Weinstein [76].

Lemma 4.17. *The second eigenvalue of* $\tilde{L}_1$ *is* 0 *and*

$$Z := \ker \tilde{L}_1 = \operatorname{span}\left\{\frac{\partial\varphi}{\partial x_j}; j = 1, \ldots, N\right\}.$$

It is out of reach in these notes to give a complete proof of Lemma 4.17. Therefore, we only give a partial proof and refer to [2] for a complete one.

Partial proof of Lemma 4.17. We remember that φ satisfies

$$-\Delta\varphi + \omega\varphi - \varphi^p = 0. \tag{4.17}$$

Differentiating (4.17) with respect to x_j gives

$$-\Delta\frac{\partial\varphi}{\partial x_j} + \omega\frac{\partial\varphi}{\partial x_j} - p\varphi^{p-1}\frac{\partial\varphi}{\partial x_j} = 0.$$

This is allowed since $\varphi \in W^{3,2}(\mathbb{R}^N)$ implies $\frac{\partial\varphi}{\partial x_j} \in H^2(\mathbb{R}^N)$. Hence 0 is an eigenvalue of $\tilde{L}_1$ and

$$\operatorname{span}\left\{\frac{\partial\varphi}{\partial x_j}; j = 1, \ldots, N\right\} \subset \ker \tilde{L}_1.$$

We admit that the reverse inclusion also holds. □

Lemma 4.18. *The space* $H^1(\mathbb{R}^N)$ *can be decomposed as* $H^1(\mathbb{R}^N) = E_1 \oplus Z \oplus E_+$, *where* $E_1 = \operatorname{span}\{e_1\}$ *and* E_+ *is the image of the spectral projection corresponding to the positive part of the spectrum of* $\tilde{L}_1$.

Note that in the direct sum $E_1 \oplus Z \oplus E_+$ the spaces are mutually orthogonal with respect to the inner product of $L^2(\mathbb{R}^N)$.

Proof of Lemma 4.18. The assertion follows immediately from Lemmas 4.14, 4.16, 4.17 and the spectral decomposition theorem (see, e.g., [43, p. 177]). □

Lemma 4.19. *For all* $u \in H^1(\mathbb{R}^N) \setminus \{0\}$ *satisfying*

$$(u, \varphi)_2 = \left(u, \frac{\partial \varphi}{\partial x_j}\right)_2 = 0 \text{ for all } j = 1, \dots, N, \tag{4.18}$$

we have

$$\langle L_1 u, u\rangle > 0. \tag{4.19}$$

Proof. Let $u \in H^1(\mathbb{R}^N)$ satisfying (4.18). We first look for a function ψ such that

$$L_1 \psi = -\omega\varphi.$$

To do this, we use the scaled functions φ_λ defined by $\varphi_\lambda(\cdot) := \lambda^{\frac{1}{p-1}}\varphi(\lambda^{\frac{1}{2}}\cdot)$ for $\lambda > 0$. The functions φ_λ satisfy

$$-\Delta\varphi_\lambda + \omega\lambda\varphi_\lambda - \varphi_\lambda^p = 0. \tag{4.20}$$

Differentiating (4.20) with respect to λ gives at $\lambda = 1$

$$-\Delta\psi + \omega\psi - p\varphi^{p-1}\psi = -\omega\varphi$$

for $\psi := \frac{\partial\varphi_\lambda}{\partial\lambda}\big|_{\lambda=1} = \frac{1}{p-1}\varphi + \frac{1}{2}x \cdot \nabla\varphi$. Note that $\psi \in H^1(\mathbb{R}^N)$ since $x \cdot \nabla\varphi \in H^1(\mathbb{R}^N)$ by Proposition 3.2. Furthermore, since φ is radial, φ is even in each x_j. Therefore, ψ is also even in each x_j whereas $\frac{\partial\varphi}{\partial x_j}$ is odd in x_j. This implies

$$\left(\psi, \frac{\partial\varphi}{\partial x_j}\right)_2 = 0 \text{ for all } j = 1, \dots, N.$$

We decompose u and ψ with respect to $H^1(\mathbb{R}^N) = E_1 \oplus Z \oplus E_+$: There exist $\alpha, \beta \in \mathbb{R}$ and $\xi, \eta \in E_+$ such that

$$\begin{aligned} u &= \alpha e_1 + \xi, \\ \psi &= \beta e_1 + \eta. \end{aligned}$$

If $\alpha = 0$ then $\xi \not\equiv 0$ since $u \not\equiv 0$, and we have

$$\langle L_1 u, u\rangle = \langle L_1\xi, \xi\rangle > 0.$$

Therefore, u satisfies (4.19). Now, we suppose $\alpha \neq 0$. We easily find:

$$\begin{aligned} \langle L_1\psi, \psi\rangle &= -\omega\left(\varphi, \frac{1}{p-1}\varphi + \frac{1}{2}x \cdot \nabla\varphi\right)_2 \\ &= -\omega\left(\frac{1}{p-1}\|\varphi\|^2_{L^2(\mathbb{R}^N)} + \frac{1}{2}\int_{\mathbb{R}^N} \varphi x \cdot \nabla\varphi dx\right). \end{aligned}$$

With integration by parts, it is easy to see that

$$\int_{\mathbb{R}^N} \varphi x \cdot \nabla\varphi dx = -N\|\varphi\|^2_{L^2(\mathbb{R}^N)} - \int_{\mathbb{R}^N} \varphi x \cdot \nabla\varphi dx.$$

Therefore, $\int_{\mathbb{R}^N} \varphi x \cdot \nabla\varphi dx = -\frac{N}{2}\|\varphi\|^2_{L^2(\mathbb{R}^N)}$ and

$$\langle L_1\psi, \psi\rangle = -\omega\left(\frac{1}{p-1} - \frac{N}{4}\right)\|\varphi\|^2_{L^2(\mathbb{R}^N)}.$$

Since $\omega > 0$ and $p < 1 + \frac{4}{N}$ this implies

$$\langle L_1\psi, \psi\rangle < 0$$

and thus $\beta \neq 0$.

On E_+, L_1 defines a positive definite quadratic form, and we have the Cauchy–Schwarz inequality

$$\langle L_1\xi, \eta\rangle^2 \leqslant \langle L_1\xi, \xi\rangle \langle L_1\eta, \eta\rangle, \quad \xi, \eta \in E_+.$$

Therefore, we find

$$\langle L_1u, u\rangle = -\alpha^2\lambda_1 + \langle L_1\xi, \xi\rangle \geqslant -\alpha^2\lambda_1 + \frac{\langle L_1\xi, \eta\rangle^2}{\langle L_1\eta, \eta\rangle}. \tag{4.21}$$

On the other hand, we get

$$0 = -\omega\,(\varphi, u)_2 = \langle L_1\psi, u\rangle = -\alpha\beta\lambda_1 + \langle L_1\xi, \eta\rangle$$

and thus obtain

$$\langle L_1\xi, \eta\rangle = \alpha\beta\lambda_1.$$

This gives

$$\begin{aligned} -\alpha^2\lambda_1 + \frac{\langle L_1\xi, \eta\rangle^2}{\langle L_1\eta, \eta\rangle} &= -\alpha^2\lambda_1 + \frac{\alpha^2\beta^2\lambda_1^2}{\langle L_1\eta, \eta\rangle} \\ &= -\alpha^2\lambda_1 + \frac{\alpha^2\beta^2\lambda_1^2}{\beta^2\lambda_1 + \langle L_1\psi, \psi\rangle} \\ &= \frac{-\alpha^2\lambda_1\,\langle L_1\psi, \psi\rangle}{\langle L_1\eta, \eta\rangle} > 0. \end{aligned}$$

Combined with (4.21), this finishes the proof. □

Proof of Lemma 4.12. The proof of Lemma 4.12 follows the same lines as the proof of Lemma 4.13. We omit the details. □

5 Instability

In this section, we will prove that the standing waves are unstable by blow-up in finite time if $1 + \frac{4}{N} \leqslant p < 1 + \frac{4}{N-2}$. More precisely, for any ground state $\varphi \in \mathcal{G}$, we will find initial data, as close to φ as we want, such that the solutions of (1.1) corresponding to these initial data will all blow up in finite time. As in Proposition 2.3, our basic tool will be the virial theorem (Proposition 2.4). However, since the energy of the ground state is nonnegative, it is not possible to argue directly as in Proposition 2.3. To overcome this difficulty, we follow the approach of [48], which is a recent improvement of the method introduced by Berestycki and Cazenave in [6, 7]. At the heart of this method is a new variational characterization of the ground states as minimizers of the action S on constraints related to the Pohozhaev identity (3.4). Compared to [6, 7], the mean feature of the approach of [48] is that we do not need to solve directly a new minimization problem; instead, we take advantage of the variational characterizations of the ground states obtained in Section 3.

Before stating our results, we give a precise definition of instability by blow-up.

Definition 5.1. Let φ be a solution of (3.1). The standing wave $e^{i\omega t}\varphi(x)$ is said to be *unstable by blow-up in finite time* if for all $\varepsilon > 0$ there exists $u_{\varepsilon,0} \in H^1(\mathbb{R}^N)$ such that

$$\|u_{\varepsilon,0} - \varphi\|_{H^1(\mathbb{R}^N)} < \varepsilon$$

but the corresponding maximal solution u_ε of (1.1) in the interval $(T_{\min}^{\varepsilon}, T_{\varepsilon}^{\max})$ satisfies $T_{\min}^{\varepsilon} > -\infty$, $T_{\varepsilon}^{\max} < +\infty$ and thus

$$\lim_{t \downarrow T_{\min}^{\varepsilon}} \|u_\varepsilon(t)\|_{H^1(\mathbb{R}^N)} = +\infty \text{ and } \lim_{t \uparrow T_{\varepsilon}^{\max}} \|u_\varepsilon(t)\|_{H^1(\mathbb{R}^N)} = +\infty.$$

We start with the simplest case $p = 1 + \frac{4}{N}$. The following result is due to Weinstein [75].

Theorem 5.2. *Let $p = 1 + \frac{4}{N}$. Then for every solution φ of* (3.1) *the standing wave $e^{i\omega t}\varphi(x)$ is unstable by blow-up in finite time.*

Proof. First, we remark that $E(v) = P(v)$ (recall that P was defined in (2.6)) for all $v \in H^1(\mathbb{R}^N)$ since $p = 1 + \frac{4}{N}$. From the Pohozhaev identity (3.4), we have

$$E(\varphi) = P(\varphi) = 0.$$

Let $u_{\varepsilon,0}$ be defined by $u_{\varepsilon,0} := (1+\varepsilon)\varphi$. Then it is easy to see that $E(u_{\varepsilon,0}) < 0$. In view of the exponential decay of φ (see Proposition 3.2), we have $|x|u_{\varepsilon,0} \in L^2(\mathbb{R}^N)$. The conclusion follows from Proposition 2.3. □

We consider now the general case.

Theorem 5.3. *Let $p > 1 + \frac{4}{N}$. For all $\varphi \in \mathcal{G}$, the standing wave $e^{i\omega t}\varphi(x)$ is unstable by blow-up in finite time.*

Remark 5.4. Theorem 5.2 asserts the instability of standing waves corresponding to *any* solution of (3.1) whereas Theorem 5.3 concerns only standing waves corresponding to ground states. It is still an open problem to prove instability by blow-up for any solution of (3.1) if $p > 1 + \frac{4}{N}$ (see [74] for a review of related results and open problems).

For the proof of Theorem 5.3 it is not possible to mimic the proof of Theorem 5.2. Indeed, for $p > 1 + \frac{4}{N}$, the identity $E(v) = P(v)$ does not hold any more. Moreover, $E(\varphi) > 0$ for $\varphi \in \mathcal{G}$, which prevents to use Proposition 2.3.

The scaling $v_\lambda(\cdot) := \lambda^{N/2} v(\lambda \cdot)$ will play an important role in the proof. In the following lemma, we investigate the behavior of different functionals under the scaling.

Lemma 5.5. *Let $v \in H^1(\mathbb{R}^N) \backslash \{0\}$ be such that $P(v) \leqslant 0$. Then there exists $\lambda_0 \in (0, 1]$ such that*

(i) $P(v_{\lambda_0}) = 0$,

(ii) $\lambda_0 = 1$ *if and only if* $P(v) = 0$,

(iii) $\frac{\partial}{\partial\lambda} S(v_\lambda) = \frac{1}{\lambda} P(v_\lambda)$,

(iv) $\frac{\partial}{\partial\lambda} S(v_\lambda) > 0$ *for* $\lambda \in (0, \lambda_0)$ *and* $\frac{\partial}{\partial\lambda} S(v_\lambda) < 0$ *for* $\lambda \in (\lambda_0, +\infty)$,

(v) $\lambda \mapsto S(v_\lambda)$ *is concave on* $(\lambda_0, +\infty)$.

Proof. A simple calculation leads to

$$P(v_\lambda) = \lambda^2 \|\nabla v\|^2_{H^1(\mathbb{R}^N)} - \lambda^{\frac{N(p-1)}{2}} \frac{N(p-1)}{2(p+1)} \|v\|^{p+1}_{L^{p+1}(\mathbb{R}^N)}.$$

Recalling that $\frac{N(p-1)}{2} > 2$ (because of $p > 1 + \frac{4}{N}$), we infer that $P(v_\lambda) > 0$ for λ small enough. Thus, by continuity of P, there must exist $\lambda_0 \in (0, 1]$ such that $P(v_{\lambda_0}) = 0$. Hence (i) is proved. If $\lambda_0 = 1$, it is clear that $P(v) = 0$. Conversely, suppose that $P(v) = 0$. Then

$$\begin{aligned} P(v_\lambda) &= \lambda^2 P(v) + \left(\lambda^2 - \lambda^{\frac{N(p-1)}{2}}\right) \frac{N(p-1)}{2(p+1)} \|v\|^{p+1}_{L^{p+1}(\mathbb{R}^N)} \\ &= \left(\lambda^2 - \lambda^{\frac{N(p-1)}{2}}\right) \frac{N(p-1)}{2(p+1)} \|v\|^{p+1}_{L^{p+1}(\mathbb{R}^N)}, \end{aligned}$$

and, since $\frac{N(p-1)}{2} > 2$, this implies that $P(v_\lambda) > 0$ for all $\lambda \in (0, 1)$. Hence (ii) follows. From a simple calculation, we obtain

$$\begin{aligned} \frac{\partial}{\partial\lambda} S(v_\lambda) &= \lambda \|\nabla v\|^2_{H^1(\mathbb{R}^N)} - \lambda^{\frac{N(p-1)}{2} - 1} \frac{N(p-1)}{2(p+1)} \|v\|^{p+1}_{L^{p+1}(\mathbb{R}^N)} \\ &= \lambda^{-1} P(v_\lambda). \end{aligned}$$

Hence (iii) is shown. To see (iv), we remark that

$$\begin{aligned} P(v_\lambda) &= \lambda^2 \lambda_0^{-2} P(v_{\lambda_0}) + \left(\lambda^2 \lambda_0^{\frac{N(p-1)}{2} - 2} - \lambda^{\frac{N(p-1)}{2}}\right) \frac{N(p-1)}{2(p+1)} \|v\|^{p+1}_{L^{p+1}(\mathbb{R}^N)} \\ &= \lambda^2 \left(\lambda_0^{\frac{N(p-1)}{2} - 2} - \lambda^{\frac{N(p-1)}{2} - 2}\right) \frac{N(p-1)}{2(p+1)} \|v\|^{p+1}_{L^{p+1}(\mathbb{R}^N)}. \end{aligned}$$

Therefore, since $\frac{N(p-1)}{2}-2>0$, $P(v_\lambda)>0$ for $\lambda<\lambda_0$ and $P(v_\lambda)<0$ for $\lambda>\lambda_0$. Combined with (iii), this gives (iv). Finally,

$$\begin{aligned}\frac{\partial^2}{\partial\lambda^2}S(v_\lambda) &= \|\nabla v\|^2_{L^2(\mathbb{R}^N)}-\lambda^{\frac{N(p-1)}{2}-2}\left(\frac{N(p-1)}{2}-1\right)\frac{N(p-1)}{2(p+1)}\|v\|^{p+1}_{L^{p+1}(\mathbb{R}^N)}\\ &= \lambda^{-2}P(v_\lambda)-\lambda^{\frac{N(p-1)}{2}-2}\left(\frac{N(p-1)}{2}-2\right)\frac{N(p-1)}{2(p+1)}\|v\|^{p+1}_{L^{p+1}(\mathbb{R}^N)},\end{aligned}$$

which implies $\frac{\partial^2}{\partial\lambda^2}S(v_\lambda)<0$ for $\lambda>\lambda_0$ since $\frac{N(p-1)}{2}>2$. Hence (v) follows. □

The proof of Theorem 5.3 runs in three steps. First, we derive a new variational characterization for the ground states of (3.1). Then we use this characterization to define a set $\mathcal{I}$, invariant under the flow of (1.1), such that any initial datum in $\mathcal{I}$ gives rise to a blowing up solution of (1.1). Finally, we prove that the ground states can be approximated by sequences of elements of $\mathcal{I}$.

We consider the minimization problem

$$d_{\mathcal{M}}:=\inf_{v\in\mathcal{M}}S(v),$$

where the constraint $\mathcal{M}$ is given by

$$\mathcal{M}:=\{v\in H^1(\mathbb{R}^N)\setminus\{0\};P(v)=0,I(v)\leqslant 0\}.$$

Recall that I was defined in Proposition 3.12.

Lemma 5.6. *The following equality holds:*

$$m=d_{\mathcal{M}},$$

where m is the least energy level defined in (3.20).

Proof. Let $\varphi\in\mathcal{G}$. Because of Lemma 3.3, we have $I(\varphi)=P(\varphi)=0$. Thus it is clear that $\varphi\in\mathcal{M}$. Therefore, $S(\varphi)\geqslant d_{\mathcal{M}}$ and thus

$$m\geqslant d_{\mathcal{M}}. \tag{5.1}$$

Now, let $v\in\mathcal{M}$. If $I(v)=0$ then $S(v)\geqslant m$ by Proposition 3.12. Suppose that $I(v)<0$. We use the scaling $v_\lambda(\cdot):=\lambda^{N/2}v(\lambda\cdot)$. Then

$$I(v_\lambda)=\lambda^2\|\nabla v\|^2_{L^2(\mathbb{R}^N)}+\omega\|v\|^2_{L^2(\mathbb{R}^N)}-\lambda^{\frac{N(p-1)}{2}}\|v\|^{p+1}_{L^{p+1}(\mathbb{R}^N)}$$

implies $\lim_{\lambda\to 0}I(v_\lambda)=\omega\|v\|^2_{L^2(\mathbb{R}^N)}>0$. By continuity of I, there exists $\lambda_1<1$ such that $I(v_{\lambda_1})=0$. Therefore, by Proposition 3.12,

$$m\leqslant S(v_{\lambda_1}). \tag{5.2}$$

From $P(v)=0$ and Lemma 5.5, we deduce that

$$S(v_{\lambda_1})<S(v). \tag{5.3}$$

Combining (5.2) and (5.3) gives $m \leqslant S(v)$, hence

$$m \leqslant d_{\mathcal{M}}. \tag{5.4}$$

The conclusion follows from (5.1) and (5.4) □

Now, we define the set $\mathcal{I}$ by

$$\mathcal{I} := \{v \in H^1(\mathbb{R}^N); I(v) < 0, P(v) < 0, S(v) < m\}$$

and use Lemma 5.6 to prove that $\mathcal{I}$ is invariant under the flow of (1.1).

Lemma 5.7. *If* $u_0 \in \mathcal{I}$ *then the corresponding maximal solution* u *in* $(T_{\min}, T^{\max})$ *of* (1.1) *satisfies* $u(t) \in \mathcal{I}$ *for all* $t \in (T_{\min}, T^{\max})$.

Proof. Let $u_0 \in \mathcal{I}$ and u be the corresponding maximal solution of (1.1) in $(T_{\min}, T^{\max})$. Since S is a conserved quantity for (1.1), we find

$$S(u(t)) = S(u_0) < m \text{ for all } t \in (T_{\min}, T^{\max}). \tag{5.5}$$

The assertion is proved by contradiction. Suppose that there exists t such that

$$I(u(t)) \geqslant 0.$$

Then, by the continuity of I and u, there exists t_0 such that

$$I(u(t_0)) = 0.$$

By Proposition 3.12, this implies

$$S(u(t_0)) \geqslant m,$$

which is in contradiction with (5.5). Therefore,

$$I(u(t)) < 0 \text{ for all } t \in (T_{\min}, T^{\max}). \tag{5.6}$$

Finally, suppose that for some $t \in (T_{\min}, T^{\max})$

$$P(u(t)) \geqslant 0.$$

Still by continuity, there exists t_1 such that

$$P(u(t_1)) = 0.$$

From (5.6), we also have $I(u(t_1)) < 0$, and thus by Lemma 5.6

$$S(u(t_1)) \geqslant m,$$

which is another contradiction. Therefore,

$$P(u(t)) < 0 \text{ for all } t \in (T_{\min}, T^{\max}),$$

and this completes the proof. □

Lemma 5.8. *Let $u_0 \in \mathcal{I}$ and u be the corresponding maximal solution of* (1.1) *in* $(T_{\min}, T^{\max})$. *Then there exists $\delta > 0$ independent of t such that $P(u(t)) < -\delta$ for all $t \in (T_{\min}, T^{\max})$.*

Proof. Let $t \in (T_{\min}, T^{\max})$ and set $v := u(t)$ and $v_\lambda(\cdot) := \lambda^{N/2} v(\lambda \cdot)$. By Lemma 5.5, there exists $\lambda_0 < 1$ such that $P(v_{\lambda_0}) = 0$. If $I(v_{\lambda_0}) \leqslant 0$, we keep λ_0, otherwise if $I(v_{\lambda_0}) > 0$ there exists $\tilde{\lambda}_0 \in (\lambda_0, 1)$ such that $I(v_{\tilde{\lambda}_0}) = 0$ and we replace λ_0 by $\tilde{\lambda}_0$. In any case, by Proposition 3.12 or Lemma 5.6,

$$S(v_{\lambda_0}) \geqslant m. \tag{5.7}$$

Now, by (v) in Lemma 5.5, we get

$$S(v) - S(v_{\lambda_0}) \geqslant (1 - \lambda_0) \frac{\partial}{\partial \lambda} S(v_\lambda)_{|\lambda=1}. \tag{5.8}$$

From (iii) in Lemma 5.5, we obtain that

$$\frac{\partial}{\partial \lambda} S(v_\lambda)_{|\lambda=1} = P(v). \tag{5.9}$$

Furthermore, $P(v) < 0$ and $\lambda_0 \in (0, 1)$ implies

$$(1 - \lambda_0) P(v) > P(v). \tag{5.10}$$

Combining (5.7)–(5.10) gives

$$S(v) - m > P(v).$$

Let $-\delta := S(v) - m$. Then $\delta > 0$ since $v \in \mathcal{I}$, and δ is independent of t since S is a conserved quantity. In conclusion, for any $t \in (T_{\min}, T^{\max})$,

$$P(u(t)) < -\delta$$

and this ends the proof. □

Lemma 5.9. *Let $u_0 \in \mathcal{I}$ be such that $|x| u_0 \in L^2(\mathbb{R}^N)$. Then the corresponding maximal solution u of* (1.1) *in* $(T_{\min}, T^{\max})$ *blows up in finite time, i.e., $T_{\min} > -\infty$, $T^{\max} < +\infty$,*

$$\lim_{t \downarrow T_{\min}} \|u(t)\|_{H^1(\mathbb{R}^N)} = +\infty \textit{ and } \lim_{t \uparrow T^{\max}} \|u(t)\|_{H^1(\mathbb{R}^N)} = +\infty.$$

Proof. By Lemma 5.8, there exists $\delta > 0$ such that

$$P(u(t)) < -\delta \text{ for all } t \in (T_{\min}, T^{\max}).$$

Remembering from Proposition 2.4 that $\frac{\partial^2}{\partial t^2} \|xu(t)\|^2_{L^2(\mathbb{R}^N)} = 8P(u(t))$, we get by integrating twice in time

$$\|xu(t)\|^2_{L^2(\mathbb{R}^N)} \leqslant -4\delta t^2 + C(t + 1).$$

As in the proof of Proposition 2.3, this leads to a contradiction for large $|t|$. Therefore, $T_{\min} > -\infty$, $T^{\max} < +\infty$ and, by Proposition 2.1,

$$\lim_{t \downarrow T_{\min}} \|u(t)\|_{H^1(\mathbb{R}^N)} = +\infty \text{ and } \lim_{t \uparrow T^{\max}} \|u(t)\|_{H^1(\mathbb{R}^N)} = +\infty.$$

□

Proof of Theorem 5.3. In view of Lemma 5.9, it remains to find a sequence in $\mathcal{I}$ converging to φ in $H^1(\mathbb{R}^N)$. We define $\varphi_\lambda(\cdot) := \lambda^{N/2}\varphi(\lambda\cdot)$. Then, by Lemma 5.5,

$$I(\varphi_\lambda) < 0,\ P(\varphi_\lambda) < 0, S(\varphi_\lambda) < m,$$

thus $\varphi_\lambda \in \mathcal{I}$ for all $0 < \lambda < 1$. Furthermore, by Proposition 3.2, φ is exponentially decaying and so is φ_λ. Therefore, $|x|\varphi_\lambda \in L^2(\mathbb{R}^N)$ for all $0 < \lambda < 1$. It is clear that $\varphi_\lambda \to \varphi$ when $\lambda \to 1$. Because of Lemma 5.9, the solution of (1.1) corresponding to the initial datum φ_λ blows up in finite time for any $0 < \lambda < 1$. This completes the proof. □

Remark 5.10. When the nonlinearity is more general, it may be impossible to use the virial theorem to obtain a result of instability by blow-up. In such cases, a good alternative to prove instability would be to follow the method introduced by Shatah and Strauss in [69] and then developed further in [34, 35]. Other methods, based on modifications of the original idea of Shatah and Strauss, are also available, see for example [21, 22, 32, 45, 61]. Note that proving instability by these methods does not necessarily provide information on the long time behavior (blow-up or global existence) of solutions starting near a standing wave.

Remark 5.11. The type of method employed here to prove instability by blow-up of standing waves is not restricted to nonlinear Schrödinger equations. See, for example, [6, 39, 53, 62, 63, 68] and the references cited therein for results on the instability by blow-up for standing waves of nonlinear Klein–Gordon equations.

6 Appendix

The appendix is devoted to the proof of Proposition 4.10.

For φ being a solution of (3.1), we define a *tubular neighborhood* of φ of size $\varepsilon > 0$ in $H^1(\mathbb{R}^N)$ by

$$U_\varepsilon(\varphi) := \{v \in H^1(\mathbb{R}^N);\ \inf_{\theta\in\mathbb{R}, y\in\mathbb{R}^N} \|e^{i\theta}v(\cdot - y) - \varphi\|_{H^1(\mathbb{R}^N)} < \varepsilon\}.$$

Before proving Proposition 4.10, some preliminaries are needed.

Lemma 6.1. *Let φ be a solution of* (3.1). *For all $\delta > 0$ there exists $\varepsilon > 0$ such that*

$$\|e^{i\theta}\varphi(\cdot - y) - \varphi\|_{L^2(\mathbb{R}^N)} < \varepsilon, \quad \theta \in \mathbb{R},\ y \in \mathbb{R}^N,$$

implies $|(\theta, y)| < \delta$. Here, $|\cdot|$ denotes the standard norm in $(\mathbb{R}/2\pi\mathbb{Z}) \times \mathbb{R}^N$.

Proof. The proof is carried out by contradiction. Let $\delta > 0$ and assume that for all $n \in \mathbb{N}$ there exists (θ_n, y_n) such that

$$\|e^{i\theta_n}\varphi(\cdot - y_n) - \varphi\|_{L^2(\mathbb{R}^N)} < \frac{1}{n} \text{ and } |(\theta_n, y_n)| > \delta.$$

Possibly for a subsequence only, we have $e^{i\theta_n}\varphi(\cdot - y_n) \to \varphi$ almost everywhere as $n \to +\infty$, and in fact everywhere by regularity of φ (see Proposition 3.2). Suppose that (y_n) is unbounded. Then, possibly for a subsequence only, $|y_n| \to +\infty$, but by exponential decay of φ (see Proposition 3.2), this implies that for all $x \in \mathbb{R}^N$,

$$e^{i\theta_n}\varphi(x + y_n) \to 0,$$

which is a contradiction with $\varphi \not\equiv 0$. Therefore, (y_n) is bounded and converges, possibly passing to a subsequence, to some $y \in \mathbb{R}^N$. Since each θ_n can be chosen in $[0, 2\pi)$, we also have $\theta_n \to \theta$ for some $\theta \in [0, 2\pi)$. By hypothesis, we have

$$|(\theta, y)| \geqslant \delta, \tag{6.1}$$

but for all $x \in \mathbb{R}^N$

$$e^{i\theta}\varphi(x + y) = \varphi(x). \tag{6.2}$$

We claim that (6.2) can hold true only if $y = \theta = 0$. Indeed, let x_0 be such that $\varphi(x_0) \neq 0$. If $y \neq 0$ then $\lim_{n\to+\infty} e^{in\theta}\varphi(x_0 + ny) = \varphi(x_0) \neq 0$, what is in contradiction with the decay of φ at infinity. Therefore, $y = 0$ and this immediately implies $\theta = 0$. This is in contradiction with (6.1) and finishes the proof. □

Lemma 6.2. *Let φ be a solution of* (3.1). *There exist $\varepsilon > 0$ and two functions $\sigma : U_\varepsilon(\varphi) \to \mathbb{R}$ and $Y : U_\varepsilon(\varphi) \to \mathbb{R}^N$ such that for all $v \in U_\varepsilon(\varphi)$*

$$\|e^{i\sigma(v)}v(\cdot - Y(v)) - \varphi\|_{L^2(\mathbb{R}^N)} = \inf_{\theta\in\mathbb{R}, y\in\mathbb{R}^N} \|e^{i\theta}v(\cdot - y) - \varphi\|_{L^2(\mathbb{R}^N)}.$$

Furthermore, the function $w := e^{i\sigma(v)}v(\cdot - Y(v))$ satisfies

$$(w, i\varphi)_2 = \left(w, \frac{\partial\varphi}{\partial x_j}\right)_2 = 0 \textit{ for all } j = 1, \dots, N. \tag{6.3}$$

Proof. For the sake of simplicity, we assume that φ is real-valued and radial. Let $\Phi : \mathbb{R} \times \mathbb{R}^N \times H^1(\mathbb{R}^N) \to \mathbb{R}$ be defined by

$$\Phi(\theta, y, v) := \frac{1}{2}\|e^{i\theta}v(\cdot - y) - \varphi\|^2_{L^2(\mathbb{R}^N)} = \frac{1}{2}\|v - e^{-i\theta}\varphi(\cdot + y)\|^2_{L^2(\mathbb{R}^N)}$$

and let $F : \mathbb{R} \times \mathbb{R}^N \times H^1(\mathbb{R}^N) \to \mathbb{R}^{N+1}$ be the derivative of Φ with respect to (θ, y):

$$F(\theta, y, v) := D_{\theta,y}\Phi(\theta, y, v) = \begin{pmatrix} (e^{i\theta}v(\cdot - y), i\varphi)_2 \\ -\left(e^{i\theta}v(\cdot - y), \frac{\partial\varphi}{\partial x_1}\right)_2 \\ \vdots \\ -\left(e^{i\theta}v(\cdot - y), \frac{\partial\varphi}{\partial x_N}\right)_2 \end{pmatrix}.$$

We have $F(0,0,\varphi)=0$, and the derivative of F is given by the diagonal matrix

$$D_{\theta,y}F(0,0,\varphi)=\begin{pmatrix} \|\varphi\|^2_{L^2(\mathbb{R}^N)} & & & 0 \\ & \|\frac{\partial\varphi}{\partial x_1}\|^2_{L^2(\mathbb{R}^N)} & & \\ & & \ddots & \\ 0 & & & \|\frac{\partial\varphi}{\partial x_N}\|^2_{L^2(\mathbb{R}^N)} \end{pmatrix}.$$

Since φ is a solution of (3.1), φ is not zero and, therefore, we have $\|\varphi\|^2_{L^2(\mathbb{R}^N)}>0$ and $\|\frac{\partial\varphi}{\partial x_j}\|^2_{L^2(\mathbb{R}^N)}>0$ for all $j=1,\dots,N$. Consequently, the matrix $D_{\theta,y}F(0,0,\varphi)$ is positive definite and by the implicit function theorem there exists $\varepsilon>0$, $\delta>0$ and $(\sigma,Y):V_\varepsilon\to\Omega$, where

$$V_\varepsilon:=\{v\in H^1(\mathbb{R}^N);\|v-\varphi\|_{H^1(\mathbb{R}^N)}<\varepsilon\}\text{ and }\Omega:=\{(\theta,y)\in\mathbb{R}\times\mathbb{R}^N;|(\theta,y)|<\delta\},$$

such that for all $v\in V_\varepsilon$

$$F(\sigma(v),Y(v),v)=0 \tag{6.4}$$

and

$$\|e^{i\sigma(v)}v(\cdot-Y(v))-\varphi\|_{L^2(\mathbb{R}^N)}=\inf_{(\theta,y)\in\Omega}\|e^{i\theta}v(\cdot-y)-\varphi\|_{L^2(\mathbb{R}^N)}. \tag{6.5}$$

Now we extend (6.5) to all $(\theta,y)\in\mathbb{R}\times\mathbb{R}^N$. Assume that there exists $(\tilde\theta,\tilde y)\in\mathbb{R}\times\mathbb{R}^N$ such that $|(\tilde\theta,\tilde y)|>\delta$ and

$$\|e^{i\tilde\theta}v(\cdot-\tilde y)-\varphi\|_{L^2(\mathbb{R}^N)}<\|e^{i\sigma(v)}v(\cdot-Y(v))-\varphi\|_{L^2(\mathbb{R}^N)}.$$

For $v\in V_\varepsilon$, we have

$$\|e^{i\tilde\theta}\varphi(\cdot-\tilde y)-\varphi\|_{L^2(\mathbb{R}^N)}<\|e^{i\tilde\theta}(\varphi(\cdot-\tilde y)-v(\cdot-\tilde y))\|_{L^2(\mathbb{R}^N)}+\|e^{i\tilde\theta}v(\cdot-\tilde y)-\varphi\|_{L^2(\mathbb{R}^N)}<2\varepsilon.$$

If ε is small enough, this implies by Lemma 6.1 that $|(\tilde\theta,\tilde y)|<\delta$, which is a contradiction. Therefore, we have, for all $v\in V_\varepsilon$,

$$\|e^{i\sigma(v)}v(\cdot-Y(v))-\varphi\|_{L^2(\mathbb{R}^N)}=\inf_{(\theta,y)\in\mathbb{R}\times\mathbb{R}^N}\|e^{i\theta}v(\cdot-y)-\varphi\|_{L^2(\mathbb{R}^N)}.$$

It remains to extend (σ,Y) to $U_\varepsilon(\varphi)$. Let $v\in U_\varepsilon(\varphi)$. Then there exists $(\theta^\star,y^\star)$ such that

$$\|e^{i\theta^\star}v(\cdot-y^\star)-\varphi\|_{L^2(\mathbb{R}^N)}<\varepsilon.$$

Define

$$\sigma(v):=\sigma(e^{i\theta^\star}v(\cdot-y^\star))+\theta^\star$$

and

$$Y(v):=Y(e^{i\theta^\star}v(\cdot-y^\star))+y^\star.$$

Then it is not hard to see that this definition is independent of the choice of $(\theta^\star,y^\star)$ and allows us to extend (σ,Y) to $U_\varepsilon(\varphi)$. The orthogonality relations in (6.3) follow from the relation (6.4). □

Recall that Q was defined in (2.1) and E in (2.2).

Lemma 6.3. *Under the hypothesis of Proposition 4.10, there exist $\varepsilon > 0$ and $C > 0$ such that for all $v \in U_\varepsilon(\varphi)$ satisfying $Q(v) = Q(\varphi)$, we have*

$$E(v) - E(\varphi) \geqslant C \inf_{(\theta,y)\in\mathbb{R}\times\mathbb{R}^N} \|e^{i\theta}v(\cdot - y) - \varphi\|^2_{H^1(\mathbb{R}^N)}.$$

Proof. For the sake of simplicity, we assume that φ is radial. Let $\varepsilon > 0$ and $v \in U_\varepsilon(\varphi)$. For ε small enough, let (σ, Y) be as in Lemma 6.2 and define

$$w := e^{i\sigma(v)}v(\cdot - Y(v)). \tag{6.6}$$

Then by Lemma 6.2 the function w satisfies

$$(w, i\varphi)_2 = \left(w, \frac{\partial\varphi}{\partial x_j}\right)_2 = 0 \text{ for all } j = 1, \dots, N. \tag{6.7}$$

Let $\lambda \in \mathbb{R}$ and $z \in H^1(\mathbb{R}^N)$ be such that

$$(z, \varphi)_2 = 0 \tag{6.8}$$

and

$$w - \varphi = \lambda\varphi + z. \tag{6.9}$$

Since φ is radial up to translations, we have

$$\left(\varphi, \frac{\partial\varphi}{\partial x_j}\right)_2 = 0 \text{ for } j = 1, \dots, N. \tag{6.10}$$

Combining (6.7), (6.9) and (6.10) we get

$$\left(z, \frac{\partial\varphi}{\partial x_j}\right)_2 = 0 \text{ for } j = 1, \dots, N. \tag{6.11}$$

Moreover, we have

$$(\varphi, i\varphi)_2 = \mathrm{Re}(-i\|\varphi\|^2_{L^2(\mathbb{R}^N)}) = 0,$$

and therefore

$$(z, i\varphi)_2 = 0. \tag{6.12}$$

From (6.8), (6.11) and (6.12), we see that z satisfies the orthogonality conditions in (4.10) and therefore

$$\langle S''(\varphi)z, z\rangle \geqslant \delta\|z\|^2_{H^1(\mathbb{R}^N)}. \tag{6.13}$$

By a Taylor expansion, we obtain

$$Q(\varphi) = Q(v) = Q(w) = Q(\varphi) + \langle Q'(\varphi), w - \varphi\rangle + O(\|w - \varphi\|^2_{H^1(\mathbb{R}^N)}).$$

But $Q'(\varphi) = \varphi$ and therefore

$$\langle Q'(\varphi), w - \varphi\rangle = (\varphi, w - \varphi)_2 = (\varphi, \lambda\varphi + z)_2 = \lambda\|\varphi\|^2_{L^2(\mathbb{R}^N)},$$

where the last equality follows from (6.8). Consequently,

$$\lambda = O(\|w-\varphi\|^2_{H^1(\mathbb{R}^N)}). \tag{6.14}$$

Now, another Taylor expansion gives

$$\begin{aligned} S(v)-S(\varphi) &= S(w)-S(\varphi) \\ &= \langle S'(\varphi), w-\varphi\rangle + \tfrac{1}{2}\langle S''(\varphi)(w-\varphi), w-\varphi\rangle + o(\|w-\varphi\|^2_{H^1(\mathbb{R}^N)}). \end{aligned} \tag{6.15}$$

Since φ is a solution of (3.1), we have $S'(\varphi)=0$ and therefore

$$\langle S'(\varphi), w-\varphi\rangle = 0. \tag{6.16}$$

Furthermore, from (6.9) we get

$$\langle S''(\varphi)(w-\varphi), w-\varphi\rangle = \lambda^2\langle S''(\varphi)\varphi,\varphi\rangle + 2\lambda \mathrm{Re}\,\langle S''(\varphi)\varphi, z\rangle + \langle S''(\varphi)z,z\rangle. \tag{6.17}$$

From (6.14) we have

$$\lambda^2\langle S''(\varphi)\varphi,\varphi\rangle = o(\|w-\varphi\|^2_{H^1(\mathbb{R}^N)}). \tag{6.18}$$

Since

$$\|z\|^2_{H^1(\mathbb{R}^N)} \leqslant 2\|w-\varphi\|^2_{H^1(\mathbb{R}^N)} + 2\lambda^2\|\varphi\|^2_{H^1(\mathbb{R}^N)},$$

we have by (6.14) that

$$\|z\|^2_{H^1(\mathbb{R}^N)} = O(\|w-\varphi\|^2_{H^1(\mathbb{R}^N)}), \tag{6.19}$$

and therefore

$$2\lambda \mathrm{Re}\,\langle S''(\varphi)\varphi, z\rangle = o(\|w-\varphi\|^2_{H^1(\mathbb{R}^N)}). \tag{6.20}$$

Combining (6.17), (6.18) and (6.20), we get

$$\langle S''(\varphi)(w-\varphi), w-\varphi\rangle = \langle S''(\varphi)z,z\rangle + o(\|w-\varphi\|^2_{H^1(\mathbb{R}^N)}).$$

Together with (6.15)–(6.16), this gives

$$S(v)-S(\varphi) \geqslant \frac{1}{2}\langle S''(\varphi)z,z\rangle + o(\|w-\varphi\|^2_{H^1(\mathbb{R}^N)}). \tag{6.21}$$

But since $Q(v)=Q(\varphi)$, we have

$$E(v)-E(\varphi) = S(v)-S(\varphi)$$

and with (6.13), (6.19) and (6.21), we obtain

$$E(v)-E(\varphi) \geqslant \frac{\delta}{2}\|w-\varphi\|^2_{H^1(\mathbb{R}^N)} + o(\|w-\varphi\|^2_{H^1(\mathbb{R}^N)}).$$

Therefore, we have for ε small enough

$$E(v)-E(\varphi) \geqslant \frac{\delta}{4}\|w-\varphi\|^2_{H^1(\mathbb{R}^N)}.$$

Setting $C := \delta/4$ and remembering how w was defined in (6.6) gives

$$E(v) - E(\varphi) \geqslant C \inf_{(\theta,y)\in\mathbb{R}\times\mathbb{R}^N} \|e^{i\theta}v(\cdot - y) - \varphi\|^2_{H^1(\mathbb{R}^N)},$$

which finishes the proof. □

Proof of Proposition 4.10. The assertion is proved by contradiction. Assume that there exists $u_{n,0}$ and $\varepsilon > 0$ such that

$$\|u_{n,0} - \varphi\|_{H^1(\mathbb{R}^N)} \to 0 \text{ as } n \to +\infty \tag{6.22}$$

but for all $n \in \mathbb{N}$

$$\sup_{t\in(T^n_{\min},T^{\max}_n)} \inf_{(\theta,y)\in\mathbb{R}\times\mathbb{R}^N} \|e^{i\theta}u_n(t, \cdot - y) - \varphi\|_{H^1(\mathbb{R}^N)} > \varepsilon$$

where u_n is the maximal solution of (1.1) in $(T^n_{\min}, T^{\max}_n)$ corresponding to $u_{0,n}$. By continuity, we can pick up the first time t_n such that

$$\inf_{(\theta,y)\in\mathbb{R}\times\mathbb{R}^N} \|e^{i\theta}u_n(t_n, \cdot - y) - \varphi\|_{H^1(\mathbb{R}^N)} = \varepsilon$$

In view of (6.22) and the conservation of charge and energy (see (2.3)), it is clear that

$$\begin{matrix} E(u_n(t_n)) = E(u_{n,0}) \to E(\varphi) \\ Q(u_n(t_n)) = Q(u_{n,0}) \to Q(\varphi) \end{matrix} \quad \text{as } n \to +\infty.$$

Let $v_n := \frac{u_n(t_n)}{\|u_n(t_n)\|_{L^2(\mathbb{R}^N)}}\|\varphi\|_{L^2(\mathbb{R}^N)}$. Then

$$Q(v_n) = Q(\varphi), E(v_n) \to E(\varphi) \text{ and } \|v_n - u_n(t_n)\|_{H^1(\mathbb{R}^N)} \to 0.$$

Choosing ε sufficiently small, we can apply Lemma 6.3 to get

$$\inf_{(\theta,y)\in\mathbb{R}\times\mathbb{R}^N} \|e^{i\theta}v_n(\cdot - y) - \varphi\|^2_{H^1(\mathbb{R}^N)} \leqslant C(E(v_n) - E(\varphi)) \to 0.$$

On the other hand, we have

$$\begin{aligned} \varepsilon &= \inf_{(\theta,y)\in\mathbb{R}\times\mathbb{R}^N} \|e^{i\theta}u_n(t_n, \cdot - y) - \varphi\|_{H^1(\mathbb{R}^N)} \\ &\leqslant \inf_{(\theta,y)\in\mathbb{R}\times\mathbb{R}^N} \|e^{i\theta}v_n(\cdot - y) - \varphi\|_{H^1(\mathbb{R}^N)} + \|u_n - v_n\|_{H^1(\mathbb{R}^N)}, \end{aligned}$$

which yields a contradiction for n large. □

Acknowledgments. The material presented in these notes is based on two lecture series given at the TU Berlin and at SISSA in 2008. The author would like to thank these two institutions for their hospitality. He is also grateful to Antonio Ambrosetti, Etienne Emmrich, and Petra Wittbold for giving him the opportunity to teach these courses. Finally, he wishes to thank Louis Jeanjean for helpful discussions regarding Section 3 and useful comments on a preliminary version of these notes, and Petra Wittbold and Etienne Emmrich for their careful reading of the manuscript and valuable suggestions.

References

[1] R. A. Adams, *Sobolev spaces*, Academic Press, New York–London, 1975.

[2] A. Ambrosetti and A. Malchiodi, *Perturbation methods and semilinear elliptic problems on* $\mathbb{R}^n$, Birkhäuser, Basel, 2006.

[3] ———, *Nonlinear analysis and semilinear elliptic problems*, Cambridge University Press, Cambridge, 2007.

[4] A. Ambrosetti and P. H. Rabinowitz, Dual variational methods in critical point theory and applications, *J. Funct. Anal.* **14** (1973), pp. 349–381.

[5] T. B. Benjamin, The stability of solitary waves, *Proc. Roy. Soc. London Ser. A* **328** (1972), pp. 153–183.

[6] H. Berestycki and T. Cazenave, Instabilité des états stationnaires dans les équations de Schrödinger et de Klein–Gordon non linéaires, *C. R. Acad. Sci. Paris* **293** (1981), pp. 489–492.

[7] ———, Instabilité des états stationnaires dans les équations de Schrödinger et de Klein–Gordon non linéaires, *Pub. Lab. Analyse Num. Université Paris VI* (1981).

[8] H. Berestycki, T. Gallouët and O. Kavian, Équations de champs scalaires euclidiens non linéaires dans le plan, *C. R. Acad. Sci. Paris* **297** (1983), pp. 307–310.

[9] H. Berestycki and P.-L. Lions, Nonlinear scalar field equations I, *Arch. Ration. Mech. Anal.* **82** (1983), pp. 313–346.

[10] ———, Nonlinear scalar field equations II, *Arch. Ration. Mech. Anal.* **82** (1983), pp. 347–375.

[11] F. A. Berezin and M. A. Shubin, *The Schrödinger equation*, Kluwer, Dordrecht, 1991.

[12] J. Byeon, L. Jeanjean and M. Mariş, Symmetry and monotonicity of least energy solutions, `arXiv:0806.0299`.

[13] T. Cazenave, Stable solutions of the logarithmic Schrödinger equation, *Nonlinear Anal.* **7** (1983), pp. 1127–1140.

[14] T. Cazenave, *Semilinear Schrödinger equations*, Courant Lecture Notes in Mathematics 10, American Mathematical Society, New York, 2003.

[15] T. Cazenave and P.-L. Lions, Orbital stability of standing waves for some nonlinear Schrödinger equations, *Comm. Math. Phys.* **85** (1982), pp. 549–561.

[16] S. Cingolani, L. Jeanjean and S. Secchi, Multi-peak solutions for magnetic NLS equations without non-degeneracy conditions, *ESAIM Control Optim. Calc. Var.*, to appear.

[17] S. Coleman, V. Glaser and A. Martin, Action minima among solutions to a class of Euclidean scalar field equations, *Comm. Math. Phys.* **58** (1978), pp. 211–221.

[18] T. Dauxois and M. Peyrard, *Physics of solitons*, Cambridge University Press, Cambridge, 2006.

[19] A. de Bouard and R. Fukuizumi, Stability of standing waves for nonlinear Schrödinger equations with inhomogeneous nonlinearities, *Ann. Henri Poincaré* **6** (2005), pp. 1157–1177.

[20] P. G. Drazin and R. S. Johnson, *Solitons: an introduction*, Cambridge University Press, Cambridge, 1989.

[21] R. Fukuizumi, Stability and instability of standing waves for the nonlinear Schrödinger equation with harmonic potential, *Discrete Contin. Dyn. Sys.* **7** (2001), pp. 525–544.

[22] R. Fukuizumi and M. Ohta, Instability of standing waves for nonlinear Schrödinger equations with potentials, *Differential Integral Equations* **16** (2003), pp. 691–706.

[23] ———, Stability of standing waves for nonlinear Schrödinger equations with potentials, *Differential Integral Equations* **16** (2003), pp. 111–128.

[24] R. Fukuizumi and M. Ohta, Instability of standing waves for nonlinear Schrödinger equations with inhomogeneous nonlinearities, *J. Math. Kyoto Univ.* **45** (2005), pp. 145–158.

[25] T. Gallay and M. Hărăguş, Orbital stability of periodic waves for the nonlinear Schrödinger equation, *J. Dynam. Differential Equations* **19** (2007), pp. 825–865.

[26] ______, Stability of small periodic waves for the nonlinear Schrödinger equation, *J. Differential Equations* **234** (2007), pp. 544–581.

[27] F. Genoud, Existence and orbital stability of standing waves for some nonlinear Schrödinger equations, perturbation of a model case, *J. Differential Equations* **246** (2009), pp. 1921–1943.

[28] F. Genoud and C. A. Stuart, Schrödinger equations with a spatially decaying nonlinearity: existence and stability of standing waves, *Discrete Contin. Dyn. Syst.* **21** (2008), pp. 137–186.

[29] B. Gidas, W. M. Ni and L. Nirenberg, Symmetry and related properties via the maximum principle, *Comm. Math. Phys.* **68** (1979), pp. 209–243.

[30] D. Gilbarg and N. S. Trudinger, *Elliptic partial differential equations of second order*, Springer, Berlin, 2001.

[31] R. T. Glassey, On the blowing up of solutions to the Cauchy problem for nonlinear Schrödinger equations, *J. Math. Phys.* **18** (1977), pp. 1794–1797.

[32] J. M. Gonçalves-Ribeiro, Instability of symmetric stationary states for some nonlinear Schrödinger equations with an external magnetic field, *Ann. Inst. H. Poincaré Phys. Théor.* **54** (1991), pp. 403–433.

[33] M. Grillakis, Linearized instability for nonlinear Schrödinger and Klein–Gordon equations, *Comm. Pure Appl. Math.* **41** (1988), pp. 747–774.

[34] M. Grillakis, J. Shatah and W. A. Strauss, Stability theory of solitary waves in the presence of symmetry I, *J. Funct. Anal.* **74** (1987), pp. 160–197.

[35] ______, Stability theory of solitary waves in the presence of symmetry II, *J. Funct. Anal.* **94** (1990), pp. 308–348.

[36] H. Hajaiej and C. A. Stuart, On the variational approach to the stability of standing waves for the nonlinear Schrödinger equation, *Adv. Nonlinear Stud.* **4** (2004), pp. 469–501.

[37] P. Hartman, *Ordinary differential equations*, SIAM, Philadelphia, 2002.

[38] L. Jeanjean, On the existence of bounded Palais-Smale sequences and application to a Landesman-Lazer-type problem set on $\mathbb{R}^N$, *Proc. Roy. Soc. Edinburgh Sect. A* **129** (1999), pp. 787–809.

[39] L. Jeanjean and S. Le Coz, Instability for standing waves of nonlinear Klein–Gordon equations via mountain-pass arguments, *Trans. Amer. Math. Soc.*, to appear.

[40] ______, An existence and stability result for standing waves of nonlinear Schrödinger equations, *Adv. Differential Equations* **11** (2006), pp. 813–840.

[41] L. Jeanjean and K. Tanaka, A note on a mountain pass characterization of least energy solutions, *Adv. Nonlinear Stud.* **3** (2003), pp. 445–455.

[42] ______, A remark on least energy solutions in $\mathbb{R}^N$, *Proc. Amer. Math. Soc.* **131** (2003), pp. 2399–2408.

[43] T. Kato, *Perturbation theory for linear operators*, Springer, Berlin – New York, 1976.

[44] H. Kikuchi, Existence and stability of standing waves for Schrödinger–Poisson–Slater equation, *Adv. Nonlinear Stud.* **7** (2007), pp. 403–437.

[45] H. Kikuchi and M. Ohta, Instability of standing waves for the Klein–Gordon–Schrödinger system, *Hokkaido Mathematical Journal* **37** (2008), pp. 735–748.

[46] ———, *Stability of standing waves for Klein–Gordon–Schrödinger equations and nonlinear Schrödinger equation with Yukawa potential*, Department of Mathematics, Saitama University, 2008, preprint.

[47] M. K. Kwong, Uniqueness of positive solutions of $\Delta u - u + u^p = 0$ in $\mathbb{R}^n$, *Arch. Ration. Mech. Anal.* **105** (1989), pp. 243–266.

[48] S. Le Coz, A note on Berestycki–Cazenave's classical instability result for nonlinear Schrödinger equations, *Adv. Nonlinear Stud.* **8** (2008), pp. 455–463.

[49] S. Le Coz, R. Fukuizumi, G. Fibich, B. Ksherim and Y. Sivan, Instability of bound states of a nonlinear Schrödinger equation with a Dirac potential, *Phys. D* **237** (2008), pp. 1103–1128.

[50] P.-L. Lions, The concentration-compactness principle in the calculus of variations. The locally compact case. I, *Ann. Inst. H. Poincaré Anal. Non Linéaire* **1** (1984), pp. 109–145.

[51] ———, The concentration-compactness principle in the calculus of variations. The locally compact case. II, *Ann. Inst. H. Poincaré Anal. Non Linéaire* **1** (1984), pp. 223–283.

[52] ———, Solutions complexes d'équations elliptiques semilinéaires dans $\mathbf{R}^N$, *C. R. Acad. Sci. Paris Sér. I Math.* **302** (1986), pp. 673–676.

[53] Y. Liu, M. Ohta and G. Todorova, Strong instability of solitary waves for nonlinear Klein–Gordon equations and generalized Boussinesq equations, *Ann. Inst. H. Poincaré Anal. Non Linéaire* **24** (2007), pp. 539–548.

[54] O. Lopes, Radial symmetry of minimizers for some translation and rotation invariant functionals, *J. Differential Equations* **124** (1996), pp. 378–388.

[55] M. Mariş, On the symmetry of minimizers, *Arch. Ration. Mech. Anal.*, to appear.

[56] J. B. McLeod, C. A. Stuart and W. C. Troy, Stability of standing waves for some nonlinear Schrödinger equations, *Differential and Integral Equations* **16** (2003), pp. 1025–1038.

[57] F. Merle and P. Raphael, On universality of blow-up profile for L^2 critical nonlinear Schrödinger equation, *Invent. Math.* **156** (2004), pp. 565–672.

[58] ———, The blow-up dynamic and upper bound on the blow-up rate for critical nonlinear Schrödinger equation, *Ann. of Math. (2)* **161** (2005), pp. 157–222.

[59] ———, On one blow up point solutions to the critical nonlinear Schrödinger equation, *J. Hyperbolic Differ. Equ.* **2** (2005), pp. 919–962.

[60] ———, On a sharp lower bound on the blow-up rate for the L^2 critical nonlinear Schrödinger equation, *J. Amer. Math. Soc.* **19** (2006), pp. 37–90.

[61] M. Ohta, Instability of standing waves for the generalized Davey–Stewartson system, *Ann. Inst. H. Poincaré Phys. Théor.* **62** (1995), pp. 69–80.

[62] M. Ohta and G. Todorova, Strong instability of standing waves for nonlinear Klein–Gordon equations, *Discrete Contin. Dyn. Syst.* **12** (2005), pp. 315–322.

[63] ———, Strong instability of standing waves for the nonlinear Klein–Gordon equation and the Klein–Gordon–Zakharov system, *SIAM J. Math. Anal.* **38** (2007), pp. 1912–1931.

[64] S. I. Pohožaev, On the eigenfunctions of the equation $\Delta u + \lambda f(u) = 0$, *Soviet Math. Dokl.* **6** (1965), pp. 1408–1411.

[65] P. H. Rabinowitz, On a class of nonlinear Schrödinger equations, *Z. Angew. Math. Phys.* **43** (1992), pp. 270–291.

[66] H. A. Rose and M. I. Weinstein, On the bound states of the nonlinear Schrödinger equation with a linear potential, *Phys. D* **30** (1988), pp. 207–218.

[67] J. S. Russell, Report on waves, *Report of the fourteenth meeting of the British Association for the Advancement of Science, York* (1844), pp. 311–390.

[68] J. Shatah, Unstable ground state of nonlinear Klein–Gordon equations, *Trans. Amer. Math. Soc.* **290** (1985), pp. 701–710.

[69] J. Shatah and W. A. Strauss, Instability of nonlinear bound states, *Comm. Math. Phys.* **100** (1985), pp. 173–190.

[70] W. A. Strauss, Existence of solitary waves in higher dimensions, *Comm. Math. Phys.* **55** (1977), pp. 149–162.

[71] M. Struwe, *Variational methods*, Springer, Berlin, 2000.

[72] C. A. Stuart, Lectures on the orbital stability of standing waves and application to the nonlinear Schrödinger equation, *Milan J. Math.*, to appear.

[73] C. Sulem and P.-L. Sulem, *The nonlinear Schrödinger equation*, Springer, New York, 1999.

[74] G. Todorova, Dynamics of non-linear wave equations, *Math. Methods Appl. Sci.* **27** (2004), pp. 1831–1841.

[75] M. I. Weinstein, Nonlinear Schrödinger equations and sharp interpolation estimates, *Comm. Math. Phys.* **87** (1983), pp. 567–576.

[76] ______, Modulational stability of ground states of nonlinear Schrödinger equations, *SIAM J. Math. Anal.* **16** (1985), pp. 472–491.

[77] M. I. Weinstein, Lyapunov stability of ground states of nonlinear dispersive evolution equations, *Comm. Pure Appl. Math.* **39** (1986), pp. 51–67.

[78] M. Willem, *Minimax theorems*, Birkhäuser, Boston, 1996.

Author information

Stefan Le Coz, SISSA, via Beirut 2–4, 34014 Trieste, Italy.
E-mail: `lecoz@sissa.it`

Multiscale methods coupling atomistic and continuum mechanics: some examples of mathematical analysis

Frédéric Legoll

Abstract. Many numerical methods coupling a discrete description of matter with a continuum one have been recently proposed in the mechanics literature. This contribution aims at introducing this field in a mathematical perspective, at an elementary level. We first review some modelling questions. We next carry on a mathematical analysis on a toy example of an atomistic to continuum coupling method. This example presents the same coupling features as some more advanced methods. On the other hand, its simplicity makes its analysis easier. Specific difficulties linked with the coupling of models at different scales are highlighted. We conclude these notes by reviewing some general questions frequently debated in the field, pointing out the main directions that have been followed at the front of research, and discussing some open problems.

Keywords. Atomistic to continuum coupling, error estimation, multiscale models, variational problems, QuasiContinuum method, materials science.

AMS classification. 65K10, 65N15, 65Z05, 70-08, 70C20, 70G75, 74G70, 74Qxx.

1 Introduction

Many numerical methods coupling a discrete description of matter with a continuum one have been proposed recently in the mechanics literature. Their developments stem from the need to go beyond standard continuum mechanics models in materials science, while still giving rise to computationally tractable methods.

This contribution, primarily written for applied mathematics students, aims at introducing this field in a mathematical perspective and pointing out the main directions that have been followed. We first review some modelling questions, focusing on the key elements. Our aim here is not to describe the most up-to-date (or most accurate) models, but to give some *qualitative* understanding of the models we manipulate. We next carry on a mathematical analysis on a toy example of such an atomistic to continuum coupling method. This example presents the same coupling features as some more advanced methods. On the other hand, its simplicity makes its analysis easier. Specific difficulties linked with the coupling of models at different scales will be highlighted along the text.

We conclude these notes by reviewing some general questions frequently debated in the field, about modelling, setting of the problem and spurious effects at the interface between two models written at different scales. We also briefly review the different methods proposed up to now (based on a coupling between discrete and continuum models or on some alternatives). We finally mention some open questions on how to

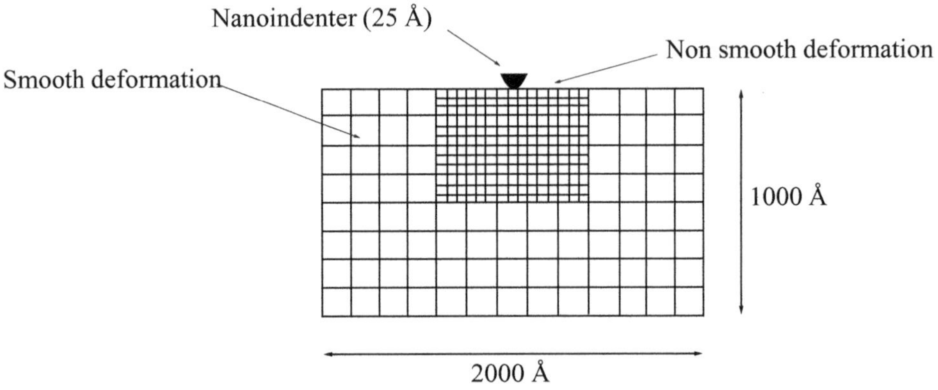

Figure 1. Schematic representation of a nanoindentation experiment: close to the stiff indenter, one expects a non-smooth deformation of the soft material, hence the need of a fine model. Further away, the deformation is smooth, and a macroscopic model, discretized on a coarse mesh (here quadrangles) provides a good enough accuracy.

take into account temperature and dynamical effects. We hope that this last section, which addresses questions at the front of research, is of interest not only for students, but also for experts in the field.

In this contribution, we only consider multiscale methods for materials science (e.g., solid mechanics). However, a lot of other applied fields have also witnessed a significant development of similar methods. See [91] for some examples of multiscale systems, in fluids mechanics, computational chemistry, ..., and the recent numerical methods to handle them. On the other hand, see [29] for a comprehensive review of mathematical results on atomistic to continuum limits for crystalline materials.

The traditional framework in solid mechanics is the continuum description. State variables are the displacement field, its gradient, the stress field, and possibly the plastic deformation field. However, there are situations for which such a model is not appropriate. A first example is nanoindentation, which consists of inserting a stiff indenter (made of a hard material, and which is often considered as non-deformable) into a piece of soft material, such as aluminium (see Figure 1). One is often interested in the force f on the indenter as a function of its depth d. When the depth is small enough, the response of the material is smooth (this is the regime of linear elasticity). However, when the depth is increased, the response becomes highly nonlinear, and eventually a singularity appears in the curve $d \mapsto f$. This corresponds to the appearance of a defect in the atomic lattice of the soft material, such as a dislocation (see [40, 79, 97, 157] for comparison between experimental results and numerical simulations of nanoindentation). Such defects cannot be described by a continuum model.

Another situation when nanoscale localized phenomena appear is crack propagation: in some materials, the crack moves because atomic bonds break [112] (see also [15]). In addition, on the fracture lips, there is some reorganization of the lattice (called surface relaxation), because of the presence of a free surface. Again, these phenomena cannot be described at the continuum scale[1].

An important literature exists on the modelling of crack propagation (and dislocations) by a continuum model. However, these theories are generally based on phenomenological arguments (for instance, to decide whether a crack appears [95], on the direction of its propagation, on how much it propagates, ...) and criteria (e.g., Griffith or Barenblatt criteria). They are also unable to describe the precise structure of the defects (the crack tip, or the dislocation core).

In all these situations, the atomistic nature of matter cannot be ignored. An appropriate model to describe the localized phenomena is an atomistic model, in which the solid is considered as a set of discrete point particles interacting through given interatomic potentials. In such a model, state variables are the positions (and momenta) of all the particles, hence the possibility to describe complex deformations such as surface relaxation, dislocation cores, ...

However, the size of materials that can be simulated by only using an atomistic model is very small in comparison to the size of the materials one is interested in. Let us first recall that the mean distance between two atoms is of the order of 10^{-10} m. Hence, there are approximately 10^{21} atoms in a volume of 1 mm^3. Simulating such a system with an atomistic model is hence computationally out of reach. This precise and expensive model can only be used in small pieces of the material. On the other hand, for some phenomena we have mentioned above, it is not possible to make accurate computations by just considering a small piece of material, because large scale or bulk effects have to be accounted for (or boundary conditions should not affect the phenomenon under study)[2].

Fortunately, it is often the case that the deformation is smooth in the main part of the solid. So, a natural idea is to try to take advantage of both models, the continuum mechanics one and the atomistic one, and to couple them. There are (at least) three central issues one should be aware of: consistency, adaptivity, and coupling conditions. Let us briefly review them.

Consistency means that the atomistic and the continuum models must be somewhat related. Indeed, for loading conditions for which a completely atomistic model leads to a smooth deformation, the continuum model should lead to a nearby solution. A related question is to build continuum models on the basis of atomistic models.

Let us now turn to adaptivity and coupling conditions. Many multiscale methods (but not all) are based on a domain decomposition paradigm: the fine scale model is used in the subdomain where non-smooth deformations are expected, while the coarse-grained model is used elsewhere (with possibly some overlapping between the two subdomains, depending on the precise method at hand). This partition of the computational domain is usually a parameter of the method. The question arises on how to choose this partition, how to adapt it to the loading conditions or the current computed solution, so as to make the best possible compromise between accuracy and numerical efficiency.

[1]See also [110] for a discussion of the smallest length scales at which classical elasticity theories hold.

[2]For instance, the stress field in a cracked material decays as $1/\sqrt{r}$ (where r is the distance to the crack tip), which is an extremely slow rate.

Once a partition is chosen, coupling conditions at the interface (or in the overlapping domain) should be chosen. Since the two models are written at different scales and thus involve different variables, some care must be taken when writing these conditions. In general, they indeed have an impact both on the accuracy of the solution of the hybrid model, and on its computational complexity.

These notes are organized as follows. In Section 2, we briefly review the key elements of the models that we manipulate later on, that is continuum mechanics models and atomistic models. As we seek for coupling these two models, we need to address the consistency question. This is the purpose of Section 3, where we show how to derive standard continuum mechanics models on the basis of atomistic models. Section 4 is dedicated first to the introduction of atomistic to continuum coupling methods, based on a domain decomposition paradigm (see Section 4.1). In the sequel of that section, we analyze a prototypical multiscale method and obtain error estimates, first in a convex setting (Section 4.2), second in a nonconvex setting (Section 4.3). The last section (Section 5) gathers some more involved discussion, both on the setting considered in this contribution, and on how to go beyond.

Before proceeding further, let us emphasize that, in these notes, we adopt a variational viewpoint. The key notion of the models that we present below is the energy. We also assume that we have at hand a fine scale model and its coarse-grained version. We look for the global minimizer of an energy, which possibly blends terms written at different scales. Hence, our aim is to compute equilibrium configurations. This corresponds to a zero temperature setting. Note that, even for such static computations, there are alternatives to looking for global minimizers of an energy: one can look for local minimizers, or critical points. Methods have also been developed in the case when no coarse-grained model is known. We also note that inserting dynamical effects or temperature effects leads to different approaches, for which there are still many open questions. We prefer to postpone all these discussions to the end of these notes (see Section 5), and present first the basic background material and one instance of a multiscale method.

2 Models

We briefly describe the continuum mechanics model and the atomistic model we work with. These models are *simplified* models in comparison to what exists in the literature. More comprehensive discussions on continuum models can be read in, e.g., [154]. See [41] for a mathematical analysis of these models, and [92] for details on the related numerical methods and their analysis (finite element methods are often the methods of choice, due to their ability to handle complex geometries). See [36, 46, 137] for a more comprehensive description of atomistic models.

2.1 A continuum mechanics model

Let us consider a deformable solid body which occupies, in the absence of any loading, the smooth bounded domain $\Omega \subset \mathbb{R}^3$ (this is the reference configuration). When submitted to body or surface forces, the solid deforms. We introduce the map $\phi : \Omega \to \mathbb{R}^3$, which is such that the material point at $x \in \Omega$ in the reference configuration moves to $\phi(x) \in \phi(\Omega)$. The map ϕ is called the *deformation* map. It is often useful to introduce the *displacement* $u(x) = \phi(x) - x$. The *deformation gradient* is the map

$$F = \nabla\phi : \Omega \to \mathbb{R}^{3\times 3}.$$

Forces inside the body are described by the *stress tensor* $\pi : \Omega \to \mathbb{R}^{3\times 3}$. By definition, $\pi(x) \cdot n$ is the force per unit area (measured in the reference configuration) that is applied on a surface located, in the reference configuration, at $x \in \Omega$, and normal to the unit vector n.[3] Consider a body submitted to body forces $f : \Omega \to \mathbb{R}^3$ (such as, e.g., gravity) and surface forces $g : \partial\Omega \to \mathbb{R}^3$ (such as, e.g., pressure on the boundary). The equilibrium equations (balance of forces) read

$$\begin{cases} -\mathrm{div}\,\pi &= f \quad \text{in } \Omega, \\ \pi \cdot n &= g \quad \text{on } \partial\Omega, \end{cases} \tag{2.1}$$

where n is the unit normal (outward-pointing) vector to $\partial\Omega$. The vector-valued equations (2.1) can be equivalently written as

$$\begin{cases} -\displaystyle\sum_{j=1}^{3} \frac{\partial \pi_{ij}}{\partial x_j}(x) &= f_i(x) \quad \text{in } \Omega, \quad 1 \le i \le 3, \\ \displaystyle\sum_{j=1}^{3} \pi_{ij}(x) n_j(x) &= g_i(x) \quad \text{on } \partial\Omega, \quad 1 \le i \le 3. \end{cases}$$

Kinematics are described by the deformation ϕ, whereas forces are described by the stress tensor π. The link between these two quantities is the *constitutive law*, which is material dependent, and can formally be written as $\pi = \mathcal{F}(x, \phi, \nabla\phi, \ldots)$.

A material is said to be *elastic* if π depends on ϕ only through $\nabla\phi$ (and not, for instance, through higher order derivatives): $\pi(x) = \mathcal{F}(x, \nabla\phi(x))$. An elastic material is said to be *hyperelastic* when there exists a function $W = W(x, M) : \Omega \times \mathbb{R}^{3\times 3} \to \mathbb{R}$, the elastic energy density, such that

$$\pi_{ij}(x) = \frac{\partial W}{\partial M_{ji}}(x, \nabla\phi(x)). \tag{2.2}$$

Then, formally, equations (2.1) are the Euler–Lagrange equations of the variational problem

$$\inf\left\{E_M(\phi);\ \phi \text{ sufficiently regular}\right\},$$

where the macroscopic energy[4] is

$$E_M(\phi) = \int_\Omega W(x, \nabla\phi(x))\,dx - \int_\Omega f(x) \cdot \phi(x)\,dx - \int_{\partial\Omega} g \cdot \phi\,d\sigma. \tag{2.3}$$

[3] We work here with the first Piola–Kirchhoff stress tensor π, so that the equilibrium equations are written in the reference configuration. We could alternatively work with the Cauchy stress tensor, defined in the current configuration.

[4] The subscript M stands for macroscopic.

Note that the variational space for ϕ should be chosen such that $E_M(\phi)$ is well-defined.

In the following, we focus on hyperelastic materials which are also homogeneous, so that W does not depend on x, but only on $\nabla\phi$.

Example 2.1. Linear elasticity corresponds to the choice

$$W(M) = \frac{1}{2}\,\varepsilon(M) : A : \varepsilon(M),$$

where A is a constant symmetric 4th-order tensor, and $\varepsilon(M)$ is the symmetric part of the tensor $M - \mathrm{Id}$. Hence, for $M = \nabla\phi$, we have $\varepsilon = \frac{1}{2}(\nabla u + \nabla u^T)$ (recall $\phi(x) = x + u(x)$, hence $\nabla\phi = \mathrm{Id} + \nabla u$). By definition of the contraction product, the density $W(M)$ reads

$$W(M) = \frac{1}{2} \sum_{i,j,k,l=1}^{3} \varepsilon_{ij}(M)\, A_{ijkl}\, \varepsilon_{lk}(M),$$

where $A_{ijkl} = A_{jikl} = A_{ijlk} = A_{klij}$. In the special case of isotropic materials, the tensor A depends only on two coefficients, the Young modulus E and the Poisson ratio ν, and the density is

$$W(M) = \frac{E}{2(1+\nu)} \left(\frac{\nu}{1-2\nu} \left(\mathrm{tr}\, \varepsilon(M)\right)^2 + \mathrm{tr}\left(\varepsilon(M)^2\right) \right).$$

When using continuum mechanics, we may face two difficulties. First, some deformations are difficult to describe, because they concern the underlying atomic lattice (and atoms are not degrees of freedom of a continuum model). Second, postulating an accurate constitutive law is not an easy task. A standard way is to introduce a parametric expression and determine the parameters on the basis of experimental results. However, materials are used in loading conditions and temperatures that are more and more demanding (think for instance of materials in nuclear power plants). Extrapolating a constitutive law that has been optimized in some loading range to another range of loads may be dangerous. In addition, it is not always possible to perform experiments at the interesting values of temperature, pressure, etc. This motivates the use of other models.

2.2 An atomistic model

At the atomistic scale, we consider the material as a set of N point particles (the atoms). Variables are the positions ϕ^i, $i = 1, \dots, N$, of the particles. The energy of the system is a function of $\{\phi^i\}_{i=1}^N$. In the following, we assume that the atoms interact through pairwise interactions, so that the energy[5] reads

$$E_\mu(\phi^1, \dots, \phi^N) = \frac{1}{2} \sum_{i=1}^{N} \sum_{j \neq i} V(\phi^j - \phi^i) \tag{2.4}$$

for a given interatomic potential V.

[5] The subscript μ stands for microscopic.

Remark 2.2. Physically reasonable potentials V only depend on the distance between atoms: $V(\phi^j - \phi^i) = \mathcal{V}(\|\phi^j - \phi^i\|)$ for some function $\mathcal{V} : \mathbb{R}^+ \to \mathbb{R}$. We also expect that $\lim_{r \to 0+} \mathcal{V}(r) = +\infty$ (bringing two atoms to the same point costs an infinite amount of energy), whereas $\lim_{r \to +\infty} \mathcal{V}(r) = 0$ (atoms do not interact when they are infinitely far away). Some of these requirements create difficulties from the mathematical viewpoint, mostly because they prevent the energy E_μ from being a convex function of ϕ. In the sequel, we will sometimes consider potentials that do not fullfill the above physical requirements, in order to make the mathematical analysis simpler.

We have only considered here pairwise interactions. More realistic models include, e.g., three-body interactions, and write

$$E_\mu(\phi^1, \dots, \phi^N) = \sum_{i<j} V_2(\phi^i, \phi^j) + \sum_{i<j<k} V_3(\phi^i, \phi^j, \phi^k).$$

In practice, more complex potentials have to be used to obtain quantitatively meaningful results. We restrain ourselves to the case (2.4) of two-body interactions.

Note that we do not consider any quantum effects. Electrons are not described in an explicit way. Actually, electronic interactions are implicitly taken into account in V. For instance, a parametric expression for V can be postulated, and parameters can be optimized so as to fit computations done with an ab initio quantum model, that involves no parameter but only physical constants.

The energy (2.4) describes the self-interaction energy of the atoms with one another. It is the analogue of the elastic energy $\int_\Omega W(\nabla\phi(x))\,dx$ in (2.3). We now wish to consider external forces applied on the atomistic system. We consider only forces that are conservative, e.g., that come from a potential energy G. When the material is submitted to this external field, its energy becomes

$$E_\mu(\phi^1, \dots, \phi^N) = \frac{1}{2}\sum_{i=1}^{N}\sum_{j\neq i} V(\phi^j - \phi^i) + \sum_{i=1}^{N} G(\phi^i). \tag{2.5}$$

Such an atomistic model is able to describe the interesting phenomena that we have mentioned above. However, as noted in the Introduction, it is incredibly expensive. Hence, we will try to couple it with a cheaper model. We explain in the next section how to obtain such a hybrid model.

3 From an atomistic model to a continuum model

In this section, we show how one can build a continuum model on the basis of an atomistic model. We hence address the issue of consistency between the models that we raised earlier. We follow the approach proposed in [26, 27], which is a pointwise approach: given a deformation, we study the limit of the atomistic energy when the lattice parameter goes to zero. Note that a variant of this approach has been proposed in [7]. An alternative to this pointwise approach is a Γ-limit approach, which amounts to looking at the limit of the atomistic variational problem itself (see [31, 45] for introductory material on Γ-convergence, and [32, 33] for some applications of this methodology to discrete problems). We refer the reader to the recent review [29] for a more comprehensive discussion on atomistic to continuum limits.

Our starting point is the expression of the atomistic energy. For the sake of clarity, we consider here the expression (2.4) rather than (2.5), e.g., we do not take into account any external loading. Taking into account this term is an easy generalization. On the basis of (2.4), we want to build a function $E_M = E_M(\phi)$, where the variable is a deformation field ϕ. We are going to make two assumptions. First, we assume

(H1) In the reference configuration, the atoms occupy positions on a perfect periodic lattice (of parameter h).

To simplify notations, we assume that this lattice is $h\mathbb{Z}^d$, where h is the lattice parameter (of order 10^{-10} m), and d is the dimension of the ambient space ($d = 1, 2$ or 3). Generalization to other lattices is straightforward. Let x^i be the position of atom i in the reference (undeformed) configuration. We hence assume that

$$x^i \in \Omega_h := h\mathbb{Z}^d \cap \Omega.$$

The energy in the reference configuration is

$$E_\mu = \frac{1}{2} \sum_{x^i \in \Omega_h} \sum_{x^j \in \Omega_h, j \neq i} V(x^j - x^i).$$

Such a material, where atoms are located on a periodic lattice, is called a *monocrystal*.

Assumption (H1) is certainly a strong assumption. Indeed, in the best possible case, a crystalline material is an aggregate of such monocrystals (one speaks of *polycrystals*), where each monocrystal corresponds to a perfect lattice (see [93] and [94] for an example of homogenization of polycrystals). Two monocrystals at least differ by the orientation of the lattice. The lattice itself may be different from one monocrystal to another. It may also be the case that, in the reference configuration, the atoms do not occupy positions on a regular lattice but, for instance, on a random perturbation of such a regular lattice. We also wish to mention that the assumption (H1) is somewhat related to the so-called Cauchy–Born rule, which is discussed in [54, 70]. Note that it is possible to derive continuum models from atomistic models without resorting to the periodicity assumption (H1). See for instance [28, 30] for such a development in a stochastic setting.

Let us now consider a *macroscopic* deformation ϕ. We want to compute the energy associated to this deformation, based on the atomistic model. We hence need to know what is the deformation at the microscopic scale, when the macroscopic deformation is prescribed to be ϕ. We make the assumption

(H2) When submitted to the macroscopic deformation ϕ, the atoms positions become $\phi(x^i)$.

Hence, at the atomistic scale, the deformation is *equal* to the macroscopic deformation. Again, (H2) is a strong assumption, which is discussed in details in [27, 66].

Under the assumptions (H1) and (H2), the atomistic energy associated to a macroscopic deformation ϕ is

$$E_\mu(\phi) = \frac{1}{2} \sum_{x^i \in \Omega_h} \sum_{x^j \in \Omega_h, j \neq i} V(\phi(x^j) - \phi(x^i)).$$

At this stage, it is important to realize that the interaction potential V actually depends on the lattice parameter h. Indeed, in the reference configuration, the interatomic distance is h, and this equilibrium distance is dictated by the properties of V. As h is a parameter that will go to zero, it is important to write explicitly how V depends on h. A simple scaling is to assume that

$$V(r) = V_0\left(\frac{r}{h}\right), \tag{3.1}$$

where V_0 does not depend on h. See [27] and Remark 4.2 below for a discussion of this scaling.

The energy E_μ is the energy of the whole system. We are going to consider a thermodynamic limit, in the sense that the number of particles in the system will go to $+\infty$. There are two equivalent ways to do this. Either we fix the lattice parameter h to some value and let the volume of the system go to $+\infty$, or we consider a system of given volume and let the lattice parameter go to zero. These two approaches are equivalent, and we follow the latter one. As the number of particles in the system diverges, we expect its energy to diverge as well (the energy is an extensive quantity). Hence, we need to consider the energy *per particle*, which is

$$e_\mu(\phi) = \frac{1}{N} E_\mu(\phi) = \frac{1}{2N} \sum_{x^i \in \Omega_h} \sum_{x^j \in \Omega_h, j \neq i} V_0\left(\frac{\phi(x^j) - \phi(x^i)}{h}\right). \tag{3.2}$$

The number of particles N, the atomic lattice parameter h and the solid volume $|\Omega|$ are linked by $|\Omega| = Nh^d$.

The following theorem, which allows to go from the atomistic energy (3.2) to a continuum mechanics energy, has been proved in [26, 27].

Theorem 3.1. *Let us assume that V_0 is defined on $\mathbb{R}^d \setminus \{0\}$, and that it is a Lipschitz function on the exterior of any ball B_R, centered at the origin and of positive radius $R > 0$. We also assume that there exist R_0, $C > 0$ and $p > d$ such that, for any $|x| \geq R_0$, we have $|V_0(x)| \leq C|x|^{-p}$. We assume that Ω is a smooth bounded domain and that ϕ is a diffeomorphism of class C^2. Then the atomistic energy* (3.2) *satisfies*

$$\lim_{h \to 0} e_\mu(\phi) = \int_\Omega W(\nabla\phi(x))\, dx,$$

where

$$W(M) = \frac{1}{2|\Omega|} \sum_{k \in \mathbb{Z}^d, k \neq 0} V_0(Mk) \quad \textit{for all } M \in \mathbb{R}^{d \times d}. \tag{3.3}$$

Note that, in view of the growth assumption on V_0, the density W is well-defined. Indeed, consider a non-singular matrix M, in the sense that there exists $c_1 > 0$ such that, for all $k \in \mathbb{Z}^d$, $|Mk| \geq c_1|k|$. Then the growth assumption on V_0 implies that $\sum_{k\in\mathbb{Z}^d, k\neq 0} V_0(Mk)$ is a converging series, hence W is well-defined.

Two remarks are in order:

- For a *regular* deformation ϕ, the atomistic energy converges to a continuum mechanics energy. The elastic energy density W is given as a function of the interatomic potential V_0. Hence, symmetries that are present in V_0 (and that correspond to crystal symmetries of the material) are naturally preserved in W.
- The quantity $\int_\Omega W(\nabla\phi(x))\,dx$ can be understood in two different ways. First, one can consider it as a continuum mechanics energy. However, depending on the physical size of the sample (which may be small), one may argue about the validity of a continuum approach. Another viewpoint is to consider $\int_\Omega W(\nabla\phi(x))\,dx$ as a good approximation of the atomistic energy $e_\mu(\phi)$, which has the advantage of being easier to evaluate. Indeed, $e_\mu(\phi)$ involves a sum over all atoms, which can be computationally intractable.

Actually, $e_\mu(\phi)$ can be considered as a function of h and expanded in series in powers of h. Theorem 3.1 provides the leading order term of the Taylor expansion. The next order terms have also been identified (see [26, 27] for more details). Note also that the same strategy has been used to build continuum mechanics *surface energies* (as opposed to *bulk energies* here) on the basis on atomistic models [22].

4 Atomistic to continuum coupling

We are interested in situations where the expected deformation is not smooth in the whole domain Ω. Hence, starting from the atomistic energy (2.4), we cannot apply Theorem 3.1 everywhere on Ω and simply work with the associated continuum model. For computational reasons, we cannot either work with an atomistic description in the whole body. We hence seek for a coupled description.

For a given deformation ϕ, let us split the domain Ω according to

$$\Omega = \Omega_M(\phi) \cup \Omega_\mu(\phi) \subset \mathbb{R}^d,$$

where $\Omega_M(\phi)$ is the domain where ϕ is regular (we expect to be able to use a macroscopic continuum model in this domain), and $\Omega_\mu(\phi)$ is the domain where ϕ is not regular[6]. Based on this partition, we define a coupled energy by

$$E_c(\phi) = \int_{\Omega_M(\phi)} W(\nabla\phi(x))\,dx + \frac{1}{2N} \sum_{x^i \in h\mathbb{Z}^d \cap \Omega_\mu(\phi)} \sum_{x^j \neq x^i} V_0\left(\frac{\phi^j - \phi^i}{h}\right), \qquad (4.1)$$

where W is defined from V_0 by (3.3). Going from (2.4) to (4.1) hence consists in introducing the scaling (3.1), considering the energy per particle (3.2), and passing to

[6] At this stage, we do not want to make more precise what we mean by regular.

the limit $h \to 0$ only in the domain $\Omega_M(\phi)$, where we expect that ϕ is sufficiently regular to allow for Theorem 3.1 to hold. In the domain $\Omega_\mu(\phi)$, we keep the original atomistic expression of the energy.

In practice, ϕ is unknown. One possibility is to define it through a variational problem, that is as the global minimizer of some energy (see Section 5.2 for a discussion of this choice, and alternatives).

The gold standard energy is the atomistic energy (2.4). We assume that the gold standard deformation is the solution[7] ϕ_μ of the minimization problem

$$\inf \left\{ e_\mu(\phi);\ \phi \in \mathbb{R}^{dN},\ \phi \in X_\mu \right\}, \tag{4.2}$$

where the variational space X_μ includes the boundary conditions imposed on the deformation ϕ (see (4.8) below for a precise definition of X_μ). The huge number of particles to consider makes (4.2) intractable. Based on the approximation (4.1) of $e_\mu(\phi)$, it is natural to approximate ϕ_μ by the solution of the variational problem

$$\inf \left\{ \begin{array}{c} E_c(\phi);\quad \phi_{|\Omega_M(\phi)} \text{ is a sufficiently regular field,} \\ \phi_{|\Omega_\mu(\phi)} \text{ is the set of discrete variables } \left\{\phi^i\right\}_{x^i \in \Omega_\mu(\phi)},\ \phi \in X_c \end{array} \right\}, \tag{4.3}$$

where the space X_c includes the boundary conditions.

Note that this approach is highly nonlinear: not only the energies V_0 and W depend nonlinearly on the unknown deformation ϕ, but also the partition of the domain Ω depends on ϕ. To fix ideas, let us choose the following naïve definition for $\Omega_M(\phi)$, assuming that $\phi \in \left(C^1(\Omega)\right)^d$:

$$\Omega_M(\phi) := \{x \in \Omega;\ \|\nabla\phi(x)\| \leq C\}$$

for some threshold C. The function $\phi \mapsto \Omega_M(\phi)$ is highly singular. Consider indeed the sequence of functions ϕ_n defined on Ω by

$$\phi_n(x) = \left(C + \frac{1}{n}\right) x, \quad x \in \Omega.$$

This sequence converges in $\left(C^1(\Omega)\right)^d$ to $\overline{\phi}(x) = Cx$. Now observe that $\Omega_M(\phi_n) = \emptyset$ whereas $\Omega_M\left(\overline{\phi}\right) = \Omega$. Hence, even if a sequence ϕ_n converges in a strong norm to some $\overline{\phi}$, it is possible that the sets $\Omega_M(\phi_n)$ and $\Omega_M\left(\overline{\phi}\right)$ remain very different[8]. Hence, studying the problem (4.3) seems very challenging from the analytical viewpoint, as well as from a numerical discretization viewpoint.

To simplify (4.3), we now remove the dependency of Ω_M with respect to ϕ. We hence a priori fix the partition $\Omega = \Omega_M \cup \Omega_\mu$, and define the coupled energy by

$$E_c(\phi, \Omega_M) = \int_{\Omega_M} W\left(\nabla\phi(x)\right) dx + \frac{1}{2N} \sum_{x^i \in h\mathbb{Z}^d \cap \Omega_\mu} \sum_{x^j \neq x^i} V_0\left(\frac{\phi^j - \phi^i}{h}\right), \tag{4.4}$$

[7] Problems of existence and uniqueness of minimizers will be addressed later on. We assume for the moment that the variational problems we consider are well-posed.

[8] Note that the same issue arises if $\Omega_M(\phi)$ is defined by $\Omega_M(\phi) := \{x \in \Omega;\ \|\nabla\phi(x)\| < C\}$, with a strict inequality on $\|\nabla\phi(x)\|$. Consider indeed the sequence $\phi_n(x) = (C - 1/n)\, x$.

where now Ω_M is a parameter, and ϕ is a mixed variable. In Ω_M, ϕ is a (sufficiently regular) field, whereas in Ω_μ, ϕ is the set of discrete variables ϕ^i, for i such that $x^i \in h\mathbb{Z}^d \cap \Omega_\mu$.

Remark 4.1. Note that the energies (2.4), (4.1) and (4.4) are all consistent with the continuum energy $\int_\Omega W(\nabla\phi(x))\,dx$.

The variational problem associated to the energy (4.4) is

$$\inf \left\{ \begin{array}{c} E_c(\phi, \Omega_M); \quad \phi_{|\Omega_M} \text{ is a sufficiently regular field,} \\ \phi_{|\Omega_\mu} \text{ is the set of discrete variables } \{\phi^i\}_{x^i \in h\mathbb{Z}^d \cap \Omega_\mu}, \ \phi \in X_c \end{array} \right\}, \tag{4.5}$$

where again boundary conditions are imposed by the constraint $\phi \in X_c$. Problem (4.5) is easier to solve than problem (4.3) since the partition is fixed. However, this parameter now has to be specified. Ideally, Ω_M is included in the domain where the solution ϕ_μ to (4.2) is "regular", and the largest possible in order to reduce the computational cost.

A possibility is to define Ω_M in an iterative way, that is:

(i) for a given Ω_M, solve (4.5);

(ii) update the partition $\Omega = \Omega_M \cup \Omega_\mu$ on the basis of the computed solution, and go back to step (i).

This is the idea followed by the QuasiContinuum method (see Section 5.3).

We now consider a simple setting that allows us to prove some estimates on the error between the solution of the atomistic problem and the solution of a coupled problem.

4.1 A simple setting

Let us consider a one-dimensional material occupying in the reference configuration the domain $\Omega = (0, L)$, submitted to an external potential G. In the atomistic model, the solid is considered as a set of $N + 1$ atoms, whose current positions are $\{\phi^i\}_{i=0}^N$. Recall that the atomistic lattice parameter h satisfies $Nh = L$. The energy per particle of the system is given by (see (2.5) and (3.1))

$$e_\mu(\phi^0, \dots, \phi^N) = \frac{1}{2N} \sum_{i=0}^N \sum_{j \neq i} V_0 \left(\frac{\phi^j - \phi^i}{h} \right) + \frac{1}{N} \sum_{i=0}^N G(\phi^i).$$

We assume that, although the deformation may be irregular, it remains small. Writing $\phi^i = x^i + u^i$, where u^i is the displacement of the atom i around the reference configuration $x^i = ih$, we assume that u^i is small. Hence, we can make the approximation

$$\begin{aligned} G(\phi^i) &\approx G(x^i) + G'(x^i)u^i \\ &= G(x^i) - G'(x^i)x^i + G'(x^i)\phi^i. \end{aligned}$$

The quantity $-G'(x)$ is identified as the force $f(x)$ imposed by the external field at the position x. In addition, the term $\sum_{i=0}^N (G(x^i) - G'(x^i)x^i)$ does not depend on ϕ^i.

Since we minimize the energy, this term plays no role, so we omit it. The atomistic energy hence becomes

$$e_\mu(\phi^0,\ldots,\phi^N) = \frac{1}{2N}\sum_{i=0}^{N}\sum_{j\neq i} V_0\left(\frac{\phi^j-\phi^i}{h}\right) - \frac{1}{N}\sum_{i=0}^{N} f(x^i)\phi^i.$$

We make the further approximation of nearest neighbour interaction. Note that this is a strong assumption (see Section 5.1 for some discussion). On the other hand, it allows us to pursue the analysis quite far. The atomistic energy hence becomes

$$e_\mu(\phi^0,\ldots,\phi^N) = \frac{1}{N}\sum_{i=0}^{N-1} V_0\left(\frac{\phi^{i+1}-\phi^i}{h}\right) - \frac{1}{N}\sum_{i=0}^{N} f(ih)\,\phi^i. \tag{4.6}$$

Remark 4.2. Assume that $V_0 = V_0(r)$ attains its minimum at a unique value, that is $r = 1$. When the material is submitted to no body forces ($f \equiv 0$) and no boundary conditions, e_μ attains its minimum for configurations ϕ such that $\phi^{i+1} - \phi^i = h$. Hence, the reference configuration corresponds to atoms located on a periodic lattice of parameter h. This shows, at least in this simple case, that introducing the scaling (3.1) is consistent with the assumption (H1).

We consider here that the microscopic equilibrium configuration is a solution of the variational problem

$$I_\mu = \inf\left\{e_\mu(\phi^0,\ldots,\phi^N);\ \phi\in X_\mu\right\}, \tag{4.7}$$

where the variational space is

$$X_\mu = \left\{\phi\in\mathbb{R}^{N+1};\ \phi^0 = 0,\ \phi^N = a\right\}. \tag{4.8}$$

Note that we have imposed fixed displacement boundary conditions. We look for a global minimizer of the energy (see Section 5.2 for some alternative approaches).

In the continuum mechanics model, the solid deformation is described by the map $\phi : \Omega = (0,L) \to \mathbb{R}$. Following the same arguments as in the proof of Theorem 3.1, one can show that, for sufficiently regular deformations ϕ and potentials V_0, the atomistic energy $e_\mu(\phi(0),\phi(h),\ldots,\phi(Nh))$ converges to the continuum mechanics expression

$$E_M(\phi) = \frac{1}{L}\int_0^L V_0(\phi'(x))\,dx - \frac{1}{L}\int_0^L f(x)\,\phi(x)\,dx. \tag{4.9}$$

We define the macroscopic configuration as a solution of the minimization problem

$$I_M = \inf\{E_M(\phi);\ \phi\in X_M\} \tag{4.10}$$

with

$$X_M = \left\{\phi\in H^1(0,L);\ \phi(0)=0,\ \phi(L)=a\right\}. \tag{4.11}$$

We assume here that the energy $E_M(\phi)$ is well-defined as soon as $\phi\in H^1(0,L)$.

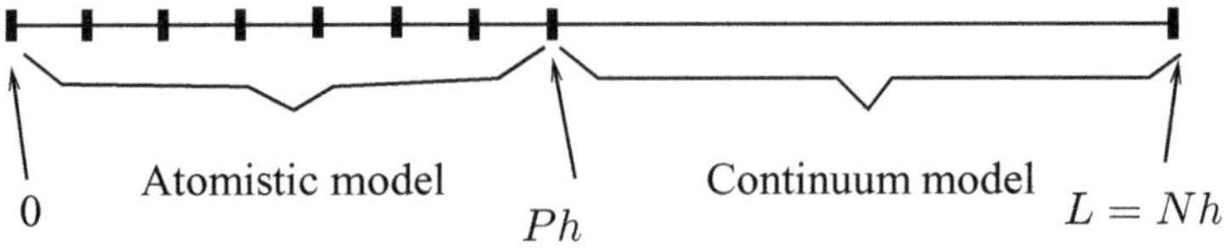

Figure 2. Partition of $\Omega = (0, L)$ according to (4.12).

We expect that solving (4.9)–(4.10) gives a good approximation of the solution of the atomistic problem when the equilibrium deformation is smooth. When non-regular deformations are expected to appear, we approximate the solution of the atomistic problem with the solution of a coupled model. In the spirit of (4.4), we introduce a partition $\Omega = (0, L) = \Omega_M \cup \Omega_\mu$ of the domain, and we consider the coupled energy

$$E_c(\phi) = \frac{1}{L} \int_{\Omega_M} [V_0(\phi'(x)) - f(x)\phi(x)] \, dx \\ + \frac{1}{N} \sum_{i;\, ih, ih+h \in \Omega_\mu} V_0\left(\frac{\phi^{i+1} - \phi^i}{h}\right) - \frac{1}{N} \sum_{i; ih \in \Omega_\mu} f(ih)\phi^i.$$

To simplify notations, we assume that Ω_M and Ω_μ are connected subdomains of Ω. To fix ideas, we make the choice

$$\Omega_\mu = (0, Ph], \quad \Omega_M = (Ph, L), \tag{4.12}$$

for some P that depends on h such that Ph is constant. Similar results are obtained without making assumption (4.12) with more technical proofs (see [23, 24]).

The coupled energy hence becomes

$$E_c(\phi) = \frac{1}{L} \int_{Ph}^{L} [V_0(\phi'(x)) - f(x)\phi(x)] \, dx \\ + \frac{1}{N} \sum_{i=0}^{P-1} V_0\left(\frac{\phi^{i+1} - \phi^i}{h}\right) - \frac{1}{N} \sum_{i=0}^{P} f(ih)\, \phi^i \tag{4.13}$$

and is defined on the variational space

$$X_c = \left\{ \begin{array}{c} \phi;\ \phi_{|\Omega_M} \in H^1(\Omega_M),\ \phi_{|\Omega_\mu} = \{\phi^i\}_{0 \le i \le P} \in \mathbb{R}^{P+1}, \\ \phi^0 = 0,\ \phi(L) = a,\ \phi^P = \phi(Ph) \end{array} \right\}. \tag{4.14}$$

The expression of the boundary conditions $\phi^0 = 0$ and $\phi(L) = a$ reflects the fact that $0 \in \partial\Omega_\mu$ and $L \in \partial\Omega_M$. The interface condition between the macroscopic and the microscopic domains reads $\phi^P = \phi(Ph)$. We hence enforce a continuity-like condition on the deformation at the boundary.

Remark 4.3. In a two-dimensional situation, writing down the interface condition is less straightforward. Actually, there are several possibilities. One can first decrease the mesh size in Ω_M down to the atomistic lattice size for mesh elements close to the interface and enforce equality of displacements of atoms and mesh nodes. Since the mesh size can only evolve smoothly (to keep good mesh qualities, hence accuracy, on the domain Ω_M), this strategy is expensive. Another possibility is to request that the positions of the atoms along (or close to) the interface are obtained from interpolation from the positions of the mesh nodes on the interface. This constraint can also be implemented in the spirit of mortar finite element methods [21].

Hence, the coupled problem we consider is

$$I_c = \inf \{E_c(\phi);\ \phi \in X_c\}. \tag{4.15}$$

The questions we address here are:

- Is the definition (4.13) of the coupled energy always the most appropriate? We require that the coupled energy is consistent with the atomistic energy, and that it leads to an efficient and accurate variational problem. The definition (4.13) is a natural choice, but it is certainly not the only possible one.
- How to (adaptively) define the partition $\Omega = \Omega_M \cup \Omega_\mu$ such that the solution of the coupled problem (4.13)–(4.15) is a good approximation of the solution of the atomistic problem (4.6)–(4.7)?
- Can error bounds be obtained?

We study both the general case of a convex energy density and a specific example of nonconvex energy, the Lennard–Jones case. See [23] and [24] for an analysis of a very similar problem. In [23] and [24], we followed a proof strategy to obtain $W^{1,\infty}$-estimates. In this contribution, we present a different proof strategy[9] and obtain H^1-estimates (see Theorem 4.15 below).

4.2 The convex case

We assume here that V_0 is a convex potential. In this case, we show that we can propose an a priori definition for the partition which is only based on properties of the body forces f. Vaguely stated, the subdomain Ω_M (in which the continuum mechanics model is used) is the part of the domain Ω where the body force f and its derivative f' are small.

With this definition, we show that the solution of the coupled problem (4.13)–(4.15) is a good approximation of the solution of the atomistic problem (4.6)–(4.7): when the atomic lattice parameter h goes to zero, the deformation and the strain given by the coupled model converge to the deformation and the strain given by the atomistic model. The main result of this section is Theorem 4.15 below.

[9] This strategy is actually the one mentioned in Remark 1.3 of [23].

We now proceed in details. We make the convexity assumption

$$\text{(H3)}\quad V_0 \in C^2(\mathbb{R}),\ V_0'(1) = 0,\ \exists \alpha, \beta \in \mathbb{R}\, \forall x \in \mathbb{R}:\ 0 < \alpha \leq V_0''(x) \leq \beta,$$

and the regularity assumption

$$\text{(H4)}\quad f \in C(\mathbb{R}).$$

Note that (H4) is required in order to give a sense to the atomistic energy (4.6). From (H3), we deduce that, for all x and y,

$$\frac{\alpha}{2}(x-1)^2 \leq V_0(x) - V_0(1), \tag{4.16}$$

$$\alpha(x-y)^2 \leq (V_0'(x) - V_0'(y))(x-y), \tag{4.17}$$

$$|V_0'(x) - V_0'(y)| \leq \beta|x-y|, \tag{4.18}$$

$$|V_0(x)| \leq |V_0(1)| + \frac{\beta}{2}|x-1|^2, \tag{4.19}$$

$$|V_0(x) - V_0(y)| \leq \beta|x-y|\,(|x| + |y| + 2)\,. \tag{4.20}$$

We first consider the macroscopic problem (4.10), where the macroscopic energy is (4.9), and the variational space is (4.11). In view of (4.19), $E_M(\phi)$ is well-defined for $\phi \in X_M$.

Theorem 4.4. *Under the assumptions (H3) and (H4), the macroscopic variational problem* (4.10) *has a unique solution* ϕ_M.

Proof. We first prove existence of a minimizer. On X_M, the energy E_M is lower-bounded. Indeed, making use of (4.16), we obtain

$$E_M(\phi) \geq V_0(1) + \frac{\alpha}{2L}\int_0^L (\phi'(x) - 1)^2\, dx - \frac{1}{L}\|f\|_{L^2(0,L)}\, \|\phi\|_{L^2(0,L)}. \tag{4.21}$$

Using a Poincaré inequality for the function $\psi : x \mapsto \phi(x) - ax/L$, which belongs to $H^1_0(0,L)$, with the Poincaré constant C_Ω, which only depends on $\Omega = (0,L)$, we find

$$\begin{aligned}
\|\phi\|^2_{H^1(0,L)} &\leq \left(\|\psi\|_{H^1(0,L)} + C\right)^2 \\
&\leq 2\,\|\psi\|^2_{H^1(0,L)} + 2C^2 \\
&\leq 2C_\Omega^2\,\|\psi'\|^2_{L^2(0,L)} + 2C^2 \\
&\leq 4C_\Omega^2\,\|\phi' - 1\|^2_{L^2(0,L)} + \widetilde{C},
\end{aligned}$$

where C is the $H^1(0,L)$-norm of the function $x \mapsto a\,x/L$. In the above bound, the constant $\widetilde{C}$ only depends on a and L. We now insert this estimate into (4.21) and obtain

$$E_M(\phi) \geq C + \frac{\alpha}{8LC_\Omega^2}\|\phi\|^2_{H^1(0,L)} - \frac{1}{L}\|f\|_{L^2(0,L)}\,\|\phi\|_{H^1(0,L)} \tag{4.22}$$

for some constant C. Since the function $t \in \mathbb{R}^+ \mapsto C + \frac{\alpha}{8LC_\Omega^2}t^2 - \frac{t}{L}\|f\|_{L^2(0,L)}$ is lower-bounded, we obtain that E_M is lower-bounded on X_M.

Let us consider a minimizing sequence of functions ϕ_n. By definition, $E_M(\phi_n)$ is bounded, hence, in view of (4.22), we obtain that $\|\phi_n\|_{H^1(0,L)}$ is bounded. Since $\Omega = (0, L)$ is bounded, we infer from Rellich's theorem (compact embedding of $H^1(0, L)$ into $L^2(0, L)$) that there exists $\overline{\phi} \in X_M$ such that, possibly for a subsequence only, ϕ_n converges to $\overline{\phi}$ weakly in $H^1(0, L)$ and strongly in $L^2(0, L)$.

The energy E_M is continuous for the strong topology of $H^1(0, L)$. Indeed, for θ_1 and θ_2 in $H^1(0, L)$, we have, using (4.20),

$$\begin{aligned}
|E_M(\theta_1) - E_M(\theta_2)| &\leq \int_0^L |V_0(\theta_1'(x)) - V_0(\theta_2'(x))|\, dx \\
&\quad + \int_0^L |f(x)|\, |\theta_1(x) - \theta_2(x)|\, dx \\
&\leq \beta \int_0^L |\theta_1'(x) - \theta_2'(x)|\, (|\theta_1'(x)| + |\theta_2'(x)| + 2)\, dx \\
&\quad + \|f\|_{L^2(0,L)}\, \|\theta_1 - \theta_2\|_{L^2(0,L)} \\
&\leq \beta \|\theta_1 - \theta_2\|_{H^1(0,L)}\, \||\theta_1'| + |\theta_2'| + 2\|_{L^2(0,L)} \\
&\quad + \|f\|_{L^2(0,L)}\, \|\theta_1 - \theta_2\|_{H^1(0,L)}.
\end{aligned}$$

This inequality yields the continuity of E_M. Since E_M is also convex, it is weakly lower semi-continuous in $H^1(0, L)$ (see, e.g., [34, Corollaire III.8]). We hence infer from $\phi_n \rightharpoonup \overline{\phi}$ in $H^1(0, L)$ that

$$E_M\left(\overline{\phi}\right) \leq \liminf E_M(\phi_n) = \lim E_M(\phi_n) = I_M.$$

Since $\overline{\phi} \in X_M$, we also have $I_M \leq E_M\left(\overline{\phi}\right)$. As a consequence, $I_M = E_M\left(\overline{\phi}\right)$, and $\overline{\phi}$ is a minimizer of (4.10).

Since V_0 is strictly convex, the energy E_M is a strictly convex function on the convex set X_M, hence the minimizer of (4.10) is unique. □

In order to write the weak formulation of (4.10), we introduce the forms

$$\begin{aligned}
A_M(\phi, \psi) &= \int_0^L V_0'(\phi'(x))\, \psi'(x)\, dx, \\
B_M(\psi) &= \int_0^L f(x)\, \psi(x)\, dx,
\end{aligned}$$

defined on $H^1(0, L) \times H^1(0, L)$ and $H^1(0, L)$, respectively. Note that A_M is nonlinear with respect to ϕ.

Lemma 4.5. *The macroscopic solution ϕ_M defined by Theorem 4.4 satisfies*

$$\forall \psi \in H_0^1(0, L), \quad A_M(\phi_M, \psi) = B_M(\psi). \tag{4.23}$$

In addition, the function $\psi_M : x \mapsto \phi_M(x) - ax/L$ *is in* $H^2(0,L)$ *and satisfies the estimate*

$$\|\psi_M\|_{H^2(0,L)} \leq \frac{C}{\alpha}\|f\|_{L^2(0,L)} \tag{4.24}$$

for some C *that only depends on the domain* $\Omega = (0,L)$.

Proof. Equation (4.23) is the weak formulation of the variational problem (4.10). It implies that, in $\mathcal{D}'(0,L)$,

$$-V_0''(\phi_M')\,\phi_M'' = -\left(V_0'(\phi_M')\right)' = f.$$

Since $V_0''(x) \geq \alpha > 0$ and since $f \in C^0(0,L) \subset L^2(0,L)$, we obtain that $\phi_M'' \in L^2(0,L)$, and

$$\|\psi_M''\|_{L^2(0,L)} = \|\phi_M''\|_{L^2(0,L)} \leq \frac{1}{\alpha}\|f\|_{L^2(0,L)}. \tag{4.25}$$

Hence ψ_M is in $H^2(0,L)$. Let us now upper-bound this function in the H^1-norm. We first note that $\psi_M \in H_0^1(0,L)$. In view of (4.17) and (4.23), we have

$$\begin{aligned}
\alpha\left\|\phi_M' - \frac{a}{L}\right\|^2_{L^2(0,L)} &\leq \int_0^L \left(V_0'(\phi_M'(x)) - V_0'\left(\frac{a}{L}\right)\right)\left(\phi_M'(x) - \frac{a}{L}\right)\,dx \\
&= \int_0^L V_0'(\phi_M'(x))\left(\phi_M'(x) - \frac{a}{L}\right)\,dx \\
&= A_M(\phi_M, \psi_M) \\
&= B_M(\psi_M) \\
&\leq \|f\|_{L^2(0,L)}\,\|\psi_M\|_{H^1(0,L)},
\end{aligned}$$

where we have used, in the second line, that $\int_0^L \left(\phi_M'(x) - \frac{a}{L}\right)\,dx = 0$ since $\psi_M(x) := \phi_M(x) - ax/L$ vanishes for $x = 0$ and $x = L$. Using a Poincaré inequality for the function ψ_M, we obtain

$$\|\psi_M\|_{H^1(0,L)} \leq \frac{C}{\alpha}\|f\|_{L^2(0,L)}$$

for a constant C that only depends on $\Omega = (0,L)$. This bound, combined with (4.25), yields (4.24). □

We now consider the microscopic problem (4.7), where the microscopic energy is (4.6) and the variational space is (4.8). The following theorem is the microscopic counterpart of Theorem 4.4.

Theorem 4.6. *Under the assumptions (H3) and (H4), the microscopic variational problem* (4.7) *has a unique solution* ϕ_μ.

We skip the proof of this theorem which follows the same lines as the proof of Theorem 4.4. In order to write the weak formulation of (4.7), we introduce the forms

$$\begin{aligned} A_\mu(\phi, \psi) &= h \sum_{i=0}^{N-1} V_0' \left(\frac{\phi^{i+1} - \phi^i}{h} \right) \frac{\psi^{i+1} - \psi^i}{h}, \\ B_\mu(\psi) &= h \sum_{i=0}^{N} f(ih)\, \psi^i, \end{aligned}$$

defined on $\mathbb{R}^{N+1} \times \mathbb{R}^{N+1}$ and $\mathbb{R}^{N+1}$, respectively, and the homogeneous microscopic space

$$X_\mu^0 = \{\psi \in \mathbb{R}^{N+1};\ \psi^0 = 0,\ \psi^N = 0\}.$$

Lemma 4.7. *The microscopic solution ϕ_μ defined by Theorem 4.6 satisfies*

$$\forall \psi \in X_\mu^0, \quad A_\mu(\phi_\mu, \psi) = B_\mu(\psi). \tag{4.26}$$

We now introduce a coupled problem based on a given partition of $\Omega = (0, L) = \Omega_M \cup \Omega_\mu$. We recall that, in order to simplify the notations, we work with the choice

$$\Omega_\mu = (0, Ph], \quad \Omega_M = (Ph, L),$$

for some P that depends on h such that Ph is constant (see (4.12) and Figure 2).

The following theorem is the coupled counterpart of Theorem 4.4.

Theorem 4.8. *Under the assumptions (H3) and (H4), the coupled variational problem* (4.15) *has a unique solution ϕ_c.*

We skip the proof of this theorem, which again follows the same lines as the proof of Theorem 4.4. In order to write the weak formulation of (4.15), we introduce the forms

$$\begin{aligned} A_c(\phi, \psi) &= h \sum_{i=0}^{P-1} V_0' \left(\frac{\phi^{i+1} - \phi^i}{h} \right) \frac{\psi^{i+1} - \psi^i}{h} + \int_{Ph}^{L} V_0'(\phi'(x))\, \psi'(x)\, dx, \\ B_c(\psi) &= h \sum_{i=0}^{P} f(ih)\, \psi^i + \int_{Ph}^{L} f(x)\, \psi(x)\, dx, \end{aligned}$$

defined on $X_c \times X_c$ and X_c, respectively, and the homogeneous coupled space

$$X_c^0 = \{\psi \in X_c;\ \psi^0 = 0,\ \psi(L) = 0\}. \tag{4.27}$$

Lemma 4.9. *The coupled solution ϕ_c defined by Theorem 4.8 satisfies*

$$\forall \psi \in X_c^0, \quad A_c(\phi_c, \psi) = B_c(\psi). \tag{4.28}$$

In addition, the function $\psi_c : x \in \Omega_M \mapsto \phi_c(x) - ax/L$ is in $H^2(\Omega_M)$ and satisfies the estimate

$$\|\psi_c\|_{H^2(\Omega_M)} \leq \frac{C}{\alpha} \|f\|_{L^2(\Omega_M)} \tag{4.29}$$

for some C that only depends on the domain Ω.

Proof. The proof of (4.28) follows the same lines as the proof of (4.23) of Lemma 4.5. The proof of (4.29) follows the same lines as the proof of (4.24): we make use of a Poincaré inequality on Ω_M, with a constant C_{Ω_M}, which is next upper-bounded by the Poincaré constant of $\Omega = (0, L)$. □

We now aim at estimating the distance between ϕ_c and ϕ_μ. We introduce an auxiliary problem, which is the discretization of the coupled problem by a finite element method, using a mesh on Ω_M of mesh size h. Of course, the resulting problem is as expensive to solve as the full microscopic problem (both problems involve as many variables), but it helps us in the numerical analysis.

We hence introduce the space

$$X_c^h = \{\theta \in X_c; \quad \theta \text{ is piecewise affine on } \Omega_M \text{ on a mesh of size } h\},$$

and its homogeneous version

$$X_c^{0h} = X_c^h \cap X_c^0.$$

We introduce the discretized coupled problem

$$\inf\{E_c(\phi);\ \phi \in X_c^h\}. \tag{4.30}$$

As for the coupled problem (4.15), the above problem has a unique solution ϕ_c^h, which satisfies the weak formulation (see Lemma 4.9)

$$\forall \psi \in X_c^{0h}, \quad A_c(\phi_c^h, \psi) = B_c(\psi). \tag{4.31}$$

To estimate the difference between the solution ϕ_c of the coupled problem and the solution ϕ_μ of the atomistic problem, we first estimate the distance between ϕ_c and ϕ_c^h, solutions of the non-discretized and discretized coupled problem, respectively. Next, we estimate the distance between ϕ_c^h and ϕ_μ.

To measure distances in X_c, we introduce the mixed semi-norm (recall $\Omega_M = (Ph, L)$)

$$|\psi|^2_{H^1(X_c)} := \int_{Ph}^{L} \psi'^2(x)\,dx + h \sum_{i=0}^{P-1} \left(\frac{\psi^{i+1} - \psi^i}{h}\right)^2.$$

We have the following estimates for the form A_c:

Lemma 4.10. *Assume that (H3) holds. For all u and g in X_c, we have the strong monotonicity*

$$A_c(u, u-g) - A_c(g, u-g) \geq \alpha |u-g|^2_{H^1(X_c)}. \tag{4.32}$$

For all u, g and θ in X_c, we have the Lipschitz-continuity

$$|A_c(u,\theta) - A_c(g,\theta)| \leq 2\beta\, |u-g|_{H^1(X_c)}\, |\theta|_{H^1(X_c)}. \tag{4.33}$$

Proof. The bound (4.32) is a direct consequence of (4.17). The bound (4.33) follows from (4.18). □

The above lemma shows that A_c fulfills the assumptions (of strong monotonicity and Lipschitz continuity) of the theorem of Zarantonello [160, Theorem 25.B], which is a generalization of the Lax-Milgram theorem to the nonlinear setting. The following lemma provides an estimate of the distance between ϕ_c and ϕ_c^h, and is a generalization of the Céa lemma.

Lemma 4.11. *The coupled solution ϕ_c, which is defined by Theorem 4.8, and its discretized counterpart ϕ_c^h, solution of (4.30), satisfy*

$$\left|\phi_c - \phi_c^h\right|_{H^1(X_c)} \leq \frac{2\beta}{\alpha} C\, h\, \|\phi_c''\|_{L^2(\Omega_M)}$$

for some C that only depends on Ω.

Proof. Using (4.32), (4.28) and (4.31), we have

$$\begin{aligned} \alpha\left|\phi_c - \phi_c^h\right|^2_{H^1(X_c)} &\leq A_c(\phi_c, \phi_c - \phi_c^h) - A_c(\phi_c^h, \phi_c - \phi_c^h) \\ &= A_c(\phi_c, \phi_c - \phi_c^h + \psi) - A_c(\phi_c^h, \phi_c - \phi_c^h + \psi) \end{aligned}$$

for any $\psi \in X_c^{0h}$. With (4.33), we therefore obtain for any $\psi \in X_c^h$

$$\begin{aligned} \alpha\left|\phi_c - \phi_c^h\right|^2_{H^1(X_c)} &\leq A_c(\phi_c, \phi_c - \psi) - A_c(\phi_c^h, \phi_c - \psi) \\ &\leq 2\beta\left|\phi_c - \phi_c^h\right|_{H^1(X_c)} \left|\phi_c - \psi\right|_{H^1(X_c)}. \end{aligned}$$

Hence,

$$\left|\phi_c - \phi_c^h\right|_{H^1(X_c)} \leq \frac{2\beta}{\alpha} \inf_{\psi \in X_c^h} \left|\phi_c - \psi\right|_{H^1(X_c)}. \tag{4.34}$$

We have shown in Lemma 4.9 that $\phi_c \in H^2(\Omega_M)$. By a standard approximation result, we thus have

$$\inf_{\psi \in X_c^h} \left|\phi_c - \psi\right|_{H^1(X_c)} \leq C\, h\, \|\phi_c''\|_{L^2(\Omega_M)}, \tag{4.35}$$

where C only depends on Ω_M and can be bounded by a constant that only depends on Ω. Collecting (4.34) and (4.35), we obtain the claimed estimate. □

We now estimate the distance between ϕ_μ and ϕ_c^h. We make use of the weak formulations of the corresponding variational problems (see Lemma 4.7 and equation (4.31)). To compare the configurations ϕ_μ and ϕ_c^h, that belong to different spaces, we introduce the *evaluation operator*

$$\begin{aligned} E : X_c &\to X_\mu \\ \psi &\mapsto E\psi \end{aligned}$$

defined by $(E\psi)^i = \psi^i$ for $0 \leq i \leq P$, and $(E\psi)^i = \psi(ih)$ for $ih \geq Ph$. We also introduce the *interpolation operator*

$$\begin{aligned} I : X_\mu &\to X_c \\ \psi &\mapsto I\psi \end{aligned}$$

defined by $(I\psi)^i = \psi^i$ for $0 \leq i \leq P$, and $I\psi_{|\Omega_M}$ is the piecewise affine interpolation of $\psi_{|\Omega_M}$. We note that $I(X_\mu) \subset X_c^h$ and

$$\forall \phi, \psi \in X_\mu, \quad A_\mu(\phi, \psi) = A_c(I\phi, I\psi). \tag{4.36}$$

Lemma 4.12. *Assume that (H3) and (H4) hold. Then*

$$\left|\phi_c^h - I\phi_\mu\right|_{H^1(X_c)} \leq \frac{\tau}{\alpha},$$

where

$$\tau = \sup_{\psi \in X_c^{0h},\ \psi \neq 0} \frac{|B_c(\psi) - B_\mu(E\psi)|}{|\psi|_{H^1(X_c)}}. \tag{4.37}$$

Proof. Let $\psi \in X_c^{0h}$. Using (4.31) and (4.26), we obtain

$$\begin{aligned} A_c(\phi_c^h, \psi) &= B_c(\psi) \\ &= B_\mu(E\psi) + B_c(\psi) - B_\mu(E\psi) \\ &= A_\mu(\phi_\mu, E\psi) + B_c(\psi) - B_\mu(E\psi) \\ &= A_c(I\phi_\mu, \psi) + B_c(\psi) - B_\mu(E\psi), \end{aligned}$$

where, in the last line, we have used (4.36) and the fact that $IE\psi = \psi$ for any $\psi \in X_c^h$. We hence obtain, for any $\psi \in X_c^{0h}$,

$$A_c(\phi_c^h, \psi) - A_c(I\phi_\mu, \psi) \leq \tau |\psi|_{H^1(X_c)},$$

where τ is defined by (4.37). We now choose $\psi = \phi_c^h - I\phi_\mu$ and use (4.32):

$$\begin{aligned} \alpha \left|\phi_c^h - I\phi_\mu\right|^2_{H^1(X_c)} &\leq A_c(\phi_c^h, \phi_c^h - I\phi_\mu) - A_c(I\phi_\mu, \phi_c^h - I\phi_\mu) \\ &\leq \tau \left|\phi_c^h - I\phi_\mu\right|_{H^1(X_c)}, \end{aligned}$$

which yields the claimed bound. □

Combining Lemma 4.11 and 4.12, we obtain the following result:

Theorem 4.13. *Assume that (H3) and (H4) hold. Then*

$$|\phi_c - I\phi_\mu|_{H^1(X_c)} \leq \frac{\tau}{\alpha} + \frac{2\beta}{\alpha} C\, h\, \|\phi_c''\|_{L^2(\Omega_M)},$$

where τ is defined by (4.37) *and C only depends on Ω.*

We now address the question of how to define the partition $\Omega = (0, L) = \Omega_M \cup \Omega_\mu$. Our aim is to upper-bound τ that appears in the estimate of Theorem 4.13. We use the following result, which can be proved by Taylor expansion:

Lemma 4.14. *Let $g \in W^{1,1}(0,L)$ with $L = Nh$. Then*

$$\left| \int_0^L g(x)\, dx - h \sum_{i=1}^{N} g(ih) \right| \leq h \|g'\|_{L^1(0,L)}.$$

Let $\psi \in X_c^{0h}$. Recall that $\Omega_M = (Ph, L)$ with $L = Nh$. Then

$$B_c(\psi) - B_\mu(E\psi) = \int_{Ph}^{Nh} f(x)\, \psi(x)\, dx - h \sum_{i=P+1}^{N} f(ih)\, \psi(ih).$$

We want to use Lemma 4.14 with $g = f\psi$. In view of assumption (H4), we have $f \in C^0\left(\overline{\Omega}_M\right)$. Since $\psi \in X_c^{0h}$, we have $\psi \in C^0\left(\overline{\Omega}_M\right)$ and $\psi' \in L^1(\Omega_M)$.

Let us now make the additional assumption that the body force f satisfies $f' \in L^1(\Omega_M)$. Then

$$\begin{aligned} |B_c(\psi) - B_\mu(E\psi)| &\leq h \|(f\psi)'\|_{L^1(\Omega_M)} \\ &\leq h \left(\|f'\psi\|_{L^1(\Omega_M)} + \|f\psi'\|_{L^1(\Omega_M)} \right) \\ &\leq h \left(\|f'\|_{L^1(\Omega_M)} \|\psi\|_{L^\infty(\Omega_M)} + \|f\|_{L^\infty(\Omega_M)} \|\psi'\|_{L^1(\Omega_M)} \right). \end{aligned}$$

We work with $\psi \in X_c^{0h}$, so that $\psi(L) = 0$. Therefore, we have $\|\psi\|_{L^\infty(\Omega_M)} \leq \|\psi'\|_{L^1(\Omega_M)} \leq \sqrt{L}\, \|\psi'\|_{L^2(\Omega_M)} \leq \sqrt{L}\, |\psi|_{H^1(X_c)}$, and find

$$|B_c(\psi) - B_\mu(E\psi)| \leq h\sqrt{L}\, \left(\|f'\|_{L^1(\Omega_M)} + \|f\|_{L^\infty(\Omega_M)} \right) |\psi|_{H^1(X_c)}.$$

In view of the definition (4.37) of τ, this yields

$$\tau \leq h\sqrt{L}\, \left(\|f'\|_{L^1(\Omega_M)} + \|f\|_{L^\infty(\Omega_M)} \right). \tag{4.38}$$

We thus obtain the following result.

Theorem 4.15. *Assume that (H3) and (H4) hold, and that the partition $\Omega = \Omega_M \cup \Omega_\mu$ is such that $f' \in L^1(\Omega_M)$. We also assume that*

$$\Omega_\mu = (0, Ph], \quad \Omega_M = (Ph, L),$$

for some P that depends on h such that Ph is constant. Then

$$|\phi_c - I\phi_\mu|_{H^1(X_c)} \leq C\, h \left(\|f'\|_{L^1(\Omega_M)} + \|f\|_{L^\infty(\Omega_M)} \right),$$

where C only depends on L, α and β.

Proof. In view of Theorem 4.13 and equation (4.38), we have

$$|\phi_c - I\phi_\mu|_{H^1(X_c)} \leq \frac{h\sqrt{L}}{\alpha} \left(\|f'\|_{L^1(\Omega_M)} + \|f\|_{L^\infty(\Omega_M)} \right) + \frac{2\beta}{\alpha} C\, h \|\phi_c''\|_{L^2(\Omega_M)},$$

where C only depends on $\Omega = (0, L)$. In view of Lemma 4.9, we also have

$$\|\phi_c''\|_{L^2(\Omega_M)} = \|\psi_c''\|_{L^2(\Omega_M)} \leq \|\psi_c\|_{H^2(\Omega_M)} \leq \frac{C}{\alpha}\|f\|_{L^2(\Omega_M)} \leq \frac{C\sqrt{L}}{\alpha}\|f\|_{L^\infty(\Omega_M)},$$

where $\psi_c(x) = \phi_c(x) - ax/L$, and where C only depends on Ω. We hence obtain the claimed bound. □

Theorem 4.15 provides a bound on the first derivatives of the deformations. This is hence a bound on strains, in a L^2-kind of norm. Using a Poincaré inequality, we deduce from it a bound on the deformations, e.g., a bound on $\|\phi_c - I\phi_\mu\|_{L^2(X_c)}$, where the norm $\|\cdot\|_{L^2(X_c)}$ is defined on X_c by

$$\forall \psi \in X_c, \quad \|\psi\|^2_{L^2(X_c)} := \int_{Ph}^{L} \psi^2(x)\, dx + h \sum_{i=0}^{P} \left(\psi^i\right)^2.$$

Remark 4.16. We observe that, if we choose the partition according to the assumptions of Theorem 4.15 with f and f' small on Ω_M (in the L^∞- and the L^1-norm, respectively), then the coupled problem is a good (and converging, when $h \to 0$) approximation of the microscopic problem. Hence, we can define an appropriate partition based on the knowledge of the body forces only. This is related to the convexity assumption. Indeed, due to the convexity of the problems, the Euler–Lagrange equations of the problems are uniformly elliptic, hence their solutions are "singular" only where the datum f is singular. In the nonconvex case, the situation is different, as is shown in Section 4.3.

Remark 4.17. The above estimates involve L^2-bounds on the deformation and its first derivative. It is also possible to obtain L^∞-bounds, as is shown in [23, 24], where a different strategy is followed. See also [42].

4.3 A nonconvex case: the Lennard–Jones case

We now consider a particular example of nonconvex potential, namely the Lennard–Jones potential

$$V_{\mathrm{LJ}}(x) = \frac{1}{x^{12}} - \frac{2}{x^6}. \tag{4.39}$$

We observe that $\lim_{x\to 0} V_{\mathrm{LJ}}(x) = +\infty$ and $\lim_{x\to+\infty} V_{\mathrm{LJ}}(x) = 0$, which are physically relevant properties (see Remark 2.2). Actually, the property that we use in the sequel is $\lim_{x\to+\infty} V_{\mathrm{LJ}}(x)/x = 0$, rather than $\lim_{x\to+\infty} V_{\mathrm{LJ}}(x) = 0$ (see Remark 4.19 for the mechanical implication of this property). Note also that V_{LJ} attains its unique minimum at $x = 1$.

Remark 4.18. The particular choice of exponents in V_{LJ} (6 and 12) has no influence on the following analysis. Similar results hold for the potential $V(x) = \frac{q}{x^p} - \frac{p}{x^q}$ with $p > q > 0$.

Remark 4.19. For a one-dimensional solid described by the continuum energy

$$E_M(\phi) = \int_0^L W(\phi'(x))\,dx,$$

the behaviour of $W(x)/x$ when $x \to +\infty$ is related to the cost of making a fracture in the material. Consider indeed a nonnegative energy density W that attains its unique minimum at $x = 1$, and for which $c = \lim_{x\to+\infty} W(x)/x$ exists, $c \in [0, +\infty]$. Let $a > L$ and $0 < \ell < L$, and consider the deformations ϕ_n defined by $\phi_n(0) = 0$ and

$$\phi_n'(x) = \begin{cases} 1 & \text{on } \left(0, \ell - \frac{1}{n}\right) \cup (\ell, L), \\ n(a-L)+1 & \text{on } \left(\ell - \frac{1}{n}, \ell\right), \end{cases}$$

which all satisfy the same boundary conditions $\phi_n(0) = 0$ and $\phi_n(L) = a$. This sequence represents configurations with a strain larger and more localized when n increases. In the limit $n \to +\infty$, it represents a fractured material. The energy of the configuration ϕ_n is

$$E_M(\phi_n) = \left(L - \frac{1}{n}\right) W(1) + \frac{1}{n} W(n(a-L)+1),$$

hence

$$\lim_{n\to+\infty} E_M(\phi_n) = LW(1) + c(a-L).$$

If $c = +\infty$, then making a fracture costs an infinite energy. If $c = 0$, then making a fracture costs no energy: $\lim_{n\to+\infty} E_M(\phi_n)$ is equal to the energy of the reference configuration $\phi(x) = x$. If $c \in (0, +\infty)$, then making a fracture costs a finite amount of energy.

In this nonconvex case, we show that expression (4.13) for the coupled energy might be inappropriate, and we describe several ways to circumvent this difficulty. The complete analysis of this case can be read in [23]. Here, we briefly summarize it (see also [155]).

We consider a material that occupies the domain $\Omega = (0, L)$ in the reference configuration and that is put in tension. We focus on the case when there are no body forces: $f \equiv 0$. Taking into account body forces does not change the qualitative conclusions of the analysis. If the material is compressed rather than extended, then the analysis is very similar to the convex case analysis (see, for instance, [23, Theorem 2.1]).

We first look at the macroscopic problem (4.10) with the macroscopic energy

$$E_M(\phi) = \frac{1}{L} \int_0^L V_{\mathrm{LJ}}(\phi'(x))\,dx$$

and the variational space

$$X_M = \left\{ \phi \in W^{1,1}(0,L);\ \frac{1}{\phi'(x)} \in L^{12}(0,L),\ \phi(0) = 0,\ \phi(L) = a \right\}. \tag{4.40}$$

We consider here the tension case: $a > L$.

Lemma 4.20. *The minimum I_M in (4.10) is equal to* $\inf V_{\mathrm{LJ}}$. *In addition, the infimum in the problem (4.10) is not attained: there exists no $\phi \in X_M$ such that $E_M(\phi) = I_M$.*

Proof. For any $\phi \in X_M$, we have that $E_M(\phi) \geq \inf V_{\mathrm{LJ}}$. Hence, E_M is lower-bounded and $I_M \geq \inf V_{\mathrm{LJ}} > -\infty$. Consider now the following sequence of continuous functions:

$$\phi_n(x) = \begin{cases} x & \text{on } \left[0, L - \frac{1}{n}\right], \\ a + n\left(a - L + \frac{1}{n}\right)(x - L) & \text{on } \left(L - \frac{1}{n}, L\right]. \end{cases}$$

The function ϕ_n is piecewise affine, satisfies the boundary conditions $\phi_n(0) = 0$, $\phi_n(L) = a$, and

$$\phi_n'(x) = 1 \text{ on } \left(0, L - \frac{1}{n}\right), \; \phi_n'(x) = n\left(a - L + \frac{1}{n}\right) \text{ elsewhere.}$$

We find

$$E_M(\phi_n) = \left(1 - \frac{1}{nL}\right) \inf V_{\mathrm{LJ}} + \frac{1}{nL} V_{\mathrm{LJ}}\left(n(a - L) + 1\right).$$

Since $\lim_{x \to +\infty} V_{\mathrm{LJ}}(x)/x = 0$, we obtain $\lim_{n \to +\infty} E_M(\phi_n) = \inf V_{\mathrm{LJ}}$. As $E_M(\phi_n) \geq I_M$, this yields $\inf V_{\mathrm{LJ}} \geq I_M$. Therefore, we get $I_M = \inf V_{\mathrm{LJ}}$.

Let us now assume that there exists $\phi \in X_M$ such that $E_M(\phi) = I_M$. This implies $\int_0^L V_{\mathrm{LJ}}(\phi'(x))\, dx = L \inf V_{\mathrm{LJ}}$ and thus $V_{\mathrm{LJ}}(\phi'(x)) = \inf V_{\mathrm{LJ}}$ almost everywhere on $(0, L)$. We conclude $\phi'(x) = 1$ almost everywhere, which is in contradiction with the boundary conditions $\phi(0) = 0$, $\phi(L) = a > L$. Hence, the infimum in the problem (4.10) is not attained. □

If the variational space is enlarged (in (4.40), replace $W^{1,1}(0, L)$ by $SBV(0, L)$, whose definition is recalled below), then the problem (4.10) has an infinite number of solutions with jumps. They are of the form

$$\phi(x) = x + \sum_{j \in J} v_j \mathcal{H}(x - y_j), \tag{4.41}$$

where $J \subset \mathbb{N}$, y_j are any points in $\Omega = (0, L)$, $\sum_j v_j = a - L$, and $\mathcal{H}$ is the Heaviside function: $\mathcal{H}(t) = 1$ if $t \geq 0$, $\mathcal{H}(t) = 0$ otherwise. This statement can be shown by arguments similar to the ones used to prove Theorem 2.1 of [23]. We recall (see [3, 5]) that the set $SBV(0, L)$ of special functions of bounded variation is

$$SBV(0, L) = \left\{ \begin{array}{c} u \in \mathcal{D}'(0, L);\; u' = D_a u + \sum_{j \in \mathbb{N}} v_j \delta_{y_j},\; D_a u \in L^1(0, L), \\ y_j \in (0, L),\; \sum_{j \in \mathbb{N}} |v_j| < +\infty \end{array} \right\},$$

where δ_{y_j} is the Dirac distribution centered at y_j (note that $\mathcal{H}'(\cdot - y_j) = \delta_{y_j}$ in the distributional sense).

Note that, with the above continuum model, neither the number nor the locations of the fractures are determined. All the configurations (4.41) have the same energy, whatever J, v_j and y_j are.

The fully atomistic model can be analyzed with methods similar to the ones used in [23, Section 2.2]. It turns out that the minimizers of this model have a *unique* fracture, whose location is *not* determined (the energy is the same whatever the fracture location is).

Hence, working with the Lennard–Jones potential and tension boundary conditions is interesting, since it leads to a deformation with a singularity (here, a jump, which represents a fracture in the one-dimensional bar). Our aim is thus to design a multi-scale method able to describe this singularity by an atomistic model, while keeping a continuum description where it is accurate enough. Since the location of the fracture is not prescribed by the atomistic model, we can afford to fix a priori the partition of Ω, with the aim to find the singularity in Ω_μ.

Let us now follow the strategy explained in Section 4.2. We consider a given partition of Ω, and, to simplify notations, we again assume that this partition is (4.12). In the spirit of the coupled energy (4.13) and the coupled space (4.14), we consider the energy

$$E_c(\phi) = \frac{1}{L}\int_{Ph}^{L} V_{\mathrm{LJ}}(\phi'(x))\,dx + \frac{1}{N}\sum_{i=0}^{P-1} V_{\mathrm{LJ}}\left(\frac{\phi^{i+1}-\phi^i}{h}\right), \tag{4.42}$$

defined on the variational space

$$X_c = \left\{ \begin{array}{c} \phi;\quad \phi_{|\Omega_M} \in SBV(\Omega_M),\ 1/\phi' \in L^{12}(\Omega_M), \\ \phi_{|\Omega_\mu} = \{\phi^i\}_{0\le i\le P} \in \mathbb{R}^{P+1},\ \phi^0 = 0,\ \phi(L) = a,\ \phi^P = \phi(Ph) \end{array} \right\}. \tag{4.43}$$

We consider the tension case: $a > L$. Following the arguments of [23, Lemma 2.3], one can show that the minimizers of

$$\inf\{E_c(\phi);\ \phi \in X_c\}$$

are of the form

$$\begin{aligned} \phi^i &= ih, \quad 0 \le i \le P, \\ \phi(x) &= x + \sum_{j\in J} v_j \mathcal{H}(x - y_j), \quad \forall x \ge Ph, \end{aligned}$$

where $J \subset \mathbb{N}$, y_j are any points in Ω_M and $\sum_j v_j = a - L$. Therefore, as in the completely macroscopic description, the material breaks. However, the fracture is *always* in the domain Ω_M, in which the deformation ϕ is assumed to be regular (since we use a macroscopic model in that domain). Hence, we do not reach our aim with the coupled method (4.42)–(4.43).

Proving the above statement is quite technical. The following calculations help to understand why the fracture systematically occurs in the continuum domain. We consider two configurations, the first one with a fracture in Ω_M and the second one with a fracture in Ω_μ, and compare their energies.

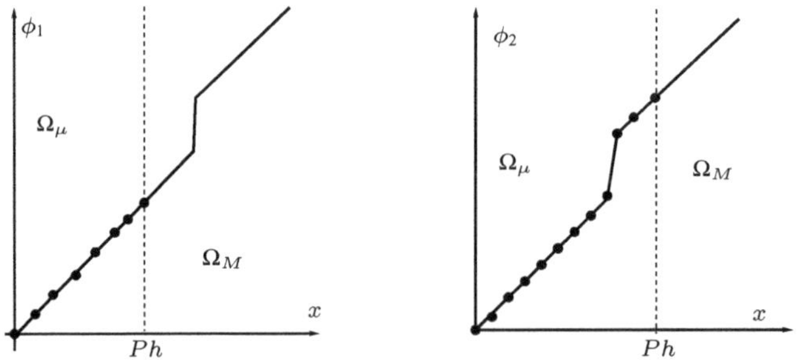

Figure 3. Configurations ϕ_1 (with a crack in Ω_M) and ϕ_2 (with a crack in Ω_μ).

Let us define $\phi_1 \in X_c$ by

$$\begin{aligned} \phi_1^i &= ih, \quad 0 \le i \le P, \\ \phi_1(x) &= x + (a-L)\mathcal{H}(x-y_0), \quad \forall x \ge Ph, \end{aligned} \tag{4.44}$$

where y_0 is any real number, $Ph < y_0 < L$ (see Figure 3). The map ϕ_1 represents a deformation with a fracture in Ω_M, located at y_0. We also consider $\phi_2 \in X_c$ defined by

$$\begin{aligned} \phi_2^i &= ih, \quad 0 \le i \le Q_0, \\ \phi_2^i &= ih + (a-L), \quad 1 + Q_0 \le i \le P, \\ \phi_2(x) &= x + (a-L), \quad \forall x \ge Ph, \end{aligned} \tag{4.45}$$

which represents a deformation with a fracture in Ω_μ on the atomic bond $(Q_0, 1+Q_0)$, with arbitrary integer $Q_0 \in [0, P-1]$.

Their energies are

$$\begin{aligned} E_c(\phi_1) &= V_{\rm LJ}(1), \\ E_c(\phi_2) &= \left(1 - \frac{1}{N}\right) V_{\rm LJ}(1) + \frac{1}{N} V_{\rm LJ}\left(\frac{a-L}{h} + 1\right) \\ &= \left(1 - \frac{1}{N}\right) V_{\rm LJ}(1) + o\left(\frac{1}{N}\right). \end{aligned}$$

Hence, $E_c(\phi_2) > E_c(\phi_1)$: a configuration with a fracture in Ω_M has a lower energy than a configuration with a fracture in Ω_μ. Actually, the energy cost for a fracture in Ω_μ is $\frac{h}{L} |V_{\rm LJ}(1)| + o\left(\frac{h}{L}\right)$, whereas the cost for a fracture in Ω_M is zero[10]. Hence, even if the difference $E_c(\phi_2) - E_c(\phi_1)$ is small, the minimization of E_c leads to locating the fracture in Ω_M.

There are several ways out of this difficulty: First, it is possible to modify the definition of the coupled energy. More precisely, the idea is to modify the elastic energy

[10] We define the cost of a fracture by the difference between the energy of a configuration with a fracture and the energy of the reference configuration.

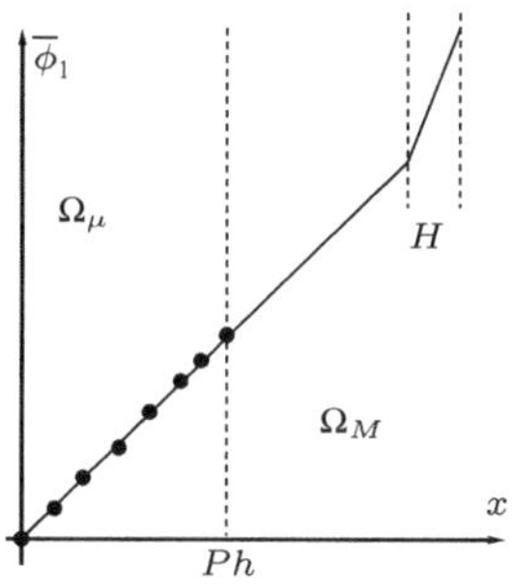

Figure 4. Configuration $\overline{\phi}_1$ (with a crack in Ω_M).

density that is used in Ω_M in such a way as it remains consistent with the atomistic model, and such that making a fracture in Ω_M costs more than making a fracture in Ω_μ. See [23] (in particular, Section 2.4) for details. Introducing this modified coupled energy is possible because we have a very detailed understanding of the difficulties associated with the coupled energy (4.42).

Second, we notice that, in practice, one does not work with functions in the space (4.43), but within a finite dimensional subset of it. Consider for instance the Galerkin approximation

$$X_c^H = \{\phi \in X_c;\ \phi \text{ is piecewise affine on } \Omega_M \text{ on a mesh of size } H\}. \qquad (4.46)$$

Let us again compare the energies of cracked configurations. We see that ϕ_2 defined by (4.45) belongs to X_c^H. However, ϕ_1 defined by (4.44) does not belong to X_c^H. Let us define its discretized version $\overline{\phi}_1 \in X_c^H$ by

$$\begin{aligned} \overline{\phi}_1^i &= ih, \quad 0 \le i \le P, \\ \overline{\phi}_1(x) &= x, \quad Ph \le x \le L-H, \\ \overline{\phi}_1(x) &= a + (x-L)\left(\frac{a-L}{H}+1\right), \quad L-H \le x \le L. \end{aligned}$$

The map $\overline{\phi}_1$ belongs to X_c^H and represents a deformation with a fracture in Ω_M, located in the last finite element (see Figure 4). We find

$$E_c\left(\overline{\phi}_1\right) = \left(1-\frac{H}{L}\right)V_{\mathrm{LJ}}(1) + \frac{H}{L}V_{\mathrm{LJ}}\left(\frac{a-L}{H}+1\right).$$

If H is chosen such that $H \gg h$ (which makes the problem defined on X_c^H cheaper to solve than the fully atomistic problem, since it involves fewer degrees of freedom), then $E_c\left(\overline{\phi}_1\right) > E_c(\phi_2)$. Actually, one can prove that the minimizers of the coupled energy E_c on X_c^H have a fracture which occurs in the atomistic domain Ω_μ.

Let us summarize our observations:

- Minimizing the coupled energy (4.42) on the space (4.43) leads to issues, since a fracture appears in the *continuum* domain.

- Introducing a (coarse enough) mesh in the continuum domain regularizes the problem: when minimizing the coupled energy (4.42) on the space (4.46), a fracture appears, which is contained in the *atomistic* domain.

The situation is hence saved by the fact that we look at a discretized version of the problem, for a sufficiently large discretization parameter H.

This discussion shows that the nonconvex case is much more challenging than the convex case, and that unexpected difficulties appear, although the general setting is extremely simple (a one-dimensional system with nearest-neighbour interactions). We expect to meet similar difficulties in more general settings (problems in higher dimension or long-range interactions).

5 Discussion

In this section, we review some general questions about the modelling and setting of the problem. We also review some methods recently proposed and discuss some open questions at the front of research.

5.1 The nearest-neighbour assumption

We gather here some remarks concerning the atomistic model (2.4) that we used:

$$E_\mu(\phi^1,\dots,\phi^N) = \frac{1}{2}\sum_{i=1}^N \sum_{j\neq i} V(\phi^j - \phi^i).$$

In general, the potential energy $V(\phi^j - \phi^i)$ tends to zero when the distance between atoms i and j goes to $+\infty$ (see Remark 2.2). Hence, one often simplifies the expression (2.4) (which is a sum of $N^2/2$ terms) by introducing a cutoff $r^{\text{cut}} > 0$ in V: we approximate $E_\mu(\phi^1,\dots,\phi^N)$ by

$$E_\mu^{\text{cut}}(\phi^1,\dots,\phi^N) = \frac{1}{2}\sum_{i=1}^N \sum_{j\neq i, \|\phi^j-\phi^i\|\leq r^{\text{cut}}} V(\phi^j - \phi^i), \tag{5.1}$$

which is cheaper to evaluate than (2.4), in general (indeed, only a finite number of atoms j are within the cutoff radius of any atom i for reasonable deformations). Analyzing a model such as (2.4) is difficult due to the long-range interactions (whose range increases when N increases!). Analyzing a model based on (5.1) is also challenging, because we do not know a priori which atoms interact with each other. Hence, for analysis purposes, a further simplification is often made: instead of considering the energy (5.1), with a finite range of interaction in terms of the *distance between atoms*, we consider an energy with a finite range of interaction in terms of *atoms indices*:

$$E_\mu^{\text{FR}}(\phi^1,\dots,\phi^N) = \frac{1}{2}\sum_{i=1}^N \sum_{j\neq i, \|j-i\|\leq I_R} V(\phi^j - \phi^i), \tag{5.2}$$

for some cutoff I_R. In a one-dimensional setting, the extreme case is $I_R = 1$, corresponding to nearest-neighbour interactions.

When making the above assumption of working with (5.2), we implicitly assume that the ordering of the atoms in the deformed configuration is the same as the ordering in the reference configuration. Consider indeed the case of nearest-neighbour interactions for a one-dimensional model:

$$E_\mu^{\mathrm{FR}}(\phi^1, \dots, \phi^N) = \sum_{i=1}^{N-1} V(\phi^{i+1} - \phi^i). \tag{5.3}$$

If the ordering in the current (deformed) configuration is not the original one, then we have one of the following situations:

- There exist indices i and j, $1 \le i, j \le N$, with $j > i+1$, such that $\phi^i \le \phi^j \le \phi^{i+1}$. The interaction between atoms i and j is not taken into account in (5.3), although they are closer to each other than atoms i and $i+1$, whose interaction is taken into account in (5.3). This does not make sense.
- There exist indices i and j, $1 \le i, j \le N$, with $j \le i-1$, such that $\phi^i \le \phi^j \le \phi^{i+1}$. Again, it does not make sense not to consider the couple $(j, i+1)$ whereas the couple $(i, i+1)$ is considered.

So, when working with (5.2), a natural assumption is that the ordering of the atoms is preserved by the deformation. A simple way to enforce this is to impose the constraint $\phi^{i+1} \ge \phi^i$ for all i in the variational space.

For pedagogical purposes, we have not included this constraint in the atomistic variational space (4.8). See [23, 24] for an analysis of the same models as in Section 4, where the variational spaces include this ordering constraint. We show there that, if the boundary value a is large enough (recall that, in (4.8), we impose $\phi^N = a$), then the constraint is not attained, and the analysis is exactly the same as the one without constraint that we perform here. On the other hand, if a is not large enough, then the variational problems do not have any solution.

5.2 Beyond global minimization

In these notes, we have adopted a variational viewpoint: we have defined the solution of the various models as the *global minimizer* of the respective energies (as we did in [23, 24]). This approach was also followed in [102], where problems similar to the one studied here were studied. More precisely, a finite one-dimensional atomistic chain is considered, with the energy (2.4), and the Lennard–Jones interaction potential (4.39) for V. There are no body forces and no boundary conditions (except a Dirichlet condition to remove translational invariance). The reference problem is hence written as

$$\inf \left\{ \frac{1}{2} \sum_{i=1}^{N} \sum_{j \ne i} V_{\mathrm{LJ}}(\phi^j - \phi^i); \quad \phi \in \mathbb{R}^N,\ \phi^1 = 0,\ \phi^{i+1} > \phi^i \right\}.$$

The influence, on the solution of the above variational problem, of introducing a cutoff radius in $V_{\rm LJ}$, as well as the influence of making a QuasiContinuum like approximation (see Section 5.3) are analyzed.

An alternative approach to global minimization is to look for *local minimizers* (see, for instance, [63, 64, 66]). A difficulty linked to this approach is that there are in general many local minimizers. In that setting, one often proves that, close to a given atomistic solution, there exists a solution of the hybrid model.

Another possibility is to try and select *one* local minimizer by some criterion. For instance, one can consider a gradient flow dynamics [128, 138], which writes, in its simplest formulation,

$$\frac{d\phi}{dt} = -\nabla E_\mu(\phi).$$

In the long-time limit, the solution of the above dynamical system converges to a configuration which is a local minimizer of the energy, and which is considered as the reference solution of the atomistic model.

Yet another option is to study *critical points* of the energy rather than local minimizers. See [129, 130] for such a choice in an atomistic to continuum setting (and also [103] for a study in a two-dimensional setting, under some simplifying assumptions, reminiscent of convexity assumptions). Note that critical points are also considered in [12, 153], where the authors go beyond standard continuum mechanics models by considering elastic densities that depend on second derivatives of the deformation, hence obtaining a continuum model with a small parameter (see Section 5.5, and energy (5.11) in particular).

5.3 The QuasiContinuum method and the ghost forces

The method that we have analyzed in Section 4 is a toy example for more advanced methods such as the QuasiContinuum method (QCM). In its initial version [150, 151], this method starts from the continuum scale, with a standard continuum mechanics model, discretized by a finite element method. The multiscale feature of the method appears when the elastic energy of an element is computed. Depending on some criteria, some elements are declared to be too heterogeneously strained for a macroscopic description to hold, and they are considered as a set of discrete particles. The energy of each element is computed according to the scale at which the element is described: either the element is too heterogeneously strained, and its energy is computed on the basis of an underlying atomistic model, or a standard continuum mechanics formula is used.

In the second version of the method [146], that we describe below, the opposite viewpoint is adopted. The starting point is a multibody atomistic energy, which is such that it is possible to define the energy $E_i(\phi)$ of the atom i when the current configuration of the atomistic system is ϕ. We start from the energy

$$E_\mu(\phi) = \sum_{i=1}^{N} E_i(\phi). \tag{5.4}$$

Let us assume here that we look for the global minimizer of (5.4), for ϕ subjected to boundary conditions BC that we do not make precise:

$$\inf\left\{E_\mu(\phi);\ \phi \in \mathbb{R}^{dN},\ \phi \text{ satisfies BC}\right\}, \tag{5.5}$$

where d is the space dimension. There are two difficulties with such a model: there are too many degrees of freedom, and the sum (5.4) involves too many terms.

To circumvent the first difficulty, a few atoms are identified as representative atoms. We denote them as the so-called *repatoms*. There are N_r of them, with $N_r \ll N$. Let i_α, $1 \le \alpha \le N_r$, denote their indices. Their current positions $\left\{\phi^{i_\alpha}\right\}_{\alpha=1}^{N_r}$ are the degrees of freedom of the reduced system, whereas the positions of the non-representative atoms are obtained by interpolation (the idea of interpolation is related to the Cauchy–Born rule; see [70] for a seminal paper on the validity of this assumption for a spring lattice system, [54] for an enhanced version of this rule, and [65] for some analysis of this rule in a dynamical setting). More precisely, a mesh is built upon the repatoms in the reference configuration. Let ϕ_0^i be the reference position of atom i, and $S_\alpha(x)$ be the piecewise affine function associated to the node α (we hence consider a P_1-finite element method). In a one-dimensional setting, we thus have

$$S_\alpha(x) = \begin{cases} \dfrac{x - \phi_0^{i_{\alpha-1}}}{\phi_0^{i_\alpha} - \phi_0^{i_{\alpha-1}}} & \text{if } \phi_0^{i_{\alpha-1}} \le x \le \phi_0^{i_\alpha}, \\[2ex] \dfrac{x - \phi_0^{i_{\alpha+1}}}{\phi_0^{i_\alpha} - \phi_0^{i_{\alpha+1}}} & \text{if } \phi_0^{i_\alpha} \le x \le \phi_0^{i_{\alpha+1}}, \\[2ex] 0 & \text{otherwise.} \end{cases}$$

The position of any atom i in the system is obtained from the positions of the repatoms by

$$\phi^i = \sum_{\alpha=1}^{N_r} S_\alpha(\phi_0^i)\, \phi^{i_\alpha}. \tag{5.6}$$

Otherwise stated, we make a Galerkin approximation of (5.5):

$$\inf\left\{E_\mu(\phi);\ \phi \in \mathbb{R}^{dN},\ \phi \text{ satisfies (5.6) and BC}\right\}. \tag{5.7}$$

The second difficulty (the number of terms in (5.4)) is solved by the fact that, due to the above reduction in the number of degrees of freedom, and due to the use of a P_1-interpolation, many atoms actually have the same energy. Assume indeed that the energy of atom i reads

$$E_i(\phi) = \sum_{j \neq i, \|\phi^j - \phi^i\| \le r^{\text{cut}}} V(\phi^j - \phi^i)$$

for some interaction potential V with some cutoff radius r^{cut}. Consider an atom i such that all the atoms j with which it interacts are in the same finite element. Then, in view of (5.6), $E_i(\phi)$ does not depend on i. Hence, all atoms i of a finite element ℓ satisfying the above condition have the same energy E_ℓ that only depends on the current positions of the repatoms (actually, only on the ones on the vertices of the finite element): $E_i(\phi) = E_\ell\left(\{\phi^{i_\alpha}\}_{\alpha=1}^{N_r}\right)$. For the other atoms, which interact with atoms belonging to different finite elements, we *approximate* their energy by the same quantity. We hence approximate (5.7) by

$$\inf\left\{\widetilde{E_\mu}(\phi);\ \phi \in \mathbb{R}^{dN},\ \phi \text{ satisfies (5.6) and BC}\right\}, \tag{5.8}$$

with $\widetilde{E_\mu}(\phi) = \sum_\ell n_\ell E_\ell\left(\{\phi^{i_\alpha}\}_{\alpha=1}^{N_r}\right)$, where n_ℓ is the number of atoms included in the finite element ℓ. The problem (5.8) can be solved in practice since it involves a reasonable number of degrees of freedom ($N_r \ll N$) and energies $\widetilde{E_\mu}(\phi)$ that actually can be computed. So, in its second version, the QuasiContinuum method somewhat consists in an efficient quadrature rule to compute (5.4). Following this viewpoint, a reportedly more stable quadrature rule has been proposed in [86], based on cluster summation rules. Some analysis of such rules can be read in [109].

The QuasiContinuum method has been applied on a number of practical examples (see, e.g., [117, 145, 152]). A review of its current status can be read in [118]. See also [6] for an application of a similar idea to another context. Note that an interesting feature of the QuasiContinuum method is the use of some criteria to introduce (or remove) repatoms. Hence, the accuracy of the model is adapted, on the fly and automatically, to the loading conditions.

An interesting application for the QuasiContinuum method is the computation of deformations with dislocations that are atomistically localized phenomena. The Frenkel–Kontorova model is a one-dimensional model that somewhat describes dislocations. Some numerical analysis of the QuasiContinuum method for this model can be read in [8, 9, 10].

Let us conclude this section by addressing the issue of ghost forces (this notion has been first introduced and discussed in [146], see also [51, 52] for another presentation and [148]). Consider a one-dimensional chain with second-nearest neighbour interactions (this is the simplest case when such ghost forces arise) of $N+1$ atoms, whose current positions are $\phi^0, \dots, \phi^N$. The energy per particle reads (see (4.6))

$$E_\mu(\phi) = \frac{1}{N}\sum_{i=0}^{N-1} V_1\left(\frac{\phi^{i+1}-\phi^i}{h}\right) + \frac{1}{N}\sum_{i=0}^{N-2} V_2\left(\frac{\phi^{i+2}-\phi^i}{h}\right)$$

for some interaction potentials V_1 and V_2, and with $Nh = L$ (we consider the case of no body forces). For all $2 \le i \le N-2$, we obtain

$$\frac{\partial E_\mu}{\partial \phi^i}(\phi) = \frac{1}{Nh}\left[-V_1'\left(\frac{\phi^{i+1}-\phi^i}{h}\right) - V_2'\left(\frac{\phi^{i+2}-\phi^i}{h}\right)\right.$$
$$\left. + V_1'\left(\frac{\phi^i-\phi^{i-1}}{h}\right) + V_2'\left(\frac{\phi^i-\phi^{i-2}}{h}\right)\right].$$

It is easy to check that any homogeneously deformed state, that is $\phi^i_{\rm HD} = a + bi$ for any a and b, satisfies $\frac{\partial E_\mu}{\partial \phi^i}(\phi_{\rm HD}) = 0$ for all $2 \leq i \leq N-2$. Hence, up to boundary effects (responsible for what is called *surface relaxation*), any homogeneously deformed state is a critical point of the energy.

Let us now consider a coupled energy, in the spirit of (4.13), but with no external body forces, and taking into account second-nearest neighbour interactions. Recall that the partition $\Omega = \Omega_M \cup \Omega_\mu$ is defined by (4.12), that the variational space X_c is defined by (4.14), and its homogeneous version X_c^0 is defined by (4.27). For any $\phi \in X_c$, we consider the coupled energy

$$E_c(\phi) = \frac{1}{L}\int_{Ph}^{L} \left[V_1(\phi'(x)) + V_2(2\phi'(x))\right] dx$$
$$+ \frac{1}{N}\sum_{i=0}^{P-1} V_1\left(\frac{\phi^{i+1}-\phi^i}{h}\right) + \frac{1}{N}\sum_{i=0}^{P-2} V_2\left(\frac{\phi^{i+2}-\phi^i}{h}\right).$$

We calculate the Gâteaux derivative of E_c for any $\psi \in X_c^0$:

$$\begin{aligned}
\langle DE_c(\phi), \psi\rangle &= \lim_{t\to 0} \frac{E_c(\phi + t\psi) - E_c(\phi)}{t} \\
&= \frac{1}{L}\int_{Ph}^{L} \left[V_1'(\phi'(x)) + 2V_2'(2\phi'(x))\right] \psi'(x)\, dx \\
&\quad + \frac{1}{N}\sum_{i=0}^{P-1} V_1'\left(\frac{\phi^{i+1}-\phi^i}{h}\right) \frac{\psi^{i+1}-\psi^i}{h} \\
&\quad + \frac{1}{N}\sum_{i=0}^{P-2} V_2'\left(\frac{\phi^{i+2}-\phi^i}{h}\right) \frac{\psi^{i+2}-\psi^i}{h}.
\end{aligned}$$

Consider now a homogeneously deformed configuration $\phi_{\rm HD}$ (e.g., there exists c such that $\phi'_{\rm HD}(x) = c$ for all $x \in (Ph, L)$ and $\phi^{i+1}_{\rm HD} - \phi^i_{\rm HD} = ch$ for $0 \leq i \leq P-1$). Denoting $C_1 = V_1'(c)$ and $C_2 = V_2'(2c)$, we find

$$\begin{aligned}
\langle DE_c(\phi_{\rm HD}), \psi\rangle &= \frac{C_1 + 2C_2}{L}\int_{Ph}^{L} \psi'(x)\, dx \\
&\quad + \frac{C_1}{N}\sum_{i=0}^{P-1} \frac{\psi^{i+1}-\psi^i}{h} + \frac{C_2}{N}\sum_{i=0}^{P-2} \frac{\psi^{i+2}-\psi^i}{h} \\
&= \frac{C_2}{L}\left(-\psi^P + \psi^{P-1} - \psi^1\right),
\end{aligned}$$

where we have used that $\psi(Ph) = \psi^P$. We see that $DE_c(\phi_{\rm HD})$ is not equal to zero, due to forces on atom 1 (this was already the case for the fully atomistic problem; here, it again gives rise to surface relaxation), and due to forces on atoms P and $P-1$, that is at the interface between the two models. Hence, some spurious forces (called ghost forces) appear at the interface between the atomistic and the continuum domains[11]. This is also the case with the QuasiContinuum method. Even up to effects on the boundary of the computational domain, a homogeneously deformed configuration is not a critical point of the hybrid energy.

Two possibilities are at hand. The first one consists in going on working with the hybrid energy and the forces derived from it. This method is for instance analyzed in [53] and also in [52], where the influence of the ghost forces is studied: estimates between the solution of the hybrid model and the solution of the atomistic model are derived, and the size of the domain (around the interface) which is impacted by these ghost forces is estimated. The second one consists in *correcting* the forces such that a homogeneously deformed configuration makes the forces zero. An extra term is hence added to the forces to compensate for the spurious effects at the micro-macro interface. As a result, these forces are not the gradient of any energy. See [51] for some numerical analysis of such a method.

5.4 Multiscale methods without a coarse-grained model

Some methods have been developed in the literature that do not need an explicit macroscopic model. The idea is to design a macroscopic-like method, which is used in the *whole* computational domain, and to resort to microscopic computations each time the constitutive law has to be used. Hence, in a continuum mechanics setting, the method consists in bypassing an analytical formula that provides the macroscopic stress as a function of the macroscopic strain by solving a fine scale problem. The macroscopic strain is used to design appropriate boundary conditions for that fine scale problem. The macroscopic stress is obtained by postprocessing its solution.

A first endeavour of such a method is the FE^2 method proposed in [67], where the fine scale model is a continuum model with some small length scales (see also [119]). See [4, 75, 76, 77, 82, 111, 113, 144] for some mathematical analysis of related methods, namely numerical homogenization methods for elliptic partial differential equations.

Note that the coarse and the fine scale models can be different in nature. See [58] for some mathematical analysis in a general setting of such a method, [114] for an application to granular media equilibrium, and [59, 60, 61, 62, 98, 99] for some application to elastodynamics, where the constitutive law is computed from molecular dynamics (in order to take into account finite temperature effects, molecular dynamics is run with a thermostatting method [37]).

5.5 The Arlequin method and other methods for equilibrium computation

In a similar spirit as the QuasiContinuum method, we mention the Arlequin method [16, 17, 18, 19], which is based on a domain decomposition idea: on each subdomain,

[11] Note that, for nearest neighbour interactions, $V_2 \equiv 0$, hence $C_2 = 0$ for any deformation rate c, and there are no ghost forces.

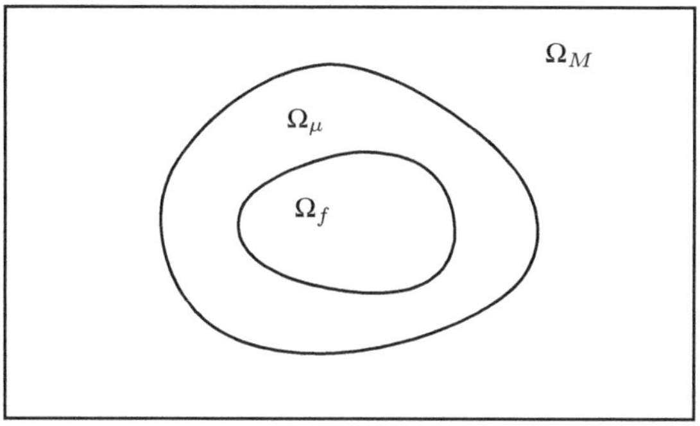

Figure 5. The three domains Ω_f, Ω_μ and Ω_M are strictly embedded one in each other: $\Omega_f \subsetneq \Omega_\mu \subsetneq \Omega_M$. The gluing domain is $\Omega_g = \Omega_\mu \setminus \Omega_f$.

different models are used. The method was originally developed to be used with different continuum mechanics models. However, it can also be used to couple continuum mechanics with a discrete model [13, 136] (see also [126, 135] for goal-oriented a posteriori error analysis in such a multiscale framework).

Let us briefly present the method when it is used to couple two different linear continuum mechanics models in dimension d. We follow the presentation of [17]. Consider a domain Ω_f, strictly embedded in the domain Ω_μ, which is itself strictly embedded in the computational domain Ω_M (see Figure 5). In Ω_f, we want to use a precise and expensive model (denoted μ-model), for instance because of the complexity of the expected deformation. On the contrary, a coarse and cheap model (denoted M-model) is enough on $\Omega_M \setminus \Omega_\mu$. The domain $\Omega_g = \Omega_\mu \setminus \Omega_f$ helps to make the transition between the two models. We assume for simplicity that we impose homogeneous Dirichlet boundary conditions on $\partial\Omega_M$.

Let us consider smooth nonnegative weight functions α_M and β_M, defined on Ω_M, and α_μ and β_μ, defined on Ω_μ. We now introduce the bilinear forms

$$\begin{aligned} a_M(u_M, v_M) &= \int_{\Omega_M} \alpha_M(x)\, \varepsilon(v_M) : \Lambda_M : \varepsilon(u_M)\, dx, \\ a_\mu(u_\mu, v_\mu) &= \int_{\Omega_\mu} \alpha_\mu(x)\, \varepsilon(v_\mu) : \Lambda_\mu : \varepsilon(u_\mu)\, dx, \end{aligned}$$

with

$$\varepsilon(u) = \frac{1}{2}\left(\nabla u + \nabla u^T\right), \tag{5.9}$$

and where Λ_M and Λ_μ are the two symmetric 4th-order tensors that correspond to the two models that we consider (see Example 2.1 for some details on linear elasticity). These two bilinear forms are defined on $X_M \times X_M$ and $X_\mu \times X_\mu$, respectively, with $X_M = (H^1_0(\Omega_M))^d$ (recall that we impose homogeneous Dirichlet boundary conditions on $\partial\Omega_M$) and $X_\mu = (H^1(\Omega_\mu))^d$.

Let $f \in (L^2(\Omega_M))^d$ be a body force and let us define on X_M and X_μ, respectively, the linear forms

$$\begin{aligned} b_M(v_M) &= \int_{\Omega_M} \beta_M(x)\, f(x) \cdot v_M(x)\, dx, \\ b_\mu(v_\mu) &= \int_{\Omega_\mu} \beta_\mu(x)\, f(x) \cdot v_\mu(x)\, dx. \end{aligned}$$

The standard problem (associated to the M-model) would be to find $u_M \in X_M$ such that

$$\forall v_M \in X_M, \quad a_M(u_M, v_M) = b_M(v_M)$$

with $\alpha_M \equiv \beta_M \equiv 1$ on Ω_M. The Arlequin method consists in finding $(u_M, u_\mu, \Phi) \in X_M \times X_\mu \times X_g$ such that, for all test functions $(v_M, v_\mu, \Psi) \in X_M \times X_\mu \times X_g$, we have

$$\begin{aligned} a_M(u_M, v_M) + a_\mu(a_\mu, v_\mu) + C(\Phi, v_M - v_\mu) &= b_M(v_M) + b_\mu(v_\mu), \\ C(\Psi, u_M - u_\mu) &= 0, \end{aligned} \tag{5.10}$$

where $X_g = (H^1(\Omega_g))^d$, and where the bilinear form C is defined on $X_g \times X_g$ by

$$C(\psi, v) = \int_{\Omega_g} (k_0 \psi(x) \cdot v(x) + k_1 \varepsilon(\psi) : \varepsilon(v))\, dx$$

for two positive parameters k_0 and k_1. Hence, the two displacements u_M and u_μ are coupled through the bilinear form C, which penalizes their difference (both in term of displacement and strain, through k_0 and k_1) in the gluing domain $\Omega_g = \Omega_\mu \setminus \Omega_f$.

The weight functions are chosen such that $\alpha_M = \beta_M = 1$ in $\Omega_M \setminus \Omega_\mu$ and $\alpha_M + \alpha_\mu = \beta_M + \beta_\mu = 1$ in Ω_μ. We can hence choose $\alpha_\mu = \beta_\mu = 0$ in $\Omega_M \setminus \Omega_\mu$. So, in the exterior domain $\Omega_M \setminus \Omega_\mu$, only the M-model is taken into account. A natural choice would be to choose $\alpha_M = \beta_M = 0$ in Ω_f. In the interior domain Ω_f, only the μ-model would be taken into account, and both models would be taken into account in the gluing domain $\Omega_g = \Omega_\mu \setminus \Omega_f$ with a smoothly varying proportion. However, it is shown in [17] that such a choice is not a good one. A better choice is to ensure that

$$\alpha_M(x) \geq \alpha_0 \quad \text{and} \quad \alpha_\mu(x) \geq \alpha_0 \quad \text{in } \Omega_f$$

for some $\alpha_0 > 0$. Hence, even in the interior domain, the M-model should be taken into account. Under these assumptions, well-posedness, stability and consistency results for the Arlequin method are proved in [17].

A particular feature of the Arlequin method is that, on the domain Ω_μ, *both models coexist* (see Section 5.6 for another example of a multiscale method, dedicated to dynamical simulations, with the same feature). Note that this is not the case with the QuasiContinuum method. See [11] for some discussion on various possibilities to couple atomistic and continuum models by using a blending zone, where both models coexist. For such methods, based on an overlapping domain decomposition idea, alternating Schwarz schemes are often the methods of choice for parallel computing.

See [133] for some numerical analysis of these schemes in an atomistic-to-continuum framework.

Let us conclude this section by noting that the methods mentioned above all start from a fine scale model, which is computationally used in different ways. An alternative to multiscale methods is to try and homogenize the fine scale model under sufficiently weak assumptions, so that the resulting macroscopic model can be used everywhere in the domain, even if the deformation is not smooth. Along these lines, we mention the articles [12, 87, 153], where a continuum mechanics model is built, in which the elastic energy depends not only on the strain, but also on higher derivatives of the displacement. For instance, a one-dimensional setting is considered in [153] with the energy

$$E_M(\phi) = \int \left[W(\phi'(x)) + h^2 W_2(\phi'(x))(\phi''(x))^2 \right] dx \tag{5.11}$$

for some functions W and W_2 and some small parameter h. In the same vein, for some justification on the basis of atomistic models of the physical foundations of a class of continuum models, namely microcontinuum theories, see [38].

The energy (5.11) is a bulk energy. Some works have also lead to the consideration, in addition to such a bulk term, of surface energy terms [47] or terms that penalize displacement discontinuities [32].

See also [131] for the derivation, from an atomistic model, of some continuum models predicting phase transitions. Note that a Γ-limit approach is often the mathematical method of choice for such challenging homogenization questions (it was employed, e.g., in [32] and [131]).

It is not possible in such notes to describe all the recently proposed coupling methods in details. We just mention here the LATIN method [89, 90], the BSM method [132, 158], and the Virtual Internal Bond method [71, 72, 83, 84, 85, 105, 156, 161]. Note that several methods have also been proposed in a fluid mechanics context, which share many features with the methods described here, see [80, 81] and [106, 124, 125] (with also an application to solid contact modelling [108]). The question of atomistic to continuum coupling also arises for magnetic forces computation [120, 143]. Here, the idea is to start from an atomistic model for magnetic forces and pass to the continuum limit, hence obtaining expressions that only depend on macroscopic variables. More details on atomistic to continuum coupling methods in materials science can be read in [29] from a mathematical perspective, and in the review article [44] and in the monographs [36, 107, 137] from a materials science perspective.

5.6 Temperature and dynamical effects in multiscale methods

Taking into account temperature (that brings in fluctuations and randomness) and dynamics (that brings in inertial effects) in an atomistic to continuum coupling method is a very challenging issue for which a lot of questions are still open.

One viewpoint to take into account temperature is to consider statistical mechanics averages in the canonical ensemble. Consider a system of N particles at positions

$\phi = (\phi^1, \dots, \phi^N) \in \mathbb{R}^{dN}$ and let $E_\mu(\phi)$ be its energy. The finite temperature thermodynamical properties of the material are obtained from canonical ensemble averages,

$$\langle A \rangle = Z^{-1} \int_{\Omega^N} A(\phi) \exp(-\beta E_\mu(\phi)) \, d\phi, \tag{5.12}$$

where $\Omega \subset \mathbb{R}^d$ is the macroscopic domain in which the positions ϕ^i vary, A is the observable of interest, $\beta = 1/(k_B T)$ is the inverse temperature (T is the physical temperature and k_B is the Boltzmann constant), and $Z = \int_{\Omega^N} \exp(-\beta E_\mu(\phi)) \, d\phi$ is the partition function [37, 46].

The difficulty for computing (5.12) comes from the N-fold integral, where N, the number of particles, is extremely large (say 10^5). One method, among others, is to compute (5.12) as a long-time average

$$\langle A \rangle = \lim_{T \to +\infty} \frac{1}{T} \int_0^T A(\phi_t) \, dt \tag{5.13}$$

along the trajectory generated by the stochastic differential equation

$$d\phi_t = -\nabla E_\mu(\phi_t) \, dt + \sqrt{2\beta^{-1}} \, dW_t, \tag{5.14}$$

where W_t is a standard dN-dimensional Brownian motion.

It is often the case that observables of interest do not depend on the positions of *all* the atoms but only on *some* of them (for instance, because these atoms are located in a region of interest). We assume that this set of interesting atoms (also called *repatoms* as in the QuasiContinuum method terminology) is given a priori, and we denote by ϕ_r their positions. We hence write

$$\phi = (\phi^1, \dots, \phi^N) = (\phi_r, \phi_c), \quad \phi_r \in \mathbb{R}^{dN_r}, \quad \phi_c \in \mathbb{R}^{dN_c}, \quad N = N_r + N_c,$$

and our aim is to compute (5.12) for such observables, that is

$$\langle A \rangle = Z^{-1} \int_{\Omega^N} A(\phi_r) \exp(-\beta E_\mu(\phi)) \, d\phi. \tag{5.15}$$

The objective is to design a computational method that is cheaper than (5.13)–(5.14), building upon the specificity of the observable.

The QuasiContinuum method has recently been extended to handle such cases [57], building upon previous works [69, 96] (see also [43] for a similar approach as well as [48, 49, 50, 159]). The approach described in [57] is based on a low temperature asymptotics. More precisely, one first performs a Taylor expansion of the energy with respect to ϕ_c around a state that is a function of the repatoms positions ϕ_r. We write

$$\phi_c = \overline{\phi}_c(\phi_r) + \xi_c$$

for some function $\overline{\phi}_c$. Assuming ξ_c is small, we find by expanding E_μ

$$\begin{aligned} E_\mu(\phi_r, \phi_c) &= E_\mu(\phi_r, \overline{\phi}_c(\phi_r) + \xi_c) \\ &\approx E_\mu(\phi_r, \overline{\phi}_c(\phi_r)) + \frac{\partial E_\mu}{\partial \phi_c}(\phi_r, \overline{\phi}_c(\phi_r)) \cdot \xi_c + \frac{1}{2} \xi_c \cdot \frac{\partial^2 E_\mu}{\partial \phi_c^2} \cdot \xi_c. \end{aligned}$$

We consider the case where the atoms vary in $\Omega = \mathbb{R}^d$. Due to the above harmonic approximation (which is reminiscent of a low temperature approximation), we can analytically compute

$$E_{\mathrm{CG}}(\phi_r) = -\frac{1}{\beta} \ln \int_{\mathbb{R}^{dN_c}} \exp(-\beta E_\mu(\phi_r, \phi_c))\, d\xi_c. \tag{5.16}$$

Hence (5.15) reads

$$\langle A \rangle = \frac{\int_{\mathbb{R}^{dN_r}} A(\phi_r) \exp(-\beta E_{\mathrm{CG}}(\phi_r))\, d\phi_r}{\int_{\mathbb{R}^{dN_r}} \exp(-\beta E_{\mathrm{CG}}(\phi_r))\, d\phi_r}. \tag{5.17}$$

This formulation is more amenable to numerical computations than the expression (5.15) since the dimension dramatically decreased from dN to $dN_r \ll dN$.

Taking a different route, namely a thermodynamic limit, an alternative method has been recently proposed in [25, 134]. It first aims at computing the limit of the ensemble averages (5.15) when N_c, the number of non-representative atoms that we want to get rid of, goes to infinity. Second, it allows to efficiently compute free energies. More precisely, when $N_c \to +\infty$, the free energy $E_{\mathrm{CG}}(\phi_r)$ defined by (5.16) diverges. This reflects the extensiveness of the free energy with respect to the number of particles that have been integrated out. Hence, the meaningful quantity is the free energy per (removed) particle, and its thermodynamic limit $\lim_{N_c \to +\infty} E_{\mathrm{CG}}(\phi_r)/N_c$, which can be efficiently computed following the method proposed in [25]. Note that the advocated approach is restricted to the one-dimensional case. In this simple setting, it provides a computational strategy that is accurate and efficient. This approach can also be considered as a first step towards the numerical analysis, in a simple setting, of more advanced methods.

To conclude on the temperature issue, we wish to mention the work [74], which aims at coarse-graining atomistic systems, in relation to thermalization and molecular dynamics. See also [68] for another work in that setting.

Dynamics seems to be an even more challenging issue than temperature. Methods based on hybrid Hamiltonians (relying on a domain decomposition paradigm with some overlap) have been proposed in [1, 2, 35] (see [140, 141, 142] for applications as well as [14]; in [100, 121, 122, 123, 127, 147] the method has been applied for the simulation of extremely large systems). They are based upon the coupling of an atomistic model with a linear continuum mechanics model, whose parameters are predetermined by fine scale computations. Let us formally describe the idea.

The dynamics of atomistic systems is often a Hamiltonian dynamics derived from an underlying Hamiltonian function. Assuming pairwise interactions for the simplicity of exposition, this microscopic energy reads

$$\mathcal{H}_\mu(\phi_\mu, p_\mu) = \frac{1}{2} \sum_{i=1}^{N} \sum_{j \neq i} V(\phi_\mu^j - \phi_\mu^i) + \sum_{i=1}^{N} \frac{(p_\mu^i)^2}{2m}. \tag{5.18}$$

The first term is the potential energy (2.4) that depends on the current positions ϕ_μ of the atoms. The second term is the kinetic energy that depends on the momenta p_μ of the particles whose mass is m. At the macroscopic scale, dynamics is often modelled by the Navier elastodynamics equation. Consider its linear version in the absence of any external loading:

$$\rho \frac{\partial^2 u}{\partial t^2} - \operatorname{div}\left(\Lambda : \varepsilon\left(u\right)\right) = 0, \tag{5.19}$$

where $u = u(t, x)$ is the displacement field, $\rho = \rho(x)$ the volumic mass, and $\Lambda = \Lambda(x)$ a 4th order tensor (recall that $\varepsilon(u)$ is defined by (5.9); see Example 2.1 for some details on linear elasticity). After a spatial discretization, we obtain a system of ordinary differential equations that form a Hamiltonian system with the Hamiltonian function

$$\mathcal{H}_M(u_M, p_M) = \frac{1}{2} u_M^T K u_M + \frac{1}{2} p_M^T M^{-1} p_M, \tag{5.20}$$

where u_M are the macroscopic displacement variables, p_M are the macroscopic momentum variables, K is the stiffness matrix and M is the mass matrix (which is often assumed to be diagonal through the so-called *lumped mass* approximation).

We observe that (5.18) and (5.20) have the same structure. The authors of [1] took opportunity of that observation to couple the two models through the definition of a single Hamiltonian function. The aim is to give rise to a method that is more efficient than using the atomistic model (5.18) in the whole domain and more accurate than using the continuum mechanics model (5.19)–(5.20) everywhere. First, the computational domain is split into two overlapping subdomains Ω_M and Ω_μ. Let $\Omega_{\rm HS} = \Omega_M \cap \Omega_\mu$ be the overlapping region (see Figure 6). The degrees of freedom of the hybrid method are as follows:

- in $\Omega_\mu \setminus \Omega_{\rm HS}$, an atomistic description is used, based on atom positions ϕ_μ and momenta p_μ;
- in $\Omega_M \setminus \Omega_{\rm HS}$, a continuum mechanics description is used. After spatial discretization, the variables are the displacements u_M and the momenta p_M at the mesh nodes;
- in the hand-shaking region $\Omega_{\rm HS} = \Omega_M \cap \Omega_\mu$, both descriptions are used and coincide: the mesh nodes correspond to the reference positions of the atoms[12]. Let χ^i be the displacement and w^i the momentum of the node i.

The method proposed in [1] includes no adaptation of the partition along the computation. Since a *linear* continuum mechanics model is used in Ω_M, this zone should be located where the reference deformation (given by the atomistic model used in the whole domain) is smooth and small. Some consistency between both models is provided by the fact that the continuum mechanics tensor Λ of (5.19) is precomputed from the atomistic model, by a molecular dynamics simulation, using (5.18).

[12] For elements further away from $\Omega_{\rm HS}$, the mesh size increases in Ω_M to reach dimensions of 4 to 8 atomistic lattice parameters.

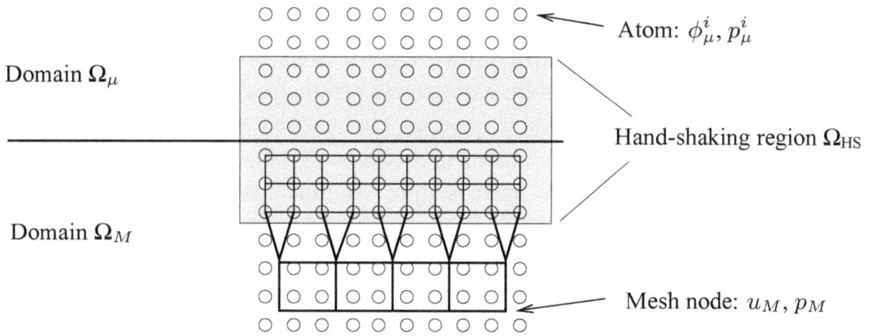

Figure 6. Degrees of freedom in the hand-shaking region Ω_{HS}.

The hybrid Hamiltonian is defined by

$$\mathcal{H}(\phi_\mu, p_\mu, u_M, p_M, \chi, w) = \frac{1}{2} \sum_{i,j \in \Omega_\mu \setminus \Omega_{\mathrm{HS}},\, i \neq j} V(\phi_\mu^j - \phi_\mu^i) + \sum_{i \in \Omega_\mu \setminus \Omega_{\mathrm{HS}}} \frac{(p_\mu^i)^2}{2m}$$

$$+ \frac{1}{2} u_M^T K u_M + \frac{1}{2} p_M^T M^{-1} p_M + V^{\mathrm{HS}}(\chi) + \sum_{i \in \Omega_{\mathrm{HS}}} \frac{(w^i)^2}{2m}, \quad (5.21)$$

where V^{HS} is some weighted average between the atomistic potential energy and the continuum mechanics one. The weight function is a parameter of the method, which is similar in spirit to the weight functions used in the Arlequin method (see Section 5.5). From (5.21), Hamiltonian equations of motion are derived for the evolution of ϕ_μ, p_μ, u_M, p_M, χ and w.

Actually, in a dynamical setting, the consistency question is already a difficult one: what is the limit of the dynamics of an increasingly large system of discrete particles? The previously described method makes the implicit assumption that this is an elastodynamics equation, but this is not clear from a rigorous mathematical viewpoint. Some analytical answer is provided in [20] for a linear system (see also [115]). More precisely, the authors work in dimension 3 and consider the Newton equations

$$m_h^i \frac{d^2 u_h^i}{dt^2} = -\nabla_{u_h^i} E_h\left(u_h^1, \ldots, u_h^N\right), \quad 1 \leq i \leq N,$$

for a system of N particles with mass m_h^i (along with initial and boundary conditions). In the above equation, u_h^i is the displacement of the particle i, E_h is the potential energy of the system, and h is a small parameter that represents the atomistic lattice parameter. The particle masses depend on h according to $m_h^i = M_i h^3$ for some M_i independent of h (note that this scaling is consistent with a finite macroscopic volumic mass ρ). The potential energy reads

$$E_h(u) = \frac{1}{2} \sum_{i,j} (u_h^i - u_h^j)^T C_h^{ij} (u_h^i - u_h^j)$$

for some symmetric matrices C_h^{ij} that again depend on h. The energy is invariant by translation, and, due to the structure of C_h^{ij}, it is also invariant by rotation. Using compactness arguments, based on a discrete Korn inequality, the authors of [20] prove that, when h goes to zero and N goes to $+\infty$ with Nh^3 fixed, $\{u_h^i\}_{i=1}^N$ converges, in some weak sense, to a function $\overline{u}$ that solves (5.19), supplemented by initial and boundary conditions, with the macroscopic volumic mass ρ and 4th order tensor Λ being derived from the microscopic masses m_h^i and matrices C_h^{ij}.

The question for nonlinear problems is much harder. The propagation of waves in a one-dimensional chain with a simple doublewell potential is analytically investigated in [39, 149]. See [55, 73, 116] for some studies with general interaction potentials, and also [139]. In [56], the problem is addressed in a somewhat different way (see also [78]). The starting point is again Newton's equation for a system of N discrete particles of mass m and potential energy E_μ:

$$m \frac{d^2 \phi^i}{dt^2} = -\nabla_{\phi^i} E_\mu \left(\phi^1, \dots, \phi^N\right), \quad 1 \leq i \leq N, \tag{5.22}$$

supplemented by initial conditions, where ϕ^i is the current position of atom i. Introduce next a nonnegative smooth window function $\chi = \chi(t,x)$, with $\int_{\mathbb{R}^4} \chi(t,x)\,dt\,dx = 1$. A natural choice is to choose χ with a compact support, which is macroscopically small and microscopically very large (hence, in that support, there are many discrete particles). For each particle i, we define

$$\chi_i(\theta,t,x) = \chi(\theta - t, \phi^i(\theta) - x), \quad \theta, t \in \mathbb{R},\ x \in \mathbb{R}^3,$$

so that $\int_{\mathbb{R}} \chi_i(\theta,t,x) d\theta$ represents the probability of finding the particle i at position x at time t. It is next natural to define the macroscopic mass density ρ and momentum density ρv by

$$\begin{aligned} \rho(t,x) &= \int_{\mathbb{R}} \sum_i m\, \chi_i(\theta,t,x)\, d\theta, \\ \rho(t,x)\, v(t,x) &= \int_{\mathbb{R}} \sum_i m\, \dot{\phi}_i(\theta)\, \chi_i(\theta,t,x)\, d\theta. \end{aligned}$$

The macroscopic energy density ρe is defined similarly. A simple computation shows that

$$\frac{\partial \rho}{\partial t} + \operatorname{div}(\rho v) = 0 \quad \frac{\partial(\rho v)}{\partial t} + \operatorname{div} \mathcal{P} = 0, \tag{5.23}$$

for some flux function $\mathcal{P}$ that depends on $\left\{\phi^i(t)\right\}_{i=1}^N$ and cannot be exactly expressed as a function of the macroscopic variables ρ, v and e (the same issue appears on the energy conservation equation). Hence, at this point, one needs to make an approximation and postulate an expression for $\mathcal{P}$ as a function of ρ, v and e (a so-called *closure relation*). In [56], several closure relations are proposed, and numerically tested: the macroscopic problem (5.23) is simulated and its results are compared with the results of the microscopic, fully atomistic, problem (5.22). It turns out that none of the tested closure approximations is fully satisfactory.

We finally mention [88] for some studies on how to transmit a pulse from an atomistic domain to a continuum domain (and back), when the continuum model is discretized by a discontinuous Galerkin scheme. See also [101, 104] for some numerical studies taking into account that, since interacting potentials are neither convex nor concave, the nonlinear wave equation, which is formally obtained from Newton's equation, can be either hyperbolic or elliptic, which creates theoretical and numerical difficulties.

Acknowledgments. This contribution gathers lecture notes of a short course given at the TU Berlin in July 2008, as part of the program "Analytical and numerical aspects of partial differential equations". This program was funded by the Luftbrückendank Foundation, and organized by E. Emmrich and P. Wittbold. I wish to thank the Luftbrückendank Foundation for its financial support, and E. Emmrich and P. Wittbold for their invitation. I wish also to thank C. Le Bris and X. Blanc for introducing me to the field of atomistic to continuum coupling methods. Finally, let me thank M. Dobson, E. Emmrich, C. Le Bris, M. Luskin, C. Patz, E. Tadmor and P. Wittbold for their suggestions on a previous version of these notes.

References

[1] F. F. Abraham, J. Q. Broughton, N. Bernstein and E. Kaxiras, Spanning the continuum to quantum length scales in a dynamic simulation of brittle fracture, *Europhys. Lett.* **44** (1998), pp. 783–787.

[2] ———, Spanning the length scales in dynamic simulation, *Computers in Physics* **12** (1998), pp. 538–546.

[3] G. Alberti and C. Mantegazza, A note on the theory of SBV functions, *Bollettino U.M.I. Sez. B* **7** (1997), pp. 375–382.

[4] G. Allaire, Homogenization and two-scale convergence, *SIAM J. Math. Anal.* **23** (1992), pp. 1482–1518.

[5] L. Ambrosio, N. Fusco and D. Pallara, *Functions of bounded variation and free discontinuity problems*, Oxford University Press, New York, 2000.

[6] M. Anitescu, D. Negrut, A. El-Azab and P. Zapol, A note on the regularity of reduced models obtained by nonlocal quasi-continuum-like approaches, *Mathematical Programming* **118** (2009), pp. 207–236.

[7] M. Arndt and M. Griebel, Derivation of higher order gradient continuum models from atomistic models for crystalline solids, *SIAM J. Multiscale Model. Simul.* **4** (2005), pp. 531–562.

[8] M. Arndt and M. Luskin, Goal-oriented atomistic-continuum adaptivity for the quasicontinuum approximation, *Int. J. Multiscale Comput. Eng.* **5** (2007), pp. 407–415.

[9] ———, Error estimation and atomistic-continuum adaptivity for the quasicontinuum approximation of a Frenkel-Kontorova model, *SIAM J. Multiscale Model. Simul.* **7** (2008), pp. 147–170.

[10] ———, Goal-oriented adaptive mesh refinement for the quasicontinuum approximation of a Frenkel-Kontorova model, *Comput. Methods Appl. Mech. Eng.* **197** (2008), pp. 4298–4306.

[11] S. Badia, M. Parks, P. Bochev, M. Gunzburger and R. Lehoucq, On atomistic-to-continuum coupling by blending, *SIAM J. Multiscale Model. Simul.* **7** (2008), pp. 381–406.

[12] S. Bardenhagen and N. Triantafyllidis, Derivation of higher order gradient continuum theories in 2, 3-D non-linear elasticity for periodic lattice models, *J. Mech. Phys. Solids* **42** (1994), pp. 111–139.

[13] P. T. Bauman, H. Ben Dhia, N. Elkhodja, J. T. Oden and S. Prudhomme, On the application of the Arlequin method to the coupling of particle and continuum models, *Computational Mechanics* **42** (2008), pp. 511–530.

[14] T. Belytschko and S. P. Xiao, Coupling methods for continuum model with molecular model, *Int. J. Multiscale Comput. Eng.* **1** (2003), pp. 115–126.

[15] T. Belytschko, S. P. Xiao, G. C. Schatz and R. S. Ruoff, Atomistic simulations of nanotube fracture, *Phys. Rev. B* **65** (2002), p. 235430.

[16] H. Ben Dhia, Problèmes mécaniques multiéchelles: la méthode Arlequin, *C.R. Acad. Sci. Paris, Série IIb* **326** (1998), pp. 899–904.

[17] ______, Further insights by theoretical investigations of the multiscale Arlequin method, *Int. J. Multiscale Comput. Eng.* **6** (2008), pp. 215–232.

[18] H. Ben Dhia and G. Rateau, Analyse mathématique de la méthode Arlequin mixte, *C.R. Acad. Sci. Paris, Série I* **332** (2001), pp. 649–654.

[19] ______, The Arlequin method as a flexible engineering design tool, *Int. J. Numer. Meth. Eng.* **62** (2005), pp. 1442–1462.

[20] M. Berezhnyy and L. Berlyand, Continuum limit for three-dimensional mass-spring networks and discrete Korn's inequality, *J. Mech. Phys. Solids* **54** (2006), pp. 635–669.

[21] C. Bernardi and Y. Maday, Mesh adaptivity in finite elements using the mortar method, *Rev. Eur. Elem. Finis* **9** (2000), pp. 451–465.

[22] X. Blanc and C. Le Bris, Définition d'énergies d'interfaces à partir de modèles atomiques, *C.R. Acad. Sci. Paris, Série I* **340** (2005), pp. 535–540.

[23] X. Blanc, C. Le Bris and F. Legoll, Analysis of a prototypical multiscale method coupling atomistic and continuum mechanics, *Math. Mod. Num. Anal.* **39** (2005), pp. 797–826.

[24] ______, Analysis of a prototypical multiscale method coupling atomistic and continuum mechanics: the convex case, *Acta Math. Appl. Sin. Engl. Ser.* **23** (2007), pp. 209–216.

[25] X. Blanc, C. Le Bris, F. Legoll and C. Patz, *Finite temperature coarse-graining of one-dimensional models: a possible computational approach*, INRIA, Report no. RR-6544, 2008, http://hal.inria.fr/inria-00282107/en/.

[26] X. Blanc, C. Le Bris and P.-L. Lions, Convergence de modèles moléculaires vers des modèles de mécanique des milieux continus, *C.R. Acad. Sci. Paris, Série I* **332** (2001), pp. 949–956.

[27] ______, From molecular models to continuum mechanics, *Arch. Rat. Mech. Anal.* **164** (2002), pp. 341–381.

[28] ______, Du discret au continu pour des modèles de réseaux aléatoires d'atomes, *C.R. Acad. Sci. Paris, Série I* **342** (2006), pp. 627–633.

[29] ______, Atomistic to continuum limits for computational materials science, *Math. Mod. Num. Anal.* **41** (2007), pp. 391–426.

[30] ______, The energy of some microscopic stochastic lattices, *Arch. Rat. Mech. Anal.* **184** (2007), pp. 303–339.

[31] A. Braides, *Gamma-convergence for beginners*, Oxford University Press, Oxford, 2002.

[32] A. Braides, G. Dal Maso and A. Garroni, Variational formulation of softening phenomena in fracture mechanics: the one-dimensional case, *Arch. Rat. Mech. Anal.* **146** (1999), pp. 23–58.

[33] A. Braides and M. S. Gelli, *From discrete systems to continuous variational problems: an introduction*, Topics on concentration phenomena and problems with multiple scales (A. Braides and V. Chiadò Piat, eds.), Lect. Notes Unione Mat. Ital. 2, Springer, Berlin, 2006, pp. 3–77.

[34] H. Brézis, *Analyse fonctionnelle*, Dunod, Paris, 1999.

[35] J. Q. Broughton, F. F. Abraham, N. Bernstein and E. Kaxiras, Concurrent coupling of length scales: methodology and application, *Phys. Rev. B.* **60** (1999), pp. 2391–2403.

[36] V. V. Bulatov, T. Diaz de la Rubia, R. Phillips, E. Kaxiras and N. Ghoniem, *Multiscale modelling of materials*, Material Research Society, 1999.

[37] E. Cancès, F. Legoll and G. Stoltz, Theoretical and numerical comparison of some sampling methods for molecular dynamics, *Math. Mod. Num. Anal.* **41** (2007), pp. 351–389.

[38] Y. Chen, J. D. Lee and A. Eskandarian, Atomistic viewpoint of the applicability of microcontinuum theories, *Int. J. Solids Struct.* **41** (2004), pp. 2085–2097.

[39] A. Cherkaev, E. Cherkaev and L. Slepyan, Transition waves in bistable structures. I. Delocalization of damage, *J. Mech. Phys. Solids* **53** (2005), pp. 383–405.

[40] Y. Choi, K. J. Van Vliet, J. Li and S. Suresh, Size effects on the onset of plastic deformation during nanoindentation of thin films and patterned lines, *J. Applied Physics* **94** (2003), pp. 6050–6058.

[41] P. G. Ciarlet, *Mathematical elasticity, vol. I: Three-dimensional elasticity*, North Holland, 1988.

[42] ———, *Basic error estimates for elliptic problems*, Handbook of Numerical Analysis (P. G. Ciarlet and J.-L. Lions, eds.), vol. II, North-Holland, 1991, pp. 17–351.

[43] S. Curtarolo and G. Ceder, Dynamics of an inhomogeneously coarse grained multiscale system, *Phys. Rev. Lett.* **88** (2002), p. 255504.

[44] W. A. Curtin and R. E. Miller, Atomistic/continuum coupling in computational materials science, *Modelling Simul. Mater. Sci. Eng.* **11** (2003), pp. 33–68.

[45] G. Dal Maso, *An introduction to* Γ*-convergence*, Birkhäuser, Boston, 1993.

[46] P. Deák, Th. Frauenheim and M. R. Pederson, *Computer simulation of materials at atomic level*, Wiley, Berlin, 2000.

[47] G. Del Piero and L. Truskinovsky, Macro- and micro-cracking in one-dimensional elasticity, *Int. J. Solids Struct.* **38** (2001), pp. 1135–1148.

[48] D. J. Diestler, Coarse-graining description of multiple scale processes in solid systems, *Phys. Rev. B* **66** (2002), p. 184104.

[49] D. J. Diestler, H. Zhou, R. Feng and X. C. Zeng, Hybrid atomistic-coarse-grained treatment of multiscale processes in heterogeneous materials: a self-consistent-field approach, *J. Chem. Phys.* **125** (2006), p. 064705.

[50] D. J. Diestler, Z.-B. Zhu and X. C. Zeng, An extension of the quasicontinuum treatment of multiscale solid systems to nonzero temperarure, *J. Chem. Phys.* **121** (2004), pp. 9279–9282.

[51] M. Dobson and M. Luskin, Analysis of a force-based quasicontinuum approximation, *Math. Mod. Num. Anal.* **42** (2008), pp. 113–139.

[52] ———, *An analysis of the effect of ghost force oscillation on quasicontinuum error*, arXiv preprint, Report no. 0811.4202, 2008.

[53] ______, Iterative solution of the quasicontinuum equilibrium equations with continuation, *Journal of Scientific Computing* **37** (2008), pp. 19–41.

[54] M. Dobson, M. Luskin, R. S. Elliott and E. B. Tadmor, A multilattice quasicontinuum for phase transforming materials: Cascading Cauchy Born kinematics, *J. Computer-Aided Mater. Des.* **14** (2007), pp. 219–237.

[55] W. Dreyer, M. Herrmann and A. Mielke, Micro-macro transition in the atomic chain via Whitham's modulation equation, *Nonlinearity* **19** (2006), pp. 471–500.

[56] W. Dreyer and M. Kunik, Cold, thermal and oscillator closure of the atomic chain, *J. Phys. A* **33** (2000), pp. 2097–2129.

[57] L. M. Dupuy, E. B. Tadmor, R. E. Miller and R. Phillips, Finite temperature quasicontinuum: Molecular dynamics without all the atoms, *Phys. Rev. Lett.* **95** (2005), p. 060202.

[58] W. E and B. Engquist, The heterogeneous multiscale methods, *Comm. Math. Sci.* **1** (2003), pp. 87–132.

[59] W. E and Z. Huang, Matching conditions in atomistic-continuum modeling of materials, *Phys. Rev. Lett.* **87** (2001), p. 135501.

[60] ______, A dynamic atomistic-continuum method for the simulation of crystalline materials, *J. Comput. Phys.* **182** (2002), pp. 234–261.

[61] W. E and X. Li, *Multiscale modeling of crystalline solids*, Handbook of Materials Modeling (S. Yip, ed.), part A, Springer, 2005, pp. 1491–1506.

[62] W. E, X. Li and E. Vanden-Eijnden, *Some recent progress in multiscale modeling*, Multiscale Modelling and Simulation (S. Attinger and P. Koumoutsakos, eds.), Lect. N. Comput. Sci. Eng. 39, Springer, 2004, pp. 3–22.

[63] W. E and P. B. Ming, Analysis of multiscale methods, *J. Comp. Math.* **22** (2004), pp. 210–219.

[64] ______, *Analysis of the local quasicontinuum method*, Frontiers and prospects of contemporary applied mathematics, 6, Higher Education Press, Beijing, 2005, pp. 18–32.

[65] ______, Cauchy-Born rule and the stability of crystalline solids: dynamics problems, *Acta Mathematicae Applicatae Sinica Engl. Ser.* **23** (2007), pp. 529–550.

[66] ______, Cauchy-Born rule and the stability of crystalline solids: static problems, *Arch. Rat. Mech. Anal.* **183** (2007), pp. 241–297.

[67] F. Feyel and J.-L. Chaboche, FE2 multiscale approach for modelling the elastoviscoplastic behaviour of long fibre SiC/Ti composite materials, *Comput. Methods Appl. Mech. Eng.* **183** (2000), pp. 309–330.

[68] J. Fish, W. Chen and R. G. Li, Generalized mathematical homogenization of atomistic media at finite temperatures in three dimensions, *Comput. Methods Appl. Mech. Eng.* **196** (2007), pp. 908–922.

[69] S. M. Foiles, Evaluation of harmonic methods for calculating the free energy of defects in solids, *Phys. Rev. B* **49** (1994), pp. 14930–14939.

[70] G. Friesecke and F. Theil, Validity and failure of the Cauchy-Born hypothesis in a two-dimensional mass-spring lattice, *J. Nonlinear Sci.* **12** (2002), pp. 445–478.

[71] H. Gao and B. Ji, Modeling fracture in nano-materials via a virtual internal bond method, *Engineering Fracture Mechanics* **70** (2003), pp. 1777–1791.

[72] H. Gao and P. Klein, Numerical simulation of crack growth in an isotropic solid with randomized internal cohesive bonds, *J. Mech. Phys. Solids* **46** (1998), pp. 187–218.

[73] J. Giannoulis, M. Herrmann and A. Mielke, *Continuum descriptions for the dynamics in discrete lattices: derivation and justification*, WIAS, Report no. 1126, Berlin, 2006.

[74] S. P. A. Gill, Z. Jia, B. Leimkuhler and A. C. F. Cocks, Rapid thermal equilibration in coarse-grained molecular dynamics, *Phys. Rev. B* **73** (2006), p. 184304.

[75] A. Gloria, An analytical framework for the numerical homogenization of monotone elliptic operators and quasiconvex energies, *SIAM J. Multiscale Model. Simul.* **5** (2006), pp. 996–1043.

[76] ______, A direct approach to numerical homogenization in nonlinear elasticity, *Networks and heterogeneous media* **1** (2006), pp. 109–141.

[77] ______, An analytical framework for numerical homogenization. Part II: Windowing and oversampling, *SIAM J. Multiscale Model. Simul.* **7** (2008), pp. 274–293.

[78] I. Goldhirsch and C. Goldenberg, On the microscopic foundations of elasticity, *Eur. Phys. J. E* **9** (2002), pp. 245–251.

[79] A. Gouldstone, K. J. Van Vliet and S. Suresh, Nanoindentation: Simulation of defect nucleation in a crystal, *Nature* **411** (2001), p. 656.

[80] N. G. Hadjiconstantinou, Combining atomistic and continuum simulations of contact-line motion, *Phys. Rev. E* **59** (1999), pp. 2475–2478.

[81] ______, Hybrid atomistic-continuum formulations and the moving contact-line problem, *J. Comput. Phys.* **154** (1999), pp. 245–265.

[82] T. Hou and X. Wu, A multiscale finite element method for elliptic problems in composite materials and porous media, *J. Comput. Phys.* **134** (1997), pp. 169–189.

[83] B. Ji and H. Gao, A study of fracture mechanisms in biological nano-composites via the virtual internal bond model, *Materials Science & Engineering A* **366** (2004), pp. 96–103.

[84] P. Klein and H. Gao, Crack nucleation and growth as strain localization in a virtual-bond continuum, *Engineering Fracture Mechanics* **61** (1998), pp. 21–48.

[85] P. A. Klein and H. Gao, *Study of crack dynamics using the virtual internal bond method*, Multiscale deformation and fracture in materials and structures: The J. R. Rice 60th Anniversary Volume (T.-J. Chuang and J. W. Rudnicki, eds.), Kluwer Academic Publishers, Dordrecht, The Netherlands, 2000, pp. 275–309.

[86] J. Knap and M. Ortiz, An analysis of the quasicontinuum method, *J. Mech. Phys. Solids* **49** (2001), pp. 1899–1923.

[87] V. Kouznetsova, M. G. D. Geers and W. A. M. Brekelmans, Multiscale constitutive modelling of heterogeneous materials with a gradient-enhanced computational homogenization scheme, *Int. J. Numer. Meth. Eng.* **54** (2002), pp. 1235–1260.

[88] B. Kraczek, D. D. Johnson and R. B. Haber, private communication.

[89] P. Ladevèze and A. Nouy, Une stratégie de calcul multiéchelle avec homogénéisation en espace et en temps, *C. R. Acad. Sci. Paris, Mécanique* **330** (2002), pp. 683–689.

[90] ______, On a multiscale computational strategy with time and space homogenization for structural mechanics, *Comput. Methods Appl. Mech. Eng.* **192** (2003), pp. 3061–3087.

[91] C. Le Bris, *Systèmes multiéchelles: modélisation et simulation*, Mathématiques et Applications 47, Springer, Berlin, 2005.

[92] P. Le Tallec, *Numerical methods for nonlinear three-dimensional elasticity*, Handbook of Numerical Analysis (P.G. Ciarlet and J.-L. Lions, eds.), vol. III, North-Holland, 1994, pp. 465–622.

[93] F. Legoll, *Molecular and multiscale methods for the numerical simulation of materials*, Ph.D. thesis, Université Paris VI, 2004, http://cermics.enpc.fr/~legoll/these_Legoll.pdf.

[94] ______, Numerical homogenization of nonlinear viscoplastic two-dimensional polycrystals, *Computational and Applied Mathematics* **23** (2004), pp. 309–325.

[95] D. Leguillon, Strength or toughness? A criterion for crack onset at a notch, *Eur. J. Mech A / Solids* **21** (2002), pp. 61–72.

[96] R. LeSar, R. Najafabadi and D. J. Srolovitz, Finite-temperature defect properties from free-energy minimization, *Phys. Rev. Lett.* **63** (1989), pp. 624–627.

[97] J. Li, K. J. Van Vliet, T. Zhu, S. Yip and S. Suresh, Atomistic mechanisms governing elastic limit and incipient plasticity in crystals, *Nature* **418** (2002), pp. 307–310.

[98] X. Li and W. E, Multiscale modeling of the dynamics of solids at finite temperature, *J. Mech. Phys. Solids* **53** (2005), pp. 1650–1685.

[99] ———, Variational boundary conditions for molecular dynamics simulations of solids at low temperature, *Commun. Comput. Phys.* **1** (2006), pp. 135–175.

[100] E. Lidorikis, M. E. Bachlechner, R. K. Kalia, A. Nakano and P. Vashishta, Coupling length scales for multiscale atomistic-continuum simulations: atomistically induced stress distribution in Si / Si_3N_4 nanopixels, *Phys. Rev. Lett.* **87** (2001), p. 086104.

[101] P. Lin, *A nonlinear wave equation of mixed type for fracture dynamics*, The National University of Singapore, Department of Mathematics, Report no. 777, 2000, http://www.math.nus.edu.sg/∼matlinp/WWW/linTR777.pdf.

[102] ———, Theoretical and numerical analysis for the quasi-continuum approximation of a material particle model, *Math. of Comput.* **72** (2003), pp. 657–675.

[103] ———, Convergence analysis of a quasi-continuum approximation for a two-dimensional material, *SIAM J. Numer. Anal.* **45** (2007), pp. 313–332.

[104] P. Lin and P. Plechac, Numerical studies of a coarse-grained approximation for dynamics of an atomic chain, *Int. J. Multiscale Comput. Eng.* **5** (2007), pp. 351–367.

[105] P. Lin and C. W. Shu, Numerical solution of a virtual internal bond model for material fracture, *Physica D* **167** (2002), pp. 101–121.

[106] J. Liu, S. Y. Chen, X. B. Nie and M. O. Robbins, A continuum-atomistic simulation of heat transfer in micro- and nano-flows, *J. Comput. Phys.* **227** (2007), pp. 279–291.

[107] W. K. Liu, D. Qian and eds. M. F. Horstmeyer, Special issue on multiple scale methods for nanoscale mechanics and materials, *Comput. Methods Appl. Mech. Eng.* **193** (2004).

[108] B. Luan, S. Hyun, J.-F. Molinari, N. Bernstein and M. O. Robbins, Multiscale modeling of two dimensional contacts, *Phys. Rev. E* **74** (2006), p. 046710.

[109] M. Luskin and C. Ortner, *An analysis of node-based cluster summation rules in the quasicontinuum method*, arXiv preprint, Report no. 0811.4214, 2008.

[110] R. Marangati and P. Sharma, Length scales at which classical elasticity breaks down for various materials, *Phys. Rev. Lett.* **98** (2007), p. 195504.

[111] P. Marcellini, Periodic solutions and homogenization of nonlinear variational problems, *Ann. Mat. Pura Appl.* **117** (1978), pp. 139–152.

[112] M. Marder, Molecular dynamics of cracks, *Computing in Science & Engineering* (September/October 1999), pp. 48–55.

[113] A.-M. Matache and Ch. Schwab, Two-scale FEM for homogenization problems, *Math. Mod. Num. Anal.* **36** (2002), pp. 537–572.

[114] C. Miehe and J. Dettmar, A framework for micro-macro transitions in periodic particle aggregates of granular materials, *Comput. Methods Appl. Mech. Eng.* **193** (2004), pp. 225–256.

[115] A. Mielke, Macroscopic behavior of microscopic oscillations in harmonic lattices via Wigner-Husimi transforms, *Arch. Rat. Mech. Anal.* **181** (2006), pp. 441–448.

[116] ———, Weak convergence methods for Hamiltonian multiscale problems, *Discrete and continuous dynamical systems A* **20** (2008), pp. 53–79.

[117] R. Miller, E. B. Tadmor, R. Phillips and M. Ortiz, Quasicontinuum simulation of fracture at the atomic scale, *Modelling Simul. Mater. Sci. Eng.* **6** (1998), pp. 607–638.

[118] R. E. Miller and E. B. Tadmor, The quasicontinuum method: Overview, applications and current directions, *J. Computer-Aided Mater. Des.* **9** (2002), pp. 203–239.

[119] S. Moorthy and S. Ghosh, Adaptivity and convergence in the Voronoi cell finite element model for analyzing heterogeneous materials, *Comput. Methods Appl. Mech. Eng.* **185** (2000), pp. 37–74.

[120] S. Müller and A. Schlömerkemper, Discrete-to-continuum limit of magnetic forces, *C.R. Acad. Sci. Paris, Série I* **335** (2002), pp. 393–398.

[121] A. Nakano, M. E. Bachlechner, R. K. Kalia, E. Lidorikis, P. Vashishta and G. Z. Voyiadjis, Multiscale simulation of nanosystems, *Computing in Science & Engineering* (July/August 2001), pp. 56–66.

[122] A. Nakano, T. J. Campbell, R. K. Kalia, S. Kodiyalam, S. Ogata, F. Shimojo, X. Su and P. Vashishta, *Scalable multiresolution algorithms for classical and quantum molecular dynamics of nanosystems*, Handbook of Numerical Analysis (P.G. Ciarlet and C. Le Bris, eds.), vol. X: Special volume: Computational chemistry, North-Holland, 2003, pp. 639–666.

[123] A. Nakano, R. K. Kalia, P. Vashishta, T. Campbell, S. Ogata, F. Shimojo and S. Saini, Scalable atomistic simulation algorithms for materials research, *Scientific Programming* **10** (2002), pp. 263–270.

[124] X. B. Nie, S. Y. Chen, W. N. E and M. O. Robbins, A continuum and molecular dynamics hybrid method for micro and nano fluid flow, *J. Fluid Mech.* **500** (2000), pp. 55–64.

[125] X. B. Nie, S. Y. Chen and M. O. Robbins, Hybrid continuum-atomistic simulation of singular corner flow, *Physics of Fluids* **16** (2004), pp. 3579–3591.

[126] J. T. Oden, S. Prudhomme, A. Romkes and P. T. Bauman, Multiscale modeling of physical phenomena: Adaptive control of models, *SIAM J. Sci. Comput.* **28** (2006), pp. 2359–2389.

[127] S. Ogata, E. Lidorikis, F. Shimojo, A. Nakano, P. Vashishta and R. K. Kalia, Hybrid finite-element / molecular-dynamics / electronic density-functional approach to materials simulations on parallel computers, *Comput. Phys. Comm.* **138** (2001), pp. 143–154.

[128] C. Ortner, Gradient flows as a selection procedure for equilibria of nonconvex energies, *SIAM J. Math. Anal.* **38** (2006), pp. 1214–1234.

[129] C. Ortner and E. Süli, *A posteriori analysis and adaptive algorithms for the quasicontinuum method in one dimension*, Oxford Numerical Analysis Group, Report no. NA-06/13, University of Oxford, 2006.

[130] ———, Analysis of a quasicontinuum method in one dimension, *Math. Mod. Num. Anal.* **42** (2008), pp. 57–91.

[131] S. Pagano and R. Paroni, A simple model for phase transitions: from the discrete to the continuum problems, *Quarterly of Applied Math.* **61** (2003), pp. 89–109.

[132] H. S. Park, E. G. Karpov, P. A. Klein and W. K. Liu, Three-dimensional bridging scale analysis of dynamic fracture, *J. Comput. Phys.* **207** (2005), pp. 588–609.

[133] M. L. Parks, P. B. Bochev and R. B. Lehoucq, Connecting atomistic-to-continuum coupling and domain decomposition, *SIAM J. Multiscale Model. Simul.* **7** (2008), pp. 362–380.

[134] C. Patz, *in preparation*, Ph.D. thesis, WIAS, Berlin.

[135] S. Prudhomme, P. T. Bauman and J. T. Oden, Error control for molecular statics problems, *Int. J. Multiscale Comput. Eng.* **4** (2006), pp. 647–662.

[136] S. Prudhomme, H. Ben Dhia, P. T. Bauman, N. Elkhodja and J. T. Oden, Computational analysis of modeling error for the coupling of particle and continuum models by the Arlequin method, *Comput. Methods Appl. Mech. Eng.* **197** (2008), pp. 3399–3409.

[137] D. Raabe, *Computational materials science*, Wiley, Weinheim, 1998.

[138] M. O. Rieger and J. Zimmer, Young measure flow as a model for damage, *Z. Angew. Math. Phys.* **60** (2009), pp. 1–32.

[139] P. Rosenau, Dynamics of dense lattices, *Phys. Rev. B* **36** (1987), pp. 5868–5876.

[140] R. E. Rudd and J. Q. Broughton, Coarse-grained molecular dynamics and the atomic limit of finite elements, *Phys. Rev. B* **58** (1998), pp. 5893–5896.

[141] ______, Concurrent coupling of length scales in solid state systems, *Phys. Stat. Sol. B* **217** (2000), pp. 251–291.

[142] ______, Coarse-grained molecular dynamics: nonlinear finite elements and finite temperature, *Phys. Rev. B* **72** (2005), p. 144104.

[143] A. Schlömerkemper, Mathematical derivation of the continuum limit of the magnetic force between two parts of a rigid crystalline material, *Arch. Rat. Mech. Anal.* **176** (2005), pp. 227–269.

[144] Ch. Schwab and A.-M. Matache, *Generalized FEM for homogenization problems*, Multiscale and Multiresolution Methods: Theory and Applications (T. J. Barth, T. F. Chan and R. Haimes, eds.), Lect. N. Comput. Sci. Eng. 20, Springer, 2001, pp. 197–238.

[145] V. B. Shenoy, R. Miller, E. B. Tadmor, R. Phillips and M. Ortiz, Quasicontinuum models of interfacial structure and deformation, *Phys. Rev. Lett.* **80** (1998), pp. 742–745.

[146] V. B. Shenoy, R. Miller, E. B. Tadmor, D. Rodney, R. Phillips and M. Ortiz, An adaptative finite element approach to atomic scale mechanics: the quasicontinuum method, *J. Mech. Phys. Solids* **47** (1999), pp. 611–642.

[147] F. Shimojo, R. K. Kalia, A. Nakano and P. Vashishta, Linear-scaling density-functional-theory calculations of electronic structure based on real-space grids: design, analysis, and scalability test of parallel algorithms, *Comput. Phys. Comm.* **140** (2001), pp. 303–314.

[148] T. Shimokawa, J. Mortensen, J. Schiotz and K. Jacobsen, Matching conditions in the quasicontinuum method: Removal of the error introduced at the interface between the coarse-grained and fully atomistic regions, *Phys. Rev. B* **69** (2004), p. 214104.

[149] L. Slepyan, A. Cherkaev and E. Cherkaev, Transition waves in bistable structures. II. Analytical solution: wave speed and energy dissipation, *J. Mech. Phys. Solids* **53** (2005), pp. 407–436.

[150] E. B. Tadmor, M. Ortiz and R. Phillips, Quasicontinuum analysis of defects in solids, *Phil. Mag. A* **73** (1996), pp. 1529–1563.

[151] E. B. Tadmor and R. Phillips, Mixed atomistic and continuum models of deformation in solids, *Langmuir* **12** (1996), pp. 4529–4534.

[152] E. B. Tadmor, G. S. Smith, N. Bernstein and E. Kaxiras, Mixed finite element and atomistic formulation for complex crystals, *Phys. Rev. B* **59** (1999), pp. 235–245.

[153] N. Triantafyllidis and S. Bardenhagen, On higher order gradient continuum theories in 1-D nonlinear elasticity. Derivation from and comparison to the corresponding discrete models, *J. of Elasticity* **33** (1993), pp. 259–293.

[154] C. Truesdell and W. Noll, *The nonlinear field theories of mechanics*, 3rd ed, Handbuch der Physik III/3, Springer, Berlin, 2004.

[155] L. Truskinovsky, *Fracture as a phase transformation*, Contemporary research in mechanics and mathematics of materials (R. Batra and M. Beatty, eds.), Ericksen's Symposium, CIMNE, Barcelona, 1996, pp. 322–332.

[156] A. Vainchtein, P. A. Klein, H. Gao and Y. Huang, *A strain-gradient virtual-internal-bond model*, Modeling and simulation-based life cycle engineering (K. P. Chong, S. Saigal, S. Thynell and H. S. Morgan, eds.), Spon Press, London, 2002, pp. 31–46.

[157] K. J. Van Vliet, J. Li, T. Zhu, S. Yip and S. Suresh, Quantifying the early stages of plasticity through nanoscale experiments and simulations, *Phys. Rev. B* **67** (2003), p. 104105.

[158] G. J. Wagner and W. K. Liu, Coupling of atomistic and continuum simulations using a bridging scale decomposition, *J. Comput. Phys.* **190** (2003), pp. 249–274.

[159] Z.-B. Wu, D. J. Diestler, R. Feng and X. C. Zeng, Coarse-graining description of solid systems at nonzero temperature, *J. Chem. Phys.* **119** (2003), pp. 8013–8023.

[160] E. Zeidler, *Nonlinear functional analysis and its applications. Vol. IIB: Nonlinear monotone operators*, Springer, New York, 1990.

[161] P. Zhang, P. A. Klein, Y. Huang, H. Gao and P. D. Wu, Numerical simulation of cohesive fracture by the virtual-internal-bond model, *Comput. Model. Eng. Sci.* **3** (2002), pp. 263–277.

Author information

Frédéric Legoll, Université Paris-Est, Institut Navier, LAMI, Ecole des Ponts, 6 et 8 avenue Blaise Pascal, 77455 Marne-la-Vallée Cedex 2 and INRIA Rocquencourt, MICMAC Team-Project, 78153 Le Chesnay Cedex, France.
E-mail: `legoll@lami.enpc.fr`

Maximal regularity and applications to PDEs

Sylvie Monniaux

Abstract. We present here a survey of results on parabolic maximal regularity and applications to partial differential equations such as uniqueness of integral solutions of the Navier–Stokes system. The subject started in the 1960's with the case of Hilbert spaces. Certain progress has been made in the late 1980's, and the real breakthrough on the theory goes back to 2000.

Keywords. Maximal regularity, singular integral, multiplier theorem, Gaussian bounds, quasilinear parabolic equation, Navier–Stokes equation, non-autonomous Cauchy problem.

AMS classification. 47D06, 35A05, 35B65, 35K55, 35K90, 34G10, 35Q30.

1 Introduction

The purpose of this lecture is to give a flavor of the concept of maximal regularity. In the last ten to fifteen years, a lot of progress has been made on this subject. The problem of parabolic maximal L^p-regularity can be stated as follows.

> Let A be an (unbounded) linear operator on a Banach space X, with domain $D(A)$. Let $p \in]1, \infty[$. Does there exist a constant $C > 0$ such that for all $f \in L^p(0, \infty; X)$, there exists a unique solution $u \in L^p(0, \infty; D(A)) \cap W^{1,p}(0, \infty; X)$ of $u' + Au = f$, $u(0) = 0$ verifying
>
> $$\|u'\|_{L^p(0,\infty;X)} + \|Au\|_{L^p(0,\infty;X)} \leq \|f\|_{L^p(0,\infty;X)}?$$

This problem, in its theoretical point of view, has been approached in different manners.

(i) If $-A$ generates a semigroup $(T(t))_{t\geq 0}$, then the solution u is given by $u(t) = \int_0^t T(t-s)f(s)\,ds$, $t \geq 0$, and therefore, the question is whether the operator R defined by

$$Rf(t) = \int_0^t AT(t-s)f(s)\,ds, \quad t \geq 0,$$

for $f \in L^p(0, \infty; X)$, is bounded in $L^p(0, \infty; X)$. In the favorable case where the semigroup is analytic, R has a convolution form with an operator-valued kernel, singular at 0. The study of boundedness of R in $L^p(0, \infty; X)$ then leads to the theory of singular integrals.

(ii) One way to treat this convolution (in t) operator is to apply the Fourier transform to it. The problem is now to decide whether $M(t) = A(isI + A)^{-1}$, $s \in \mathbb{R}$, is a Fourier multiplier. This has been studied by L. Weis in [41] who gave an equivalent property to maximal regularity of A in terms of bounds of the resolvent of A.

(iii) Another approach is to see this problem as the invertibility of the sum of two operators $A + B$ where B is the derivative in time. G. Dore and A. Venni in [16] followed this idea, using imaginary powers of A and B..

All these characterizations are not always easy to deal with when concrete examples are concerned. To verify that a precise operator has the maximal L^p-regularity property needs other results. This has been the case for operators with Gaussian estimates (see [21] and [12]) and more recently for operator with generalized Gaussian estimates (see [24]). Among a very large literature, let us mention the surveys by W. Arendt [4] and P. C. Kunstmann and L.Weis [25] where the theory of maximal regularity is largely covered.

The first part of these notes is dedicated to the study of this problem from a theoretical point of view. In a second part, we will give examples of operators having the maximal L^p-property: generators of contraction semigroups in $L^p(\Omega)$, generators of semigroups having Gaussian estimates or generalized Gaussian estimates. Applications to partial differential equations, such as the semilinear heat equation or the incompressible Navier–Stokes equations, are given in a third part. Finally, in a last part, we give some results on the non-autonomous maximal L^p-regularity problem. Many results proved in the autonomous case are also true in the non-autonomous case provided we assume enough regularity (in t) on the operators $A(t)$. This condition may be removed if the operators $A(t)$ have the same domain D, and in that case, only continuity is required. This non-autonomous maximal L^p-regularity is far from being understood, but is nonetheless important for applications to quasilinear evolution problems.

2 Theoretical point of view

2.1 Statement of the problem

Let X be a Banach space and A be a closed (unbounded) linear operator with domain $D(A)$ dense in X. Let $f : [0, \infty[\to X$ be a measurable function. We consider the problem of existence and regularity of solutions to the problem

$$\begin{cases} u'(t) + Au(t) &= f(t), \quad t \geq 0, \\ u(0) &= 0. \end{cases} \tag{2.1}$$

Definition 2.1. Let $p \in]1, \infty[$. An operator A is said to have the (parabolic) maximal L^p-regularity property if there exists a constant $C > 0$ such that for all $f \in L^p(0, \infty; X)$, there is a unique $u \in L^p(0, \infty; D(A))$ with $u' \in L^p(0, \infty; X)$ satisfying (2.1) for almost every $t \in]0, \infty[$ and

$$\|u\|_{L^p(0,\infty;X)} + \|u'\|_{L^p(0,\infty;X)} + \|Au\|_{L^p(0,\infty;X)} \leq C\|f\|_{L^p(0,\infty;X)}. \tag{2.2}$$

Not all operators have this property. In particular, there holds

Proposition 2.2 (Analytic semigroup). *Let A be an operator on a Banach space X with the maximal L^p-regularity property for one $p \in]1,\infty[$. Then $-A$ generates a bounded analytic semigroup on X.*

Proof. Let $z \in \mathbb{C}$ with $\operatorname{Re}(z) > 0$. Define $f_z \in L^p(0,\infty;\mathbb{C})$ by

$$f_z(t) = e^{zt} \quad \text{if} \quad 0 \le t \le \frac{1}{\operatorname{Re}(z)} \qquad \text{and} \qquad f_z(t) = 0 \quad \text{if} \quad t > \frac{1}{\operatorname{Re}(z)}.$$

Step 1. Let $x \in X$ and denote by u_z the solution of (2.1) for $f = f_z \otimes x$. Define then

$$R_z x = \operatorname{Re}(z) \int_0^\infty e^{-zt} u_z(t)\, dt.$$

Then the following estimates hold:

$$\begin{aligned} \|R_z x\|_X &\le \operatorname{Re}(z) \|u_z\|_{L^p(0,\infty;X)} \|t \mapsto e^{-zt}\|_{L^{p'}(0,\infty)} \\ &\le \operatorname{Re}(z)\, C\, \|f_z\|_{L^p(0,\infty;\mathbb{C})} \|x\|_X \|t \mapsto e^{-zt}\|_{L^{p'}(0,\infty)} \\ &\le C\, \frac{(e^p - 1)^{\frac{1}{p}}}{p'^{\frac{1}{p'}}} \|x\|_X. \end{aligned}$$

The first estimate comes from the Hölder inequality applied to the integral form of $R_z x$, p' denoting the conjugate exponent of p: $\frac{1}{p} + \frac{1}{p'} = 1$. The inequality (2) is obtained by estimating u_z by f via the maximal L^p-regularity property of A; note that $\|f\|_{L^p(0,\infty;X)} = \|f_z\|_{L^p(0,\infty;\mathbb{C})} \|x\|_X$. The last estimate comes from the calculations of the different norms of the previous line. By writing $R_z x$ as

$$R_z x = \frac{\operatorname{Re}(z)}{z} \int_0^\infty e^{-zt} u_z'(t)\, dt$$

(by performing an integration by parts), the same arguments as before give the estimate

$$\|R_z x\|_X \le \frac{1}{|z|}\, C\, \frac{(e^p - 1)^{\frac{1}{p}}}{p'^{\frac{1}{p'}}} \|x\|_X.$$

Therefore, we get

$$\|R_z x\|_X \le \frac{M}{1 + |z|} \|x\|_X \tag{2.3}$$

with $M = C\, \frac{(e^p-1)^{\frac{1}{p}}}{p'^{\frac{1}{p'}}}$.

Step 2. Let now $x \in D(A)$. We have

$$\begin{aligned} R_z(zI + A)x &= zR_z x + R_z Ax \\ &= \operatorname{Re}(z)\int_0^\infty e^{-zt}u_z'(t)\,dt + \operatorname{Re}(z)\int_0^\infty e^{-zt}Au_z(t)\,dt \\ &= \operatorname{Re}(z)\int_0^\infty e^{-zt}f_z(t)x\,dt \\ &= x. \end{aligned}$$

The first equality is straightforward. The first term of the second equality comes from the integration by parts as in Step 1, whereas the second term comes from the fact that $Au_z \in L^p(0,\infty;X)$ by the maximal L^p-regularity property of A. The third equality follows from (2.1) and equality (4) is obtained by a simple calculation, reminding that $f = f_z \otimes x$.
Step 3. The equality $R_z(zI+A)x = x$ for all $x \in D(A)$ together with (2.3) ensure that R_z is the resolvent of $-A$ in z. Therefore, the spectrum $\sigma(A)$ of A is included in $\mathbb{C}_+ = \{z \in \mathbb{C}; \operatorname{Re}(z) \geq 0\}$ and there exists $M > 0$ such that for all $z \in \mathbb{C}$ with $\operatorname{Re}(z) > 0$, we have (2.3). This implies that $-A$ generates a bounded analytic semigroup in X. □

Let us consider for a moment a slightly different problem. We might ask what happens if the initial condition in (2.1) is not equal to zero.

Remark 2.3. Once we know that $-A$ generates a semigroup $(T(t))_{t\geq 0}$ on the Banach space X, we can study the initial value problem

$$\begin{cases} u'(t) + Au(t) = 0, & t \geq 0, \\ u(0) = u_0. \end{cases} \tag{2.4}$$

It is known that the solution u is given by $u(t) = T(t)u_0$, $t \geq 0$. This solution u belongs to $L^p(0,\infty;X)$ if and only if (see [30, Chapter 1]) the initial value u_0 belongs to the real interpolation space $(X, D(A))_{\frac{1}{p'},p}$.

Proposition 2.4 (Independence with respect to p). *Let A be an operator on a Banach space X with the maximal L^p-regularity property for one $p \in]1,\infty[$. Then A has the maximal L^q-regularity property for all $q \in]1,\infty[$.*

To prove this fact, we need the following auxiliary theorem due to A. Benedek, A. P. Calderón and R. Panzone.

Theorem 2.5 (Theorem 2 in [7]). *Let X be a Banach space and let $p \in]1,\infty[$. Let $k : \mathbb{R} \to \mathscr{L}(X)$ be measurable, $k \in L^1_{\mathrm{loc}}(\mathbb{R} \setminus \{0\}; \mathscr{L}(X))$. Let $S \in \mathscr{L}(L^p(\mathbb{R};X))$ be the convolution operator with kernel k, i.e., for all $f \in L^\infty(\mathbb{R};X)$ with compact support, one can write Sf for all $t \notin \operatorname{supp} f$ as*

$$Sf(t) = \int_{\mathbb{R}} k(t-s)f(s)\,ds. \tag{2.5}$$

Assume that there exists a constant $c > 0$ such that the Hörmander-type condition

$$\int_{|t|>2|s|} \|k(t-s) - k(t)\|_{\mathscr{L}(X)} dt \le c \quad \forall\, s \in \mathbb{R}, \tag{2.6}$$

is fulfilled. Then $S \in \mathscr{L}(L^q(\mathbb{R}; X))$ for all $q \in]1, \infty[$.

Proof. We prove that S is bounded from $L^1(\mathbb{R}; X)$ to $L^1_w(\mathbb{R}; X)$ where L^1_w stands for L^1-weak and is defined as the space

$$L^1_w(\mathbb{R}; X) = \left\{ f : \mathbb{R} \to X \text{ measurable} \,;\, \sup_{\alpha>0} \alpha \cdot \Big| \{t \in \mathbb{R}; \|f(t)\|_X > \alpha\} \Big| < \infty \right\},$$

endowed with the norm

$$\|f\|_{L^1_w(\mathbb{R};X)} = \sup_{\alpha>0} \alpha \cdot \Big| \{t \in \mathbb{R}; \|f(t)\|_X > \alpha\} \Big|.$$

Let $f \in L^1(\mathbb{R}; X)$ and fix $\lambda > 0$. By the Calderón–Zygmund decomposition applied to $\mathbb{R} \ni t \mapsto \|f(t)\|_X$ (see Theorem A.8 below, in the case $n = 1$), we may decompose f into a "good" part g and a "bad" part $b = \sum_k b_k$. We then have $Sf = Sg + \sum_k Sb_k$ and therefore

$$\Big\{t \in \mathbb{R}; \|Sf(t)\|_X > \lambda\Big\} \subset \Big\{t \in \mathbb{R}; \|Sg(t)\|_X > \frac{\lambda}{2}\Big\} \cup \Big\{t \in \mathbb{R}; \|Sb(t)\|_X > \frac{\lambda}{2}\Big\}.$$

The measure of the first set is easy to estimate. Since $g \in L^1(\mathbb{R}; X) \cap L^\infty(\mathbb{R}; X)$, we have $g \in L^p(\mathbb{R}; X)$ and since S is bounded in $L^p(\mathbb{R}; X)$, we have

$$\begin{aligned}
\left|\Big\{t \in \mathbb{R}; \|Sg(t)\|_X > \frac{\lambda}{2}\Big\}\right| &\le \frac{\|Sg\|^p_{L^p(\mathbb{R};X)}}{(\frac{\lambda}{2})^p} \\
&\le \frac{2^p \|S\|^p_{\mathscr{L}(L^p(\mathbb{R};X))}}{\lambda^p} \|g\|^p_{L^p(\mathbb{R};X)} \\
&\le \frac{2^p \|S\|^p_{\mathscr{L}(L^p(\mathbb{R};X))}}{\lambda^p} \left(\|g\|^{\frac{1}{p}}_{L^1(\mathbb{R};X)} \|g\|^{1-\frac{1}{p}}_{L^\infty(\mathbb{R};X)} \right)^p \\
&\le \frac{2^p \|S\|^p_{\mathscr{L}(L^p(\mathbb{R};X))}}{\lambda^p} \|f\|_{L^1(\mathbb{R};X)} (2\lambda)^{p-1}
\end{aligned}$$

and therefore

$$\lambda \left|\Big\{t \in \mathbb{R}; \|Sg(t)\|_X > \frac{\lambda}{2}\Big\}\right| \le 4^p \|S\|^p_{\mathscr{L}(L^p(\mathbb{R};X))} \|f\|_{L^1(\mathbb{R};X)}. \tag{2.7}$$

The first inequality is obvious. The second one follows from the fact that S is bounded in $L^p(\mathbb{R}; X)$ by hypothesis. The third inequality results from the Hölder inequality. The last inequality comes from the fact that $\|g\|_{L^1(\mathbb{R};X)} \le \|f\|_{L^1(\mathbb{R};X)}$ and $\|g\|_{L^\infty(\mathbb{R};X)} \le 2\lambda$

by construction in the Calderón–Zygmund decomposition. It remains now to estimate the quantity

$$\left|\left\{t \in \mathbb{R}; \|Sb(t)\|_X > \frac{\lambda}{2}\right\}\right|.$$

We decompose the set as

$$\left\{t \in \mathbb{R}; \|Sb(t)\|_X > \frac{\lambda}{2}\right\} \subset E \cup \left\{t \in \mathbb{R} \setminus E; \|Sb(t)\|_X > \frac{\lambda}{2}\right\}$$

where $E = \bigcup_{k\in\mathbb{N}} \tilde{Q}_k$ with $\tilde{Q}_k$ being the double of the cube Q_k arising in the Calderón–Zygmund decomposition. We already have, by the Calderón–Zygmund decomposition, $|E| \le 2\lambda^{-1}\|f\|_{L^1(\mathbb{R};X)}$. The last term that remains to be estimated is the measure of the set $\{t \in \mathbb{R} \setminus E; \|Sb(t)\|_X > \frac{\lambda}{2}\}$, and that is where the assumption (2.6) comes in. Denoting by s_k the center of Q_k, $k \in \mathbb{N}$, we have

$$\begin{aligned}
\int_{\mathbb{R}\setminus E} \|Sb(t)\|_X dt &\le \sum_{k\in\mathbb{N}} \int_{\mathbb{R}\setminus E} \left\| \int_{Q_k} k(t-s)b_k(s)\,ds \right\|_X dt \\
&\le \sum_{k\in\mathbb{N}} \int_{\mathbb{R}\setminus E} \left\| \int_{Q_k} (k(t-s)-k(t-s_k))b_k(s)\,ds \right\|_X dt \\
&\le \sum_{k\in\mathbb{N}} \int_{\mathbb{R}\setminus \tilde{Q}_k} \int_{Q_k} \|k(t-s)-k(t-s_k)\|_{\mathscr{L}(X)} \|b_k(s)\|_X ds\,dt \\
&\le \sum_{k\in\mathbb{N}} c \int_{Q_k} \|b_k(s)\|_X ds \\
&\le 2c\,\|f\|_{L^1(\mathbb{R};X)}.
\end{aligned}$$

The first inequality comes from (2.5) since $t \in \mathbb{R} \setminus E$ and thus $t \notin \operatorname{supp} b_k = Q_k$. The second inequality is in fact an equality since $\int_{Q_k} b_k ds = 0$ by construction of the b_k's. The third inequality is obvious and the last but one inequality comes from the fact that, for $t \in \mathbb{R}\setminus\tilde{Q}_k$ and $s \in Q_k$, we have $|(t-s_k)| > 2|(t-s)-(t-s_k)|$. We can then apply (2.6). The last inequality is obvious taking the Calderón–Zygmund decomposition into account. Therefore, we have

$$\begin{aligned}
\left|\left\{t \in \mathbb{R}; \|Sb(t)\|_X > \frac{\lambda}{2}\right\}\right| &\le |E| + \left|\left\{t \in \mathbb{R}\setminus E; \|Sb(t)\|_X > \frac{\lambda}{2}\right\}\right| \\
&\le \frac{2}{\lambda}\|f\|_{L^1(\mathbb{R};X)} + \frac{2}{\lambda}\int_{\mathbb{R}\setminus E} \|Sb(t)\|_X dt \\
&\le \frac{2(1+2c)}{\lambda}\|f\|_{L^1(\mathbb{R};X)}.
\end{aligned}$$

Together with (2.7), this gives

$$\lambda\left|\left\{t \in \mathbb{R}; \|Sf(t)\|_X > \lambda\right\}\right| \le C\|f\|_{L^1(\mathbb{R};X)}, \tag{2.8}$$

where $C = 4^p \|S\|^p_{\mathscr{L}(L^p(\mathbb{R};X))} + 2(1+2c)$, which means that S is of weak type $(1,1)$ (see Definition A.5 below). By the Marcinkiewicz interpolation theorem (see Theorem A.6 below), the operator S is of strong type (q,q) (see again Definition A.5 below) for all $q \in]1,p[$. Moreover, it is easy to see that S', the adjoint operator of S, is of the same form as S: S' is bounded in $L^{p'}(\mathbb{R};X)$ (where $\frac{1}{p} + \frac{1}{p'} = 1$) and of weak type $(1,1)$ by the same arguments as before, which implies by the Marcinkiewicz interpolation theorem that S' is of strong type (q,q) for all $q \in]1,p'[$, and therefore, by duality, S is of strong type (q,q) for all $q \in]p,\infty[$. We now have proved that S is a bounded operator in $L^q(\mathbb{R};X)$ for all $q \in]1,\infty[$. □

Proof of Proposition 2.4. Let A be an unbounded operator on a Banach space X such that $-A$ generates an analytic semigroup $(T(t))_{t \geq 0}$. Define $k : \mathbb{R} \to \mathscr{L}(X)$ by

$$k(t) = AT(t) \text{ if } t > 0 \quad \text{and} \quad k(t) = 0 \text{ if } t \leq 0.$$

Then k is measurable with $k \in L^1_{\text{loc}}(\mathbb{R} \setminus \{0\}; \mathscr{L}(X))$. For any $s \in \mathbb{R}$, we have

$$\begin{aligned}
\int_{|t|>2|s|} \|k(t-s) - k(t)\|_{\mathscr{L}(X)} dt &= \int_{t>2|s|} \left\| \int_t^{t-s} A^2 T(\tau)\, d\tau \right\|_{\mathscr{L}(X)} dt \\
&\leq C \int_{t>2|s|} \left| \int_t^{t-s} \frac{1}{\tau^2}\, d\tau \right| dt \\
&= C \int_{t>2|s|} \left| \frac{1}{t-s} - \frac{1}{t} \right| dt \\
&\leq C \ln 2.
\end{aligned}$$

The first equality comes from the fact that k vanishes on $]-\infty, 0[$ and that an analytic semigroup is differentiable on $]0,+\infty[$, its derivative at a point $t > 0$ being equal to $AT(t)$. The second inequality is due to the property of analytic semigroups that, for all $n \in \mathbb{N}$, there holds $\sup_{t>0} \|t^n A^n T(t)\|_{\mathscr{L}(X)} < \infty$. As for the third equality, it is obtained by calculating the integral $\int_t^{t-s} \frac{1}{\tau^2}\, d\tau$. The last inequality comes from the exact integration of $\int_{t>2|s|} |\frac{1}{t-s} - \frac{1}{t}|\, dt$, which gives $\ln 2$ if $s > 0$, 0 if $s = 0$ and $\ln \frac{3}{2}$ if $s < 0$. Therefore, we can apply Theorem 2.5 to conclude that the property of maximal L^p-regularity is independent of $p \in]1,\infty[$. □

2.2 Maximal regularity in Hilbert spaces

In the special case where the Banach space X is actually a Hilbert space, the reverse statement of Proposition 2.2 is true.

Theorem 2.6 (de Simon, 1964). *Let $-A$ be the generator of a bounded analytic semigroup in a Hilbert space H. Then A has the maximal L^p-regularity property for all $p \in]1,\infty[$.*

Proof. This theorem is due to de Simon [15]. Denote by $(T(t))_{t\geq 0}$ the semigroup generated by $-A$ and let $f \in L^2(0,\infty; D(A))$. Then it is easy to see that u given by

$$u(t) = \int_0^t T(t-s)f(s)\,ds, \quad t \geq 0, \tag{2.9}$$

is the solution of (2.1), and Au has the form

$$Au(t) = \int_{\mathbb{R}} k(t-s)f(s)\,ds, \quad t \geq 0,$$

where we have extended f by 0 on $]-\infty, 0[$ and $k(t) = AT(t)$ if $t > 0$, $k(t) = 0$ if $t \leq 0$. Applying the Fourier transform $\mathscr{F}$ (in t), we obtain for all $x \in \mathbb{R}$

$$\begin{aligned}
\mathscr{F}(Au)(\sigma) &= \int_{\mathbb{R}} e^{-it\sigma} Au(t)\,dt \\
&= \int_{\mathbb{R}}\int_{\mathbb{R}} e^{-it\sigma} k(t-s)f(s)\,ds\,dt \\
&= \int_{\mathbb{R}}\int_{\mathbb{R}} e^{-i(t+s)\sigma} k(t)f(s)\,ds\,dt \\
&= \int_0^\infty e^{-it\sigma} T(t)\Big(\int_{\mathbb{R}} e^{-is\sigma} Af(s)\,ds\Big)\,dt \\
&= (i\sigma + A)^{-1} A\mathscr{F}(f)(\sigma) \\
&= A(i\sigma + A)^{-1}\mathscr{F}(f)(\sigma).
\end{aligned}$$

Since $-A$ generates a bounded analytic semigroup, we have

$$\sup_{\sigma\in\mathbb{R}} \|A(i\sigma + A)^{-1}\|_{\mathscr{L}(H)} < \infty,$$

and therefore

$$\|\mathscr{F}(Au)\|_{L^2(\mathbb{R};H)} \leq c\,\|\mathscr{F}(f)\|_{L^2(\mathbb{R};H)}.$$

Since $\mathscr{F}$ is an isomorphism on $L^2(\mathbb{R}; H)$, this implies that

$$\|Au\|_{L^2(\mathbb{R};H)} \leq c\,\|f\|_{L^2(\mathbb{R};H)}.$$

This proves that A has the maximal L^2-regularity property. By Proposition 2.4, the proof is complete. □

The question now arises, whether every negative generator of a bounded analytic semigroup in any Banach space X has the property of maximal L^p-regularity. This question, posed by H. Brézis, was first partially answered by T. Coulhon and D. Lamberton in [13]. To describe their result, we need to define the notion of UMD-space (a space in which martingale differences are unconditional). Actually, we give here a property of UMD-spaces equivalent to the original definition. For more details on this subject, see [9] and [10]. The Hilbert transform $\mathcal{H}f$ of a measurable function f is, whenever it exists, the limit as $\varepsilon \to 0^+$ and $T \to +\infty$ of

$$\mathcal{H}_{\varepsilon,T}f(t) = \frac{1}{\pi}\int_{\varepsilon\le|s|\le T} \frac{f(t-s)}{s}\,ds, \quad t \in \mathbb{R}.$$

Definition 2.7. A Banach space X is said to be of class UMD if the Hilbert transform $\mathcal{H}$ is bounded in $L^p(\mathbb{R};X)$ for all (or equivalently for one) $p \in]1,\infty[$.

Example 2.8. (i) A Hilbert space is in the class UMD.

(ii) If X is a Banach space in the UMD-class, then $L^p(\Omega;X)$, for $\Omega \subset \mathbb{R}^d$ and $p \in]1,\infty[$, is also in the UMD-class.

Theorem 2.9 (Coulhon–Lamberton, 1986). *If the negative generator of the Poisson semigroup on $L^2(\mathbb{R};X)$ (see (2.10) below) has the maximal L^p-property, then the Hilbert transform is bounded in $L^2(\mathbb{R};X)$.*

This theorem implies that if X is a Banach space with the property that every negative generator of a bounded analytic semigroup has the maximal L^p-property, then necessarily X is of class UMD. The converse was an open problem until the work of N. Kalton and G. Lancien [22] where it was proved that such a Banach space is "essentially" a Hilbert space.

Theorem 2.10 (Kalton–Lancien, 2000). *On every Banach lattice which is not isomorphic to a Hilbert space, there are generators of analytic semigroups without the maximal L^p-regularity property.*

Proof of Theorem 2.9. The Poisson semigroup $(P(t))_{t\ge 0}$ on $L^2(\mathbb{R};X)$ is defined as

$$(P(t)f)(x) = \int_{\mathbb{R}} \frac{t}{\pi(y^2+t^2)} f(x-y)\,dy, \quad x \in \mathbb{R},\ t>0, \quad f \in L^2(\mathbb{R};X). \tag{2.10}$$

This semigroup is bounded analytic and its generator $-A$ satisfies

$$(AP(t)f)(x) = \int_{\mathbb{R}} \frac{t^2-y^2}{\pi(y^2+t^2)^2} f(x-y)\,dy, \quad x \in \mathbb{R},\ t>0, \quad f \in L^2(\mathbb{R};X).$$

This relation is obtained by taking the derivative in t of $P(t)f$. As we have already seen, the assumption that A has the maximal L^p-regularity property in $L^2(\mathbb{R};X)$ implies that the operator

$$f \mapsto \Big( ]0,\infty[\times\mathbb{R} \ni (t,x) \mapsto \int_0^t \Big(\int_{\mathbb{R}} \frac{s^2-y^2}{\pi(y^2+s^2)^2} f(t-s,x-y)\,dy\Big)\,ds\Big)$$

is bounded in $L^2(0,\infty;L^2(\mathbb{R};X)) = L^2(]0,\infty[\times\mathbb{R};X)$. By the change of variables $y-s=u$, $y+s=v$, $x-t=u'$ and $x+t=v'$, and the change of the function $F(u,v) = f(\frac{v-u}{2},\frac{v+u}{2})$, we see that the operator K, defined by

$$(KF)(u',v') = \int_{\mathbb{R}} \Big(\int_u^{+\infty} \frac{-uv}{(v^2+u^2)^2} F(u'-u,v'-v)\,dv\Big)\,du, \quad (u',v') \in E,$$

is bounded in $L^2(E;X)$, where $E = \{(u,v) \in \mathbb{R}^2; v > u\}$. This means that there exists a constant $C > 0$ such that

$$\int_{\mathbb{R}} \Big(\int_{u'}^{+\infty} \|KF(u',v')\|_X^2 dv' \Big)\, du' \le C \int_{\mathbb{R}} \Big(\int_{u}^{+\infty} \|F(u,v)\|_X^2 dv \Big)\, du, \qquad (2.11)$$

for all $F \in L^2(E;X)$. Let now $a > 0$ and $\phi \in L^2(\mathbb{R};X)$ and take

$$F(u,v) = \phi(u)\chi_{0<v-u<1}.$$

Then we have

$$\int_{\mathbb{R}} \Big(\int_{u}^{+\infty} \|F(u,v)\|_X^2 dv \Big)\, du = \|\phi\|_{L^2(\mathbb{R};X)}. \qquad (2.12)$$

Computing KF, we get

$$KF(u',v') = \int_{\mathbb{R}} \Big(\int_{\max\{u,u+(v'-u')-1\}}^{v'-u'+u} \frac{-uv}{(v^2+u^2)^2}\, dv \Big) \phi(u'-u)\, du$$

and therefore, if $0 < v'-u' < 1$, we have

$$\begin{aligned} KF(u',v') &= \int_{\mathbb{R}} \Big(\frac{u}{2[(v'-u'+u)^2+u^2]^2} - \frac{1}{4u} \Big) \phi(u'-u)\, du \\ &= \Big(\int_{\mathbb{R}} \frac{u}{2[(v'-u'+u)^2+u^2]^2}\, \phi(u'-u)\, du \Big) - \frac{\pi}{4} \mathcal{H}(\phi)(u'), \end{aligned}$$

or equivalently if $0 < v'-u' < 1$

$$\mathcal{H}(\phi)(u') = -\frac{4}{\pi} KF(u',v') + \Big(\int_{\mathbb{R}} \frac{2u}{\pi[(v'-u'+u)^2+u^2]^2}\, \phi(u'-u)\, du \Big).$$

We take the norm in X, square it and integrate then with respect to v' from $u'+\frac{1}{2}$ to $u'+1$. We then obtain

$$\frac{1}{2} \int_{\mathbb{R}} \|\mathcal{H}(\phi)(u')\|_X^2 du' \le 2 \left(\frac{16}{\pi^2} \|KF\|_{L^2(E;X)}^2 + c^2 \|\phi\|_{L^2(\mathbb{R};X)}^2 \right) \qquad (2.13)$$

since

$$\int_{\mathbb{R}} \frac{2u}{\pi[(v'-u'+u)^2+u^2]^2}\, \phi(u'-u)\, du = (a_{v'-u'} * \phi)(u'),$$

where

$$a_w(u) = \frac{2u}{\pi[(w+u)^2+u^2]^2}, \quad u \in \mathbb{R},$$

with $a_w \in L^1(\mathbb{R})$ for all $w \in [\frac{1}{2}, 1]$. Putting (2.11), (2.12) and (2.13) together, we conclude that the Hilbert transform is bounded in $L^2(\mathbb{R};X)$ and therefore X is of UMD-class. □

2.3 R-boundedness

All the details about the following results can be found in [41]. For a Banach space X, let $\mathscr{S}(\mathbb{R};X)$ be the space of rapidly decreasing functions mapping $\mathbb{R}$ into X. As before, $\mathscr{F}$ denotes the Fourier transform (in t).

Definition 2.11. Let X and Y be Banach spaces. A function $M:\mathbb{R}\setminus\{0\}\to\mathscr{L}(X,Y)$ is said to be a Fourier multiplier on $L^p(\mathbb{R};X)$ if the expression $Rf=\mathscr{F}^{-1}(M\mathscr{F}(f))$ is well-defined for $f\in\mathscr{S}(\mathbb{R};X)$ and if the mapping R extends to a bounded operator $R:L^p(\mathbb{R};X)\to L^p(\mathbb{R};Y)$.

It has been observed by G. Pisier that the converse of Theorem A.7 is true: if $X=Y$ and all differentiable functions $M=M(t):\mathbb{R}\setminus\{0\}\to\mathscr{L}(X,Y)$ satisfying for some constant $C>0$ the estimates

$$\|M(t)\|_{\mathscr{L}(X,Y)}\le C \quad \text{and} \quad \|tM'(t)\|_{\mathscr{L}(X,Y)}\le C \quad \text{for all } t\in\mathbb{R}\setminus\{0\}$$

are Fourier multipliers in $L^2(\mathbb{R};X)$, then X is isomorphic to a Hilbert space. Therefore, to decide whether a particular M is a Fourier multiplier, some additional assumptions are needed. Such assumptions were studied by L. Weis in [41] in 2001. His result gives also an equivalent property to maximal regularity in terms of the so-called R-boundedness of the resolvent of the operator.

Definition 2.12. A set $\tau\subset\mathscr{L}(X,Y)$ is called R-bounded if there is a constant $C>0$ such that for all $n\in\mathbb{N}$, $T_1,\dots,T_n\in\tau$ and $x_1,\dots,x_n\in X$

$$\int_0^1\Big\|\sum_{j=1}^n r_j(s)T_jx_j\Big\|_Y ds\le C\int_0^1\Big\|\sum_{j=1}^n r_j(s)x_j\Big\|_X ds, \tag{2.14}$$

where $(r_j)_{j=1,\dots,n}$ is a sequence of independent $\{-1,1\}$-valued random variables on $[0,1]$; for example, the Rademacher functions $r_j(t)=\operatorname{sign}(\sin(2^j\pi t))$. The R-bound of τ is

$$R(\tau)=\inf\{C>0;\text{(2.14) holds}\}.$$

Remark 2.13. If X and Y are Hilbert spaces, a set $\tau\subset\mathscr{L}(X,Y)$ is R-bounded if and only if it is bounded.

We have already seen that the maximal L^p-regularity property of an operator A on a Banach space X is equivalent to the boundedness in $L^p(0,\infty;X)$ of the operator R given by

$$Rf(t)=\int_0^t AT(t-s)f(s)\,ds,\quad t\in[0,\infty[,\quad f\in L^p(0,\infty;X). \tag{2.15}$$

When taking (formally) the Fourier transform of Rf, we get

$$\mathscr{F}(Rf)(\sigma)=A(i\sigma I+A)^{-1}\mathscr{F}(f)(\sigma),\quad \sigma\in\mathbb{R}.$$

Therefore, if M denotes the operator-valued function $M(\sigma)=A(i\sigma I+A)^{-1}$, our problem is now to find conditions on M (and therefore on A and its resolvent) assuring that M is a Fourier multiplier in $L^p(\mathbb{R};X)$.

Theorem 2.14 (L. Weis, 2001). *Let X and Y be UMD-Banach spaces. Let*

$$M : \mathbb{R} \setminus \{0\} \to \mathscr{L}(X,Y)$$

be a differentiable function such that the sets

$$\Big\{M(t); t \in \mathbb{R} \setminus \{0\}\Big\} \quad \text{and} \quad \Big\{tM'(t); t \in \mathbb{R} \setminus \{0\}\Big\}$$

are R-bounded. Then M is a Fourier multiplier on $L^p(\mathbb{R};X)$ for all $p \in]1,\infty[$.

References for the proof. This theorem can be found in [41, Theorem 3.4]. □

Applying this theorem to our problem of maximal L^p-regularity, we get the following result.

Corollary 2.15 (L. Weis, 2001). *Let X be a UMD-Banach space and A be the negative generator of an analytic semigroup on X. Then A has the maximal L^p-regularity property if and only if the set $\{i\sigma(i\sigma I + A)^{-1};\ \sigma \in \mathbb{R}\}$ is R-bounded.*

Idea of the proof. This result can be found in [41, Corollary 4.4]. Denoting as before

$$M(\sigma) = A(i\sigma I + A)^{-1}, \quad \sigma \in \mathbb{R},$$

we have $M(\sigma) = i\sigma(i\sigma I + A)^{-1} - I$. Therefore, if

$$M_0 : \sigma \mapsto i\sigma(i\sigma I + A)^{-1}$$

is a Fourier multiplier, so is M. The first part of Theorem 2.14 applied to M_0 holds by the assumption that

$$\Big\{i\sigma(i\sigma I + A)^{-1};\ \sigma \in \mathbb{R}\Big\}$$

is R-bounded. For the second part, we must show that $\{\sigma M_0'(\sigma); \sigma \in \mathbb{R}\}$ is also R-bounded. For that purpose, note that $\sigma M_0'(\sigma) = M_0(\sigma)(I - M_0(\sigma))$. □

3 Classes of operators with maximal regularity property

In this section, we present three classes of operators in L^q-spaces having the maximal L^p-regularity property. We consider analytic semigroups in $L^2(\Omega)$, which can be extended to $L^p(\Omega)$ (with an underlying measure space (Ω,μ)) for p in a subinterval of $]1,\infty[$ including the case $p = 2$. In the sequel, if there is no ambiguity, we always denote the $L^p(\Omega)$-norm by $\|\cdot\|_p$.

3.1 Contraction semigroups

The result presented here is due to D. Lamberton [28].

Theorem 3.1 (D. Lamberton, 1987). *Let* (Ω, μ) *be a measure space and* A *the negative generator of an analytic semigroup of contractions* $(T(t))_{t\geq 0}$ *in* $L^2(\Omega)$. *Assume that for all* $q \in [1, \infty]$, *there holds the estimate*

$$\|T(t)f\|_q \leq \|f\|_q \quad \text{for all } t \geq 0 \text{ and all } f \in L^2(\Omega) \cap L^q(\Omega).$$

Then the operator A *has the maximal* L^p*-regularity property in* $L^q(\Omega)$.

Reference for the proof. The proof of this theorem can be found in [28]. The idea is first to observe that A has the maximal L^p-regularity property in $L^2(\Omega)$ by Theorem 2.6. The strategy then is to show that the convolution operator R defined by

$$Rf(t) = \int_0^t AT(t-s)f(s)\,ds, \quad t \geq 0,\ f \in L^p(0,\infty; L^p(\Omega)) = L^p(]0,\infty[\times\Omega),$$

is bounded in $L^p(]0,\infty[\times\Omega)$. We already know that this is true for $p = 2$. To prove the result for other values of $p \in]1,\infty[$, D. Lamberton uses the so-called Coifman–Weiss transference principle (this is the core of the proof). Once this is proved, by Theorem 2.4, we conclude that A has the maximal L^p-regularity property in $L^q(\Omega)$. □

This theorem can be applied to show that certain operators have the maximal L^p-regularity property, such as the Laplacian in $L^p(\Omega)$ ($\Omega \subset \mathbb{R}^n$ sufficiently regular) with Dirichlet, Neumann or Robin boundary conditions.

Proposition 3.2. *Let* $\Omega \subset \mathbb{R}^n$ *be a domain such that the Stokes formula (integration by parts) applies. We denote by* ν *the outer unit normal at* $\partial\Omega$. *Let* A_j *(*$j = D, N$ *or* R*) be the unbounded operator defined in* $L^2(\Omega)$ *by*

$$\begin{aligned} D(A_j) &= \{u \in H^1(\Omega); \Delta u \in L^2(\Omega) \text{ and } b_j(u) = 0 \text{ on } \partial\Omega\} \\ A_j u &= -\Delta u, \end{aligned}$$

where $b_D(u) = u$, $b_N(u) = \partial_\nu u$ *and* $b_R(u) = \alpha u + \partial_\nu u$ *for* $\alpha \geq 0$.

Then the operators A_j *(*$j = D, N$ *or* R*) have the maximal* L^p*-regularity property in* $L^q(\Omega)$ *for all* $p, q \in]1, \infty[$.

Proof. Thanks to Theorem 3.1, we only need to show that $-A_j$ generates an analytic semigroup $(T_j(t))_{t\geq 0}$ in $L^2(\Omega)$ and that this semigroup satisfies the estimate

$$\|T_j(t)f\|_q \leq \|f\|_q, \quad t \geq 0,\ f \in L^2(\Omega) \cap L^q(\Omega). \tag{3.1}$$

Case $j = D$. This case corresponds to Dirichlet boundary conditions. The first assumption to verify is that $-A_D$ generates an analytic semigroup $(T_D(t))_{t\geq 0}$ in $L^2(\Omega)$. Let a_D be the sesquilinear form

$$a_D(u,v) = \int_\Omega \nabla u \cdot \overline{\nabla v}\,dx, \quad u, v \in H_0^1(\Omega; \mathbb{C}).$$

This form a_D is continuous and coercive. It is easy to show that A_D is associated to the form a_D and therefore generates a bounded analytic semigroup. It remains to show that (3.1) holds for $j = D$. Let $q \in [1, \infty[$ and let

$$u(t) = T_D(t)f,\ t \geq 0,$$

be the solution of the Cauchy problem

$$u'(t) + A_D u(t) = 0,\ u(0) = f \in L^2(\Omega) \cap L^q(\Omega).$$

We first consider the case $1 < q \leq 2$. Multiplying the equation by $v = |u|^{q-2} u \chi_{u \neq 0}$ and integrating over Ω, we obtain for all $t \geq 0$,

$$\begin{aligned}
0 &= \int_\Omega u'(t) v(t)\, dx - \int_\Omega v(t) \Delta u(t)\, dx \\
&= \frac{1}{q} \frac{d}{dt} \Big(\int_\Omega |u(t)|^q\, dx \Big) + \int_\Omega |u(t)|^{q-2} |\nabla u(t)|^2 \chi_{u \neq 0}\, dx \\
&\quad + \int_\Omega u \nabla \big(|u(t)|^{q-2} \big) \cdot \nabla u \chi_{u \neq 0}\, dx \\
&= \frac{1}{q} \frac{d}{dt} \Big(\|u\|_q^q \Big)(t) + (q-1) \int_\Omega |u(t)|^{q-2} |\nabla u(t)|^2 \chi_{u \neq 0}\, dx
\end{aligned}$$

and therefore

$$\frac{d}{dt} \Big(\|u\|_q^q \Big)(t) \leq 0$$

which implies that $\|u(t)\|_q \leq \|f\|_q$. As the operator A_D is self-adjoint, we obtain by duality the same result also for the case $2 \leq q < \infty$. Passing with q to ∞, we obtain $\|u(t)\|_\infty \leq \|f\|_\infty$. This shows (3.1) for $j = D$.

Case $j = N$. This case corresponds to Neumann boundary conditions. It goes more or less as the previous case. The integrations by parts can be performed and give 0 for the boundary terms since $\partial_\nu u = 0$ at the boundary. As before, the first assumption to verify is that $-A_N$ generates an analytic semigroup $(T_N(t))_{t \geq 0}$ in $L^2(\Omega)$. Let a_N be the sesquilinear form

$$a_N(u, v) = \int_\Omega \nabla u \cdot \overline{\nabla v}\, dx, \quad u, v \in H^1(\Omega; \mathbb{C}).$$

This form a_N is continuous and coercive. It is easy to show that A_N is associated to the form a_N and therefore generates a bounded analytic semigroup. It remains to show that (3.1) holds for $j = N$. As in the previous case, for

$$u(t) = T_N(t)f,\ t \geq 0,$$

being the solution of the Cauchy problem

$$u'(t) + A_N u(t) = 0,\ u(0) = f \in L^2(\Omega) \cap L^q(\Omega),$$

we have

$$\begin{aligned}
0 &= \int_\Omega u'(t)v(t)\,dx - \int_\Omega v(t)\Delta u(t)\,dx \\
&= \frac{1}{q}\frac{d}{dt}\Big(\int_\Omega |u(t)|^q\,dx\Big) + \int_\Omega |u(t)|^{q-2}|\nabla u(t)|^2\chi_{u\neq 0}\,dx \\
&\quad + \int_\Omega u\nabla(|u(t)|^{q-2})\cdot\nabla u\chi_{u\neq 0}\,dx \\
&= \frac{1}{q}\frac{d}{dt}\Big(\|u\|_q^q\Big)(t) + (q-1)\int_\Omega |u(t)|^{q-2}|\nabla u(t)|^2\chi_{u\neq 0}\,dx
\end{aligned}$$

and therefore

$$\frac{d}{dt}\Big(\|u\|_q^q\Big)(t) \le 0,$$

which implies that $\|u(t)\|_q \le \|f\|_q$. Passing with q to ∞, we obtain $\|u(t)\|_\infty \le \|f\|_\infty$. This shows (3.1) for $j = N$.

Case $j = R$. This case corresponds to Robin boundary conditions. The first assumption to verify is that $-A_R$ generates an analytic semigroup $(T_R(t))_{t\ge 0}$ in $L^2(\Omega)$. Let a_R be the sesquilinear form

$$a(u,v) = \int_\Omega \nabla u\cdot\overline{\nabla v}\,dx + \int_{\partial\Omega} \alpha\, u\bar{v}\,d\sigma, \quad u,v\in H^1(\Omega;\mathbb{C}).$$

This form a_R is continuous and coercive. It is easy to show that A_R is associated to the form a_R and therefore generates a bounded analytic semigroup. It remains to show that (3.1) holds for $j = R$. As in the two previous cases, for

$$u(t) = T_R(t)f,\ t\ge 0,$$

being the solution of the Cauchy problem

$$u'(t) + A_R u(t) = 0,\ u(0) = f\in L^2(\Omega)\cap L^q(\Omega),$$

we have

$$\begin{aligned}
0 &= \int_\Omega u'(t)v(t)\,dx - \int_\Omega v(t)\Delta u(t)\,dx \\
&= \frac{1}{q}\frac{d}{dt}\Big(\int_\Omega |u(t)|^q\,dx\Big) + \int_\Omega |u(t)|^{q-2}|\nabla u(t)|^2\chi_{u\neq 0}\,dx \\
&\quad + \int_\Omega u\nabla(|u(t)|^{q-2})\cdot\nabla u\chi_{u\neq 0}\,dx - \int_{\partial\Omega} \partial_\nu u(t)|u(t)|^{q-2}u(t)\,d\sigma \\
&= \frac{1}{q}\frac{d}{dt}\Big(\|u\|_q^q\Big)(t) + (q-1)\int_\Omega |u(t)|^{q-2}|\nabla u(t)|^2\chi_{u\neq 0}\,dx \\
&\quad + \int_{\partial\Omega} \alpha\,|u(t)|^q\,d\sigma
\end{aligned}$$

and therefore

$$\frac{d}{dt}\left(\|u\|_q^q\right)(t) \leq 0,$$

which implies that $\|u(t)\|_q \leq \|f\|_q$. Passing with q to ∞, we obtain $\|u(t)\|_\infty \leq \|f\|_\infty$. This shows (3.1) for $j = R$.

It suffices now to apply Theorem 3.1 to obtain that the operators A_j ($j = D, N$ or R) have the maximal L^p-regularity property in $L^q(\Omega)$ for all $p, q \in]1, \infty[$ (Proposition 3.2).

□

3.2 Gaussian bounds with pointwise estimates

The result presented here is due first to M. Hieber and J. Prüss [21] and was somewhat extended by T. Coulhon and X. T. Duong [12]. The theorem below is adapted to semigroups with Gaussian estimates (so, not stated in the full generality).

Theorem 3.3 (Hieber–Prüss 1997, Coulhon–Duong 2000). *Let $\Omega \subset \mathbb{R}^n$. Assume that $(T(t))_{t\geq 0}$ is an analytic semigroup in $L^2(\Omega)$ with the representation for all $f \in L^2(\Omega)$ and $z \in \mathbb{C}$ with $|\arg z| < \varepsilon$ by*

$$T(z)f(x) = \int_\Omega p(z, x, y) f(y)\, dy, \quad x \in \Omega,$$

where the kernel p, for $t > 0$, satisfies the estimate

$$|p(t, x, y)| \leq c g(bt, x, y), \quad x, y \in \Omega, \tag{3.2}$$

with $c, b > 0$. Here $g(t, x, y) = (4\pi t)^{-\frac{n}{2}} e^{-\frac{|x-y|^2}{4t}}$. Then the semigroup $(T(t))_{t\geq 0}$ can be extended to an analytic semigroup in $L^q(\Omega)$ for all $q \in]1, \infty[$ and its negative generator A_q has the maximal L^p-regularity property for all $p \in]1, \infty[$.

Lemma 3.4. *Under the assumptions of Theorem 3.3 there exist $\theta \in]0, \frac{\pi}{2}]$ and constants $c_1, b_1 > 0$ such that*

$$|p(z, x, y)| \leq c_1 g(b_1 \mathrm{Re}(z), x, y), \quad x, y \in \Omega, \ z \in \mathbb{C} \text{ with } |\arg z| < \varepsilon \tag{3.3}$$

and, consequently, there are two constants $c_2, b_2 > 0$ such that

$$\left|\frac{\partial p}{\partial t}(t, x, y)\right| \leq \frac{c_2}{t} g(b_2 t, x, y), \quad t > 0, \ x, y \in \Omega. \tag{3.4}$$

Proof. This result is well known (see, e.g., Davies' book [14]). The estimate (3.4) follows from (3.3) using the Cauchy integral formula for the holomorphic function $z \mapsto p(z, x, y)$. □

Idea of the proof of Theorem 3.3. Let $Q = [0, \infty[\times \Omega$. The space (Q, μ, d), where μ is the Lebesgue measure on $\mathbb{R}^{n+1}$ and d is the quasi-metric defined by

$$d((t, x), (s, y)) = |x - y|^2 + |t - s|,$$

is of homogeneous type (has the doubling property: there exists a constant $C > 0$ such that $\mu(B(\xi, 2r)) \leq c\mu(B(\xi, r))$ where $B(\xi, r) = \{\eta \in Q; d(\xi, \eta) < r^2\}$). Let K be the operator with kernel

$$k((t,x),(s,y)) = \frac{\partial p(t-s,x,y)}{\partial t}, \quad t,s > 0,\ x,y \in \Omega.$$

We know that the operator K defined by

$$Kf(\xi) = \int_Q k(\xi,\eta) f(\eta)\, d\mu(\eta), \quad \text{a.e. in } Q \ni \xi = (t,x),$$

is bounded on $L^2(Q)$. This is only a reformulation of the maximal L^2-regularity property on $L^2(\Omega)$. The strategy is to prove that K is of weak-type $(1,1)$. By interpolation, we can then prove that K is a bounded operator on $L^q(Q)$ for all $q \in]1,2]$. A duality argument is then used to prove that K is bounded on $L^{q'}(Q)$ (where $\frac{1}{q} + \frac{1}{q'} = 1$ with $q' \in [2,\infty[$). Therefore, using the p-independence of the maximal L^p-regularity property (see Theorem 2.4), we finish the proof. Of course, the core of the proof is to show that K is of weak-type $(1,1)$. For that purpose, since the kernel k has a behavior like (3.4), we have to study a singular integral. We will use a (regularized) Calderón–Zygmund decomposition (see Theorem A.8) adapted to the problem. For any $f \in L^1(Q)$ and $\alpha > 0$, there exist $g, b_i \in L^1(Q)$ with the properties

(i) $|g(\xi)| \leq \kappa\alpha$ a.e. in $Q \ni \xi$;

(ii) there exist balls $B_i = B(\xi_i, r_i) \subset Q$ (i.e., $(t,x) \in B(\xi_i, r_i)$ if $|x - x_i|^2 + |t - t_i| < r_i^2$ and $\xi_i = (t_i, x_i)$) such that $\operatorname{supp} b_i \subset B_i$ and

$$\int_Q |b_i(\xi)|\, d\mu(\xi) \leq \kappa\alpha\mu(B_i);$$

(iii) $\displaystyle\sum_{i=1}^{\infty} \mu(B_i) \leq \frac{\kappa}{\alpha}\|f\|_1$;

(iv) any $\xi \in Q$ belongs to at most N_0 balls B_i,

where κ and N_0 only depend on the dimension n.

We "regularize" the functions b_i by applying an operator R_i defined by a kernel

$$\rho_i : Q \to \mathbb{R},\ \rho_i(\xi,\eta) = \varphi_i(t-s)\chi_{[(t-r_i)_+,t]}(s) k_{r_i}(x,y), \qquad \text{where } \xi = (t,x), \eta = (s,y)$$

and where $\varphi_i(\sigma) = \frac{1}{r_i}\frac{e}{2(e-1)} e^{-\frac{|\sigma|}{r_i}}$. This idea was first applied by X. T. Duong and A. McIntosh in [17]. We can show that the norm of $\sum_{i=1}^{\infty} R_i b_i \in L^2(Q)$ is controlled by $\alpha^{\frac{1}{2}}\|f\|_1$. Therefore, if we write

$$Kf = Kg + \sum_{i=1}^{\infty} KR_i b_i + \sum_{i=1}^{\infty} (K - KR_i) b_i,$$

only the last term is still to be investigated, the first two coming directly from the fact that K is bounded in $L^2(Q)$. It remains to show that

$$\mu\Big(\Big\{\xi \in Q; \Big|\sum_{i=1}^{\infty}(K - KR_i)b_i(\xi)\Big| > \alpha\Big\}\Big) \le C\,\alpha\|f\|_1.$$

This can be done once we prove that

$$\int_{d(\xi,\eta)\ge cr_i} |k(\xi,\eta) - k_i(\xi,\eta)|d\mu(\xi) \le C,$$

where $k_i(\xi,\eta) = \int_Q k(\xi,\zeta)\rho_i(\zeta,\eta)d\mu(\zeta)$ is the kernel of KR_i. Indeed, the proof at this step is very much like the proof of Theorem 2.4, using this last estimate. □

Example 3.5. Consider the elliptic differential operator $L = -\mathrm{div}A\nabla$ of second order in divergence form with Dirichlet boundary conditions in a domain $\Omega \subset \mathbb{R}^n$ with $A \in L^\infty(\Omega;\mathbb{C}^{n\times n})$ with antisymmetric imaginary part. Then it has been proved by E. M. Ouhabaz (see [36] and [37]) that $-L$ generates an analytic semigroup in $L^2(\Omega)$ with Gaussian estimates. Therefore, L has the maximal L^p-regularity property on $L^q(\Omega)$ for all $q \in]1,\infty[$, by Theorem 3.3.

3.3 Generalized Gaussian bounds

It is sometimes not clear, or not true, whether a semigroup in $L^2(\Omega)$ has Gaussian estimates of the type (3.2). However, it is sometimes possible to prove a weaker form, namely a local integrated bound of the following form. To make the notations shorter, we will use

$$\|\cdot\|_{\mathscr{L}(L^{q_0}(\Omega),L^{q_1}(\Omega))} = \|\cdot\|_{q_0\to q_1}$$

and

$$A(x,\rho,k) = B(x,(k+1)\rho)\setminus B(x,k\rho), \quad x\in\Omega,\ \rho>0,\ k\in\mathbb{N}.$$

Definition 3.6. Let $\Omega \subset \mathbb{R}^n$ be a domain. Let A be the negative generator of an analytic semigroup $(T(t))_{t\ge 0}$ in $L^{q_0}(\Omega)$. We say that A has generalized Gaussian estimates (q_0,q_1) (where $1 < q_0 \le q_1 < \infty$) if one of the following properties holds:

(1) the semigroup $(T(t))_{t\ge0}$ satisfies

$$\|\chi_{B(x,t)}T(t)\chi_{A(x,t,k)}\|_{q_0\to q_1} \le |B(x,t)|^{-(\frac{1}{q_0}-\frac{1}{q_1})}h(k) \tag{3.5}$$

for $t>0$, $x\in\Omega$, $k\in\mathbb{N}$, and $(h(k))_{k\ge1}$ satisfying

$$h(k) \le c_\delta(k+1)^{-\delta} \text{ for some } \delta > \frac{n}{q_0} + \frac{1}{q_0'}$$

(2) the resolvent of A satisfies

$$\|\chi_{B(x,t)}(I+zA)^{-1}\chi_{A(x,t,k)}\|_{q_0\to q_1} \le |B(x,t)|^{-(\frac{1}{q_0}-\frac{1}{q_1})}h(k) \tag{3.6}$$

for $z \in \Sigma_\theta = \{w \in \mathbb{C} \setminus \{0\}; |\arg(w)| < \pi - \theta\}$ with $\theta \in [0, \frac{\pi}{2}[$, $t = |z|^{-\frac{1}{2}}$, $x \in \Omega$, $k \in \mathbb{N}$ and $(h(k))_{k\geq 1}$ satisfying

$$h(k) \leq c_\delta (k+1)^{-\delta} \text{ for some } \delta > \frac{n}{q_0} + \frac{1}{q_0'}.$$

Remark 3.7. A semigroup satisfying the Gaussian estimates (3.2) satisfies the two bounds above for all $1 < q_0 \leq q_1 < \infty$.

These kinds of bounds for $q_0 = 2$ are easier to prove than the pointwise Gaussian estimates (3.2), in particular, the second one (3.6). Indeed, it is, for instance for divergence form elliptic operators, only a matter of partial integration, as we will see below for the Navier–Lamé operator (see Theorem 3.9).

Theorem 3.8 (Kunstmann, 2008). *Let $\Omega \subset \mathbb{R}^n$ be a domain. Let A be the negative generator of an analytic semigroup $(T(t))_{t\geq 0}$ in $L^2(\Omega)$ satisfying* (3.6) *with $q_1 > q_0 = 2$. Then A has the maximal L^p-regularity property in $L^q(\Omega)$ for all $q \in [2, q_1[$.*

References for the proof. This result is due to P. C. Kunstmann [24]. The original statement uses (3.5) instead of (3.6), and is presented in a more general context: instead of $\Omega \subset \mathbb{R}^n$, Ω is assumed to be a space of homogeneous type. □

In the following, we study the Navier–Lamé operator

$$L = -\mu\Delta - (\lambda + \mu)\nabla \operatorname{div}, \quad \mu > 0,\ \lambda + \mu \geq 0,$$

supplemented by homogeneous Dirichlet boundary conditions, which appears in linear elasticity theory. To be precise, let L be the operator in $L^2(\Omega; \mathbb{R}^n)$ associated with the sesquilinear form

$$\ell(u, v) = \mu \int_\Omega \nabla u \cdot \overline{\nabla v}\, dx + (\lambda + \mu) \int_\Omega \operatorname{div} u\, \overline{\operatorname{div} v}\, dx, \quad u, v \in H_0^1(\Omega; \mathbb{R}^3).$$

Since ℓ is symmetric, continuous and coercive, the operator L is self-adjoint and generates a bounded analytic semigroup in $L^2(\Omega; \mathbb{R}^n)$.

Theorem 3.9. *Let Ω be a Lipschitz domain in $\mathbb{R}^n$ ($n \geq 3$). Then the Navier–Lamé operator with Dirichlet boundary conditions has the maximal L^p-regularity property in $L^q(\Omega)$ for all $q \in]\frac{2n}{n+2}, \frac{2n}{n-2}[$.*

Proof. Fix an arbitrary point $x \in \Omega$,

$$z \in \Sigma_\theta = \{w \in \mathbb{C} \setminus \{0\}; |\arg(w)| < \pi - \theta\},$$

$t = |z|^{-\frac{1}{2}}$ and an arbitrary partition of unity $\{\eta_j; j \in \mathbb{N}\}$ of $\mathbb{R}^n$ such that

$$\begin{aligned} &\eta_0 \in \mathscr{C}_c^\infty(B(x, 2t); \mathbb{R}), \quad \eta_j \in \mathscr{C}_c^\infty\Big(B(x, 2^{j+1}t) \setminus B(x, 2^{j-1}t); \mathbb{R}\Big), \\ &0 \leq \eta_j \leq 1, \quad |\nabla \eta_j| \leq \tfrac{1}{2^{j-1}t}, \quad \textstyle\sum_{j=0}^\infty \eta_j = 1, \end{aligned} \tag{3.7}$$

where $B(x,r)$ is the ball in $\mathbb{R}^n$ with center at $x \in \mathbb{R}^3$ and radius $r > 0$, decompose $f \in L^2(\Omega, \mathbb{R}^n)$ as

$$f = \sum_{j=0}^{\infty} f_j, \quad f_j = \eta_j f \quad \text{and set } u = \sum_{j=0}^{\infty} u_j, \quad u_j = (zI + L)^{-1} f_j \in D(L). \tag{3.8}$$

We will prove that for all $p \in [2, \frac{2n}{n-2}]$, there exist two constants $C, c > 0$ such that

$$|z| \Big[\int_{\Omega \cap B(x,t)} |u_j|^p dy \Big]^{\frac{1}{p}} \le C e^{-c2^j} t^{n(\frac{1}{p}-\frac{1}{2})} \Big[\int_\Omega |f_j|^2 dy \Big]^{\frac{1}{2}} \quad \forall j \in \mathbb{N}. \tag{3.9}$$

This will be done in three steps.
Step 1. Pick a new family of functions $(\xi_j)_{j\ge1}$ such that $\xi_j \in \mathscr{C}_c^\infty(B(x, 2^{j-1}t); \mathbb{R})$. Taking the L^2-pairing of $\xi_j^2 \overline{u_j}$ with both sides of $zu_j + Lu_j = f_j$, and keeping in mind that $\xi_j f_j = 0$ for each $j \ge 1$, we may write, based on integration by parts,

$$\begin{aligned} z \int_\Omega \xi_j^2 |u_j|^2 dy + \mu \int_\Omega \xi_j^2 |\nabla u_j|^2 dy + (\lambda + \mu) \int_\Omega \xi_j^2 |\operatorname{div} u_j|^2 dy \\ = \int_\Omega \mathcal{O}\Big(|\nabla \xi_j| |u_j| |\xi_j| \Big[\mu |\nabla u_j| + (\lambda+\mu) |\operatorname{div} u_j| \Big] \Big) dy. \end{aligned} \tag{3.10}$$

From this, via the Cauchy–Schwarz inequality and a standard trick that allows us to absorb similar terms with small coefficients in the left-hand side, we get

$$|z| \int_\Omega \xi_j^2 |u_j|^2 dy \le C \int_\Omega |\nabla \xi_j|^2 |u_j|^2 dy \tag{3.11}$$

and, since $\lambda + \mu \ge 0$,

$$\int_\Omega \xi_j^2 |\nabla u_j|^2 dy \le C \int_\Omega |\nabla \xi_j|^2 |\nabla u_j|^2 dy. \tag{3.12}$$

Step 2. Similarly as in [6], we now replace the cutoff function ξ_j in (3.11) by another cutoff function $e^{\alpha_j \xi_j} - 1$ (which has the same properties as ξ_j), with

$$\alpha_j = \frac{\sqrt{|z|}}{2\sqrt{C} \|\nabla \xi_j\|_\infty}, \ j \ge 2.$$

In a first stage, this yields

$$\int_\Omega |u_j|^2 |e^{\alpha_j \xi_j} - 1|^2 dy \le \frac{1}{4} \int_\Omega |u_j|^2 |e^{\alpha_j \xi_j}|^2 dy,$$

then further

$$\int_\Omega |u_j|^2 |e^{\alpha_j \xi_j}|^2 dy \le 4 \int_\Omega |u_j|^2 dy, \tag{3.13}$$

in view of the generic, elementary observation

$$\|f-g\| \le \frac{1}{2}\|f\| \quad \text{implies} \quad \|f\| \le 2\|g\|.$$

If we now assume that the original cutoff functions $(\xi_j)_{j\ge 2}$ also satisfy

$$0 \le \xi_j \le 1, \quad \xi_j = 1 \text{ on } B(x,t) \quad \text{and} \quad \|\nabla \xi_j\|_\infty \le \frac{\kappa}{2^j t},$$

it follows from the definition of α_j that $\alpha_j \ge c2^j$ and from (3.13) that

$$|e^{\alpha_j}|^2 \int_{\Omega\cap B(x,t)} |u_j|^2 dy \le 4\int_\Omega |u_j|^2 dy \le \frac{\mathrm{C}}{|z|^2}\int_\Omega |f_j|^2 dy,$$

the second inequality coming from the fact that $-L$ generates an analytic semigroup. This gives then

$$|z|^2 \int_{\Omega\cap B(x,t)} |u_j|^2 dy \le Ce^{-c2^j}\int_\Omega |f_j|^2 dy. \tag{3.14}$$

The same procedure allows to estimate ∇u on $B(x,t)$ using (3.12) as follows

$$|z| \int_{\Omega\cap B(x,t)} |\nabla u_j|^2 dy \le Ce^{-c2^j}\int_\Omega |f_j|^2 dy. \tag{3.15}$$

Those two estimates are also valid if $j = 0$ since the resolvent of L is bounded in $L^2(\Omega;\mathbb{R}^n)$.

Step 3. Let $p = 2^* = \frac{2n}{n-2}$ and $D \subset \mathbb{R}^n$ $(n \ge 3)$ be a Lipschitz domain of diameter $R > 0$. The continuous embedding of $H^1(D)$ into $L^p(D)$ then reads, after rescaling, as

$$R^{n(\frac{1}{2}-\frac{1}{p})}\Big(\int_D |u|^p dy\Big)^{\frac{1}{p}} \le C\Big[\Big(\int_D |u|^2 dy\Big)^{\frac{1}{2}} + R\Big(\int_D |\nabla u|^2 dy\Big)^{\frac{1}{2}}\Big]. \tag{3.16}$$

Combining this inequality with (3.14) and (3.15), and keeping in mind that $|z| = \frac{1}{t^2}$, we have for all $j \in \mathbb{N}$

$$|z|\Big(\int_{\Omega\cap B(x,t)} |u_j|^{\frac{2n}{n-2}} dy\Big)^{\frac{n-2}{2n}} \le Ct^{-1}e^{-c2^j}\Big(\int_\Omega |f_j|^2 dy\Big)^{\frac{1}{2}}. \tag{3.17}$$

By interpolation, using (3.14) and (3.17), the generalized Gaussian bound (3.9) is proved for all $2 \le p \le \frac{2n}{n-2}$.

By Theorem 3.8, we may conclude that L has the maximal L^p-regularity property in the space $L^q(\Omega;\mathbb{R}^n)$ for all $q \in [2, \frac{2n}{n-2}[$. By duality (since L is self-adjoint), we can prove that L has the maximal L^p-regularity property in $L^q(\Omega;\mathbb{R}^n)$ also for all $q \in]\frac{2n}{n+2}, 2]$, which proves Theorem 3.9. □

4 Applications to partial differential equations

In this section, we will apply maximal L^p-regularity results to show the uniqueness of solutions of certain partial differential equations. We start with a toy problem, studied by F. Weissler [42]. This will show the way to prove uniqueness of mild solutions of the incompressible Navier–Stokes system.

4.1 Existence and uniqueness for a semilinear heat equation

We are interested in the initial-boundary value problem

$$\begin{array}{rcll} \dfrac{\partial u}{\partial t} - \Delta u &=& u^2 & \text{in} \quad]0,T[\times\Omega, \\ u(t,x) &=& 0 & \text{on} \quad]0,T[\times\partial\Omega, \\ u(0) &=& u_0 & \text{in} \quad \Omega, \end{array} \tag{4.1}$$

where $T > 0$ and $\Omega \subset \mathbb{R}^n$ is a domain with no particular regularity at the boundary. We assume that $n \geq 4$. The critical space where we are looking for solutions is $L^p(\Omega)$ with $p = \frac{n}{2}$. This space is critical in the sense that if $p > \frac{n}{2}$, then the nonlinearity u^2 is a "small" perturbation of the linear part $-\Delta u$ and if $p < \frac{n}{2}$, then the nonlinearity "wins" and the methods applied here are not appropriate. Our purpose is to show that a solution $u \in \mathscr{C}([0,T]; L^{\frac{n}{2}}(\Omega))$ of (4.1) (in an integral sense defined below) is unique in the space $\mathscr{C}([0,T]; L^{\frac{n}{2}}(\Omega))$.

Definition 4.1. A function u is said to be an integral solution of (4.1) with $u_0 \in L^{\frac{n}{2}}(\Omega)$ on $[0,\tau]$ if $u \in \mathscr{C}([0,\tau]; L^{\frac{n}{2}}(\Omega))$ and if u satisfies

$$u(t) = T(t)u_0 + \int_0^t T(t-s)(u(s)^2)\, ds, \quad t \in [0,\tau],$$

where $(T(t))_{t\geq 0}$ is the semigroup generated by the Dirichlet–Laplacian in $L^{\frac{n}{4}}(\Omega)$, which we denote by $-A$.

We first study the question of existence.

Theorem 4.2 (F. Weissler, 1981). *Let $n \geq 4$. For any initial value $u_0 \in L^{\frac{n}{2}}(\Omega)$, there exists $\tau \in]0,T]$ and an integral solution $u \in \mathscr{C}([0,\tau]; L^{\frac{n}{2}}(\Omega))$ of (4.1) in $[0,\tau]$. If $\|u_0\|_{\frac{n}{2}}$ is sufficiently small, then $\tau = T$.*

Proof. We will show the local existence of an integral solution of (4.1) via a fixed point method. We reformulate the problem as to find a Banach space $\mathcal{E}_T$ containing $\mathscr{C}([0,T]; L^{\frac{n}{2}}(\Omega))$ such that $a = T(\cdot)u_0 \in \mathcal{E}_T$ and there exists $u \in \mathcal{E}_T$ verifying

$$u = a + B(u,u),$$

where B is the bilinear operator defined by

$$B(u,v)(t) = \int_0^t T(t-s)(u(s)v(s))\, ds, \quad t \in [0,T].$$

We need B to be continuous on $\mathcal{E}_T \times \mathcal{E}_T$. We choose

$$\mathcal{E}_T = \Big\{ u \in \mathscr{C}([0,T]; L^{\frac{n}{2}}(\Omega)); t \mapsto t^{\frac{1}{4}} u(t) \in \mathscr{C}([0,T]; L^{\frac{2n}{3}}(\Omega)) \Big\},$$

and we define the norm in this space to be

$$\|u\|_{\mathcal{E}_T} = \sup_{0<t<T} \|u(t)\|_{\frac{n}{2}} + \sup_{0<t<T} t^{\frac{1}{4}} \|u(t)\|_{\frac{2n}{3}} \quad u \in \mathcal{E}_T.$$

This space $\mathcal{E}_T$ endowed with its norm is a Banach space. We remark first that $(T(t))_{t\geq 0}$ is a bounded analytic semigroup in $L^p(\Omega)$ for all $p \in]1,\infty[$, which implies, in particular, that for all $\alpha \geq 0$, there exists a constant $c_{p,\alpha} > 0$ such that

$$\|t^\alpha A^\alpha T(t)\|_{\mathscr{L}(L^p(\Omega))} \leq c_{p,\alpha}, \quad t > 0.$$

Therefore, it is easy to check that $a \in \mathcal{E}_T$. We will show next that

$$B : \mathcal{E}_T \times \mathcal{E}_T \to \mathcal{E}_T$$

is continuous. Let $u, v \in \mathcal{E}_T$. Then we have

$$t \mapsto t^{\frac{1}{2}} u(t) v(t) \in \mathscr{C}([0,T]; L^{\frac{n}{3}}(\Omega))$$

with norm bounded by $\|u\|_{\mathcal{E}_T} \|v\|_{\mathcal{E}_T}$. Therefore, we obtain

$$\|t^{\frac{1}{2}} A^{-\frac{1}{2}}(u(t)v(t))\|_{\frac{n}{2}} \leq c \|u\|_{\mathcal{E}_T} \|v\|_{\mathcal{E}_T}$$

for all $t \in [0,T]$ because of the continuous embedding $W^{1,\frac{n}{3}}(\Omega) \subset L^{\frac{n}{2}}(\Omega)$. We then have

$$\begin{aligned} B(u,v)(t) &= \int_0^t T(t-s)(u(s)v(s))\, ds \\ &= \int_0^t \sqrt{t-s}\, A^{\frac{1}{2}} T(t-s) A^{-\frac{1}{2}}(\sqrt{s}\, u(s)v(s)) \frac{1}{\sqrt{t-s}\sqrt{s}}\, ds \end{aligned}$$

which gives, for all $t \in [0,T]$, the estimate

$$\begin{aligned} \|B(u,v)(t)\|_{\frac{n}{2}} &\leq c\, c_{\frac{n}{2},\frac{1}{2}} \|u\|_{\mathcal{E}_T} \|v\|_{\mathcal{E}_T} \Big(\int_0^t \frac{1}{\sqrt{t-s}\sqrt{s}}\, ds \Big) \\ &\leq \pi c\, c_{\frac{n}{2},\frac{1}{2}} \|u\|_{\mathcal{E}_T} \|v\|_{\mathcal{E}_T} \end{aligned}$$

since

$$\int_0^t \frac{1}{\sqrt{t-s}\sqrt{s}}\, ds = \int_0^1 \frac{1}{\sqrt{1-\sigma}\sqrt{\sigma}}\, d\sigma = \int_{-\frac{1}{2}}^{\frac{1}{2}} \frac{2}{\sqrt{1-4r^2}}\, dr = \pi.$$

The same arguments are used to estimate $t^{\frac{1}{4}} \|B(u,v)(t)\|_{\frac{n}{3}}$. For $u, v \in \mathcal{E}_T$, we have

$$\|t^{\frac{1}{2}} A^{-\frac{3}{4}}(u(t)v(t))\|_{\frac{2n}{3}} \leq c \|u\|_{\mathcal{E}_T} \|v\|_{\mathcal{E}_T}$$

because of $t^{\frac{1}{2}}A^{-\frac{1}{2}}(u(t)v(t)) \in L^{\frac{n}{2}}(\Omega)$ and the continuous embedding $W^{\frac{1}{2},\frac{n}{2}}(\Omega) \subset L^{\frac{2n}{3}}(\Omega)$ in dimension n. Therefore, we have

$$\begin{aligned}\|t^{\frac{1}{4}}B(u,v)(t)\|_{\frac{2n}{3}} &\leq c\,c_{\frac{2n}{3},\frac{3}{4}}\|u\|_{\mathcal{E}_T}\|v\|_{\mathcal{E}_T}\Big(t^{\frac{1}{4}}\int_0^t (t-s)^{-\frac{3}{4}}s^{-\frac{1}{2}}\,ds\Big)\\ &\leq c\,c_{\frac{n}{2},\frac{1}{2}}\|u\|_{\mathcal{E}_T}\|v\|_{\mathcal{E}_T}\Big(\int_0^1 (1-\sigma)^{-\frac{3}{4}}\sigma^{-\frac{1}{2}}\,d\sigma\Big).\end{aligned}$$

We can finish the proof of existence by applying the Picard fixed point theorem (see Theorem A.1 below) as long as $\|a\|_{\mathcal{E}_T} \leq \frac{1}{4\|B\|}$. This is the case if $\|u_0\|_{\frac{n}{2}}$ is small enough. The argument must be adapted a little if $\|u_0\|_{\frac{n}{2}}$ is not small by adjusting T so that the result remains true. □

We now turn to the question of uniqueness.

Theorem 4.3 (F. Weissler, 1981). *Assume that $n \geq 5$. Let $u_1, u_2 \in \mathscr{C}([0,T]; L^{\frac{n}{2}}(\Omega))$ be two integral solutions of (4.1) for the same initial value $u_0 \in L^{\frac{n}{2}}(\Omega)$ on some interval $[0,T]$ of local existence. Then $u_1 = u_2$ on $[0,T]$.*

The proof of the foregoing theorem will be prepared by the following result.

Lemma 4.4. *The bilinear operator*

$$B : L^q(0,T; L^{\frac{n}{2}}(\Omega)) \times \mathscr{C}([0,T]; L^{\frac{n}{2}}(\Omega)) \to L^q(0,T; L^{\frac{n}{2}}(\Omega))$$

is bounded for all $q \in]1,\infty[$.

Proof. For $u \in L^q(0,T; L^{\frac{n}{2}}(\Omega))$ and $v \in \mathscr{C}([0,T]; L^{\frac{n}{2}}(\Omega))$ the product uv is in $L^q(0,T; L^{\frac{n}{4}}(\Omega))$. Therefore,

$$f : t \mapsto A^{-1}(u(t)v(t)) \in L^q(0,T; L^{\frac{n}{2}}(\Omega)).$$

Since the Dirichlet–Laplacian enjoys the maximal L^q-regularity in $L^{\frac{n}{2}}(\Omega)$ (see Proposition 3.2), we have

$$\begin{aligned}\|B(u,v)\|_{L^q(0,T;L^{\frac{n}{2}}(\Omega))} &= \Big\|t \mapsto A\int_0^t AT(t-s)f(s)\,ds\Big\|_{L^q(0,T;L^{\frac{n}{2}}(\Omega))}\\ &\leq C\|u\|_{L^q(0,T;L^{\frac{n}{2}}(\Omega))}\|v\|_{\mathscr{C}([0,T];L^{\frac{n}{2}}(\Omega))},\end{aligned}$$

the constant C coming from the maximal L^q-regularity property of A in $L^{\frac{n}{2}}(\Omega)$. □

Proof of Theorem 4.3. With the same notations as in the previous proof, u_1 and u_2 are both solutions of the equation

$$u = a + B(u,u).$$

If we denote by v the difference between u_1 and u_2, then v must satisfy the equation

$$v = B(v, u_1 + u_2),$$

with $u_1, u_2, v \in \mathscr{C}([0,T]; L^{\frac{n}{2}}(\Omega))$ and $u_1(0) = u_2(0) = u_0$, $v(0) = 0$. To prove that $v = 0$ on a small interval $[0,\tau]$ (which implies then that $v = 0$ on the whole interval $[0,T]$), we need the following auxiliary lemma, which proof lies below, and this is where we use the maximal regularity property for the Dirichlet–Laplacian in $L^{\frac{n}{2}}(\Omega)$.

Let $\varepsilon > 0$ be fixed. Choose $u_{0,\varepsilon} \in \mathscr{C}_c^\infty(\Omega)$ such that

$$\|u_0 - u_{0,\varepsilon}\|_{\frac{n}{2}} < \varepsilon.$$

We decompose $B(v, u_1 + u_2)$ into three parts:

$$B(v, u_1 + u_2) = B(v, u_1 + u_2 - 2u_0) + 2B(v, u_0 - u_{0,\varepsilon}) + 2B(v, u_{0,\varepsilon}).$$

We can estimate the first two parts thanks to Lemma 4.4. This gives then for any $\tau \in]0,T]$

$$\begin{aligned} &\|B(v, u_1 + u_2 - 2u_0)\|_{L^q(0,\tau;L^{\frac{n}{2}}(\Omega))} \\ \leq\ & C\|v\|_{L^q(0,\tau;L^{\frac{n}{2}}(\Omega))}\Big(\|u_1 - u_0\|_{\mathscr{C}([0,\tau];L^{\frac{n}{2}}(\Omega))} + \|u_2 - u_0\|_{\mathscr{C}([0,\tau];L^{\frac{n}{2}}(\Omega))}\Big) \end{aligned} \tag{4.2}$$

and

$$\|B(v, u_0 - u_{0,\varepsilon})\|_{L^q(0,\tau;L^{\frac{n}{2}}(\Omega))} \leq C\varepsilon\|v\|_{L^q(0,\tau;L^{\frac{n}{2}}(\Omega))}. \tag{4.3}$$

Since $\|u_i - u_0\|_{\mathscr{C}([0,\tau];L^{\frac{n}{2}}(\Omega))} \to 0$ as $\tau \to 0$ for $i = 1, 2$, it remains to estimate the last part $B(v, u_{0,\varepsilon})$. Since $u_{0,\varepsilon} \in \mathscr{C}_c^\infty(\Omega)$, we have $vu_0 \in L^q(0,\tau; L^{\frac{n}{2}}(\Omega))$ and

$$\|B(v, u_{0,\varepsilon})(t)\|_{\frac{n}{2}} \leq Mt^{1-\frac{1}{q}}\|v\|_{L^q(0,\tau;L^{\frac{n}{2}}(\Omega))}\|u_{0,\varepsilon}\|_\infty.$$

It is now obvious that $\|B(v, u_{0,\varepsilon})\|_{L^q(0,\tau;L^{\frac{n}{2}}(\Omega))} \to 0$ as $\tau \to 0$. Combining this last result with the estimates (4.2) and (4.3), we obtain that for ε and τ small enough,

$$\|B(v, u_1 + u_2)\|_{L^q(0,\tau;L^{\frac{n}{2}}(\Omega))} \leq \frac{1}{2}\|v\|_{L^q(0,\tau;L^{\frac{n}{2}}(\Omega))}.$$

Since v is solution of $v = B(v, u_1 + u_2)$ on $[0,T]$, and then in particular on $[0,\tau]$, this implies that $v = 0$ almost everywhere on $[0,\tau]$. Since moreover v is continuous on $[0,\tau]$, we conclude that $v = 0$ (everywhere) on $[0,\tau]$. These arguments show that $\{t \in [0,T]; v = 0\}$ is an open set in $[0,T]$ and by continuity of v, this set is also closed. Since it is not empty, it is necessarily equal to the whole interval $[0,T]$ by connectedness. □

4.2 Uniqueness for the incompressible Navier–Stokes system

The incompressible Navier–Stokes equations in the whole space $\mathbb{R}^n$ reads as

$$\begin{aligned} \frac{\partial u}{\partial t} - \Delta u + \nabla\pi + (u \cdot \nabla)u &= 0 \quad \text{in} \quad]0,T[\times\mathbb{R}^n, \\ \operatorname{div} u &= 0 \quad \text{in} \quad]0,T[\times\mathbb{R}^n, \\ u(0) &= u_0 \quad \text{in} \quad \mathbb{R}^n, \end{aligned} \tag{4.4}$$

where

$$u : [0,T] \times \mathbb{R}^n \to \mathbb{R}^n \quad \text{and} \quad \pi : [0,T] \times \mathbb{R}^n \to \mathbb{R}$$

denote the velocity of a fluid and its pressure. The notation $(u \cdot \nabla)v$ for u and v vector fields stands for $\sum_{i=1}^n u_i \partial_i v$. We assume that there is no external force. Throughout this section, we also assume that $n \geq 3$.

We can reformulate this system in a functional analytic setting as follows: Let A be the operator with domain $W^{2,p}(\mathbb{R}^n;\mathbb{R}^n)$ (for a $p \in]1,\infty[$), given by

$$Au = \mathbb{P}(-\Delta u) \text{ where } \mathbb{P} = I + \nabla(-\Delta)^{-1}\text{div},$$

$\mathbb{P}$ is the Leray projection and is bounded in $L^p(\mathbb{R}^n;\mathbb{R}^n)$ since the Riesz projections are bounded. As in the case of the semilinear heat equation above, there is a critical space for (4.4). In the scale of Lebesgue spaces, the critical space here is $L^n(\mathbb{R}^n;\mathbb{R}^n)$, which means that if $p > n$, then the nonlinearity $(u \cdot \nabla)u$ appears as a "small" perturbation of the linear part $\mathbb{P}(-\Delta u)$ and if $p < n$, then the nonlinearity "wins".

The existence of integral solutions of (4.4) (see Definition 4.5 below) for an initial condition $u_0 \in L^n(\mathbb{R}^n;\mathbb{R}^n)$ has been proved by T. Kato in 1984 in [23]. The proof is similar to the proof of the existence of solutions for the semilinear heat equation (Theorem 4.2). A good reference for this problem is the book by P. G. Lemarié–Rieusset [29].

Definition 4.5. A function u is said to be an integral solution of (4.4) on $[0,\tau]$ for $u_0 \in L^n(\mathbb{R}^n;\mathbb{R}^n)$ with $\operatorname{div} u_0 = 0$ if $u \in \mathscr{C}([0,\tau];L^n(\mathbb{R}^n;\mathbb{R}^n))$ and if u satisfies

$$u(t) = e^{t\Delta}u_0 - \mathbb{P}\int_0^t e^{(t-s)\Delta}\nabla \cdot (u(s) \otimes u(s))\, ds, \quad t \in [0,\tau],$$

where $(e^{t\Delta})_{t\geq 0}$ is the semigroup generated by the Laplacian Δ.

Remark 4.6. (i) Since u is a divergence-free vector field, we have

$$(u \cdot \nabla)u = \sum_{i=1}^n u_i \partial_i u = \sum_{i=1}^n \partial_i(u_i u) = \nabla \cdot (u \otimes u),$$

which explains the form of u in Definition 4.5.

(ii) In the previous definition, we want the semigroup $(e^{t\Delta})_{t\geq 0}$ to act on a distribution $\nabla(u \otimes u) \in W^{-1,\frac{n}{2}}$. This makes sense since in the case of the whole space $\mathbb{R}^n$, the heat semigroup acts on all $W^{s,p}(\mathbb{R}^n;\mathbb{R}^n)$, $s \in \mathbb{R}$, $p \in]1,\infty[$.

Theorem 4.7 (T. Kato, 1984). *Let $u_0 \in L^n(\mathbb{R}^n;\mathbb{R}^n)$ satisfy $\operatorname{div} u_0 = 0$. Then there exists $T > 0$ and an integral solution $u \in \mathscr{C}([0,T];L^n(\mathbb{R}^n;\mathbb{R}^n))$ of (4.4). If $\|u_0\|_n$ is small enough, then the solution u is global (i.e., we can take $T = \infty$).*

Idea of the proof. The proof follows the line of the proof of Theorem 4.2 by working on the space

$$\begin{aligned} \mathcal{E}_T \;=\; \Big\{ & u \in \mathscr{C}([0,T];L^n(\mathbb{R}^n;\mathbb{R}^n));\ \operatorname{div} u = 0 \text{ in } \mathbb{R}^n \text{ and} \\ & t \mapsto \sqrt{t}\nabla u(t) \in \mathscr{C}([0,T];L^n(\mathbb{R}^n;\mathbb{R}^{n\times n})\Big\}, \end{aligned}$$

endowed with the norm

$$\|u\|_{\mathcal{E}_T} = \sup_{0<t<T} \|u(t)\|_n + \sup_{0<t<T} \sqrt{t}\|\nabla u(t)\|_n.$$

We are looking for $u \in \mathcal{E}_T$ solution of $u = a + B(u,u)$ where B is defined by

$$B(u,v)(t) = -\mathbb{P}\int_0^t e^{(t-s)\Delta}\Big(\frac{1}{2}\nabla\cdot(u(s)\otimes v(s)+v(s)\otimes u(s))\Big)\,ds, \quad t\in[0,T], \tag{4.5}$$

and $a(t) = e^{t\Delta}u_0$, $t \in [0,T]$. □

Theorem 4.8 (G. Furioli, P. G. Lemarié–Rieusset, E. Terraneo, 2000). *Let u_1, u_2 be two integral solutions of* (4.4) *for the same initial value $u_0 \in L^n(\mathbb{R}^n;\mathbb{R}^n)$ satisfying* $\operatorname{div} u_0 = 0$. *Then $u_1 = u_2$ on $[0,T]$.*

Proof. This result was first proved by G. Furioli, P. G. Lemarié–Rieusset and E. Terraneo in [18] (see also the very nice review on the subject by M. Cannone [11]). The proof presented here is based on [33] and of the same spirit as the proof of Theorem 4.3. The basic idea is to reformulate the problem of uniqueness as to show that $u = u_1 - u_2$ is equal to zero on the interval $[0,T]$. The function u satisfies the equation $u = B(u, u_1 + u_2)$ where B is defined by (4.5) above. As shown by F. Oru [35], the bilinear operator

$$B : \mathscr{C}([0,T]; L^n(\mathbb{R}^n;\mathbb{R}^n)) \times \mathscr{C}([0,T]; L^n(\mathbb{R}^n;\mathbb{R}^n)) \to \mathscr{C}([0,T]; L^n(\mathbb{R}^n;\mathbb{R}^n))$$

is not bounded. Had it been continuous, the proof of uniqueness of an integral solution of the Navier–Stokes system would have been straightforward. The idea of [18], in dimension $n = 3$, was then to lower the regularity of the space $L^3(\mathbb{R}^3;\mathbb{R}^3)$ and to consider a Besov space E instead (or, as shown by Y. Meyer in [31], the weak L^3 space, namely $L^{3,\infty}(\mathbb{R}^3;\mathbb{R}^3)$) to obtain a bounded bilinear operator B in $\mathscr{C}([0,T];E) \times \mathscr{C}([0,T];E)$.

The proof of [33] relies on a slightly different idea: instead of weaken the regularity of the space in the x-variable, we consider a Lebesgue space L^p in time instead of the space of continuous functions in time. As in the proof of Theorem 4.3, we write

$$u = B(u, u_1 - u_0) + B(u, u_2 - u_0) + 2B(u, u_0 - u_{0,\varepsilon}) + 2B(u, u_{0,\varepsilon}),$$

where $u_{0,\varepsilon}$ is chosen in $\mathscr{C}_c^\infty(\mathbb{R}^n;\mathbb{R}^n)$ close to u_0 in the L^n-norm. To be able to show that $u = 0$ on a small interval $[0,\tau]$ ($\tau > 0$) with the same method as in the proof of Theorem 4.3, we only need a result of the same kind as Lemma 4.4, see Lemma 4.9 below. At that point, the argument goes exactly as before. □

Lemma 4.9. *The bilinear operator*

$$B : L^p(0,T; L^n(\mathbb{R}^n;\mathbb{R}^n)) \times \mathscr{C}([0,T]; L^n(\mathbb{R}^n;\mathbb{R}^n)) \to L^p(0,T; L^n(\mathbb{R}^n;\mathbb{R}^n))$$

is bounded for all $p \in]1,\infty[$. More precisely, for all $p \in]1,\infty[$, there exists a constant $c_p > 0$ such that for all $u \in L^p(0,T;L^n(\mathbb{R}^n;\mathbb{R}^n))$ and all $v \in \mathscr{C}([0,T];L^n(\mathbb{R}^n;\mathbb{R}^n))$, we have

$$\|B(u,v)\|_{L^p(0,T;L^n(\mathbb{R}^n;\mathbb{R}^n))} \le c_p \|u\|_{L^p(0,T;L^n(\mathbb{R}^n;\mathbb{R}^n))} \|v\|_{\mathscr{C}([0,T];L^n(\mathbb{R}^n;\mathbb{R}^n))}.$$

Proof. To prove this lemma is exactly where the maximal L^p-regularity property of the Laplacian comes in. We rewrite $B(u,v)$ as

$$B(u,v)(t) = \mathbb{P}(-\Delta)\int_0^t e^{-(t-s)(-\Delta)}(-\Delta)^{-1}f(s)\,ds,$$

where $f = -\frac{1}{2}\nabla\cdot(u\otimes v + v\otimes u)$. What we only have to prove then is that we can estimate $(-\Delta)^{-1}f$ in the $L^p(0,T;L^n(\mathbb{R}^n;\mathbb{R}^n))$-norm with respect to the norm of u in $L^p(0,T;L^n(\mathbb{R}^n;\mathbb{R}^n))$ and the norm of v in $\mathscr{C}([0,T];L^n(\mathbb{R}^n;\mathbb{R}^n))$. Indeed, since we know that $-\Delta$ has the maximal L^p-regularity property, this would imply the result. Employing the continuous embedding $W^{1,\frac{n}{2}}(\mathbb{R}^n;\mathbb{R}^n)\subset L^n(\mathbb{R}^n;\mathbb{R}^n)$, we find

$$\begin{aligned}\|(-\Delta)^{-1}f\|_{L^p(0,T;L^n(\mathbb{R}^n;\mathbb{R}^n))} &\le C\|(-\Delta)^{-1}f\|_{L^p(0,T;W^{1,\frac{n}{2}}(\mathbb{R}^n;\mathbb{R}^n))}\\ &\le C\|f\|_{L^p(0,T;W^{-1,\frac{n}{2}}(\mathbb{R}^n;\mathbb{R}^n))}\\ &\le C\|(u\otimes v+v\otimes u)\|_{L^p(0,T;L^{\frac{n}{2}}(\mathbb{R}^n;\mathbb{R}^n))}\\ &\le C\|u\|_{L^p(L^n)}\|v\|_{\mathscr{C}([0,T];L^n(\mathbb{R}^n;\mathbb{R}^n))}\end{aligned}$$

and therefore we have proved Lemma 4.9. □

5 Maximal regularity for non-autonomous evolution problems

In this section, we consider non-autonomous problems of the form

$$\begin{aligned} u'(t)+A(t)u(t) &= f(t), \quad t\ge s,\\ u(s) &= u_s, \end{aligned} \tag{5.1}$$

for some $s\ge 0$. Compared with the problems studied before, the difference is now that the operator A itself depends on the time t. The main consequence is that the operators $\frac{d}{dt}$ and A do not commute anymore. We will present here results without proofs (references for the proofs are however given).

5.1 Operator families depending smoothly on time

We will here assume that the operators $(A(t))_{t\in[0,T]}$ defined on a Banach space X are uniformly (in $t\in[0,T]$) sectorial for all $t\in[0,T]$ (see Definition A.10 below). This implies in particular that $-A(t)$ is the generator of an analytic semigroup in X for all $t\in[0,T]$. Moreover, we assume the so-called Acquistapace–Terreni condition: there exist constants $c>0$, $\alpha\in[0,1[$ and $\delta\in]0,1]$ such that

$$\left\|A(t)(\lambda I + A(t))^{-1}[A(t)^{-1}-A(s)^{-1}]\right\|_{\mathscr{L}(X)} \le \frac{c|t-s|^\delta}{1+|\lambda|^{1-\alpha}} \tag{5.2}$$

holds for all $t, s \in [0,T]$ and $\lambda \in \Sigma_\theta = \{z \in \mathbb{C} \setminus \{0\}; |\arg(z)| < \pi - \theta\}$ for some $\theta \in [0, \frac{\pi}{2}[$. This condition implies in particular Hölder continuity in time of $t \mapsto A(t)^{-1}$. Let us point out that the Acquistapace–Terreni condition allows however the domains $D(A(t))$ to depend on t. This condition has been first used by P. Acquistapace and B. Terreni in [1] (and in a somewhat more abstract form by R. Labbas and B. Terreni in [26] and [27]) to prove the existence of an evolution family $(U(t,s))_{t\geq s,\, s,t\in[0,T]}$ (for all $t \geq s \geq 0$, $U(t,s)$ is a bounded operator in X) so that $u(t) = U(t,s)u_s$, $t \geq s$, is the solution of (5.1) with $f = 0$. The general form of a solution of (5.1) is then given by the formula

$$u(t) = U(t,s)u_s + \int_s^t U(t,r)f(r)\,dr, \quad t \geq s\,.$$

Definition 5.1. The family of operators $(A(t))_{t\in[0,T]}$ is said to have the maximal L^p-regularity property if, for all $f \in L^p(0,T;X)$, there exists a unique solution u of (5.1) (with $s = 0$ and $u_0 = 0$) satisfying $u' \in L^p(0,T;X)$ and

$$u(t) \in D(A(t)) \text{ a.e. in } [0,T] \ni t \quad \text{and} \quad t \mapsto A(t)u(t) \in L^p(0,T;X).$$

As in the autonomous case (see Proposition 2.4), the property of maximal L^p-regularity is independent of $p \in]1,\infty[$ under the condition (5.2).

Theorem 5.2. *Let X be a Banach space. Let $\mathscr{A} = (A(t))_{t\in[0,T]}$ be a family of uniformly sectorial operators on X satisfying the Acquistapace–Terreni condition* (5.2). *Assume that $\mathscr{A}$ enjoys the maximal L^p-regularity property for one $p \in]1,\infty[$. Then $\mathscr{A}$ has the maximal L^q-regularity property for all $q \in]1,\infty[$.*

Reference for the proof. The proof of this theorem is due to M. Hieber and S. Monniaux and can be found in [20, Theorem 3.1]. The idea of the proof is to show that the condition (5.2) implies a Hörmander-type condition for the operator S defined by

$$Sf(t) = \int_0^t A(t)e^{-(t-s)A(t)}f(s)\,ds, \quad t \in [0,T],$$

which is the singular part of the operator $f \mapsto A(\cdot)u(\cdot)$ for u being the solution of (5.1) with $s = 0$ and $u_0 = 0$. Therefore, we can apply Theorem 2.5. Since S is bounded in $L^p(0,T;X)$ for one $p \in]1,\infty[$, it is also bounded in $L^q(0,T;X)$ for all $q \in]1,\infty[$. □

In the case of Hilbert spaces, we have the following result.

Theorem 5.3. *Let $X = H$ be a Hilbert space. Let $\mathscr{A} = (A(t))_{t\in[0,T]}$ be a family of uniformly sectorial operators on H satisfying the Acquistapace–Terreni condition* (5.2). *Then $\mathscr{A}$ enjoys the maximal L^p-regularity property for all $p \in]1,\infty[$.*

Reference for the proof. The proof is due to M. Hieber and S. Monniaux and can be found in [20, Theorem 3.2]. It appears as a corollary of Theorem 5.2 and Theorem 5.4 below with the symbol a defined by

$$a(t,\tau) = \begin{cases} A(0)(i\tau I + A(0))^{-1}, & t < 0,\\ A(t)(i\tau I + A(t))^{-1}, & t \in [0,T],\\ A(T)(i\tau I + A(T))^{-1}, & t > T.\end{cases}$$

Indeed, it suffices to show that a satisfies the conditions of Theorem 5.4 to get the maximal L^2-regularity property of $\mathscr{A}$ and it remains to apply Theorem 5.2 to obtain the maximal L^p-regularity property for all $p \in]1,\infty[$. □

Theorem 5.4. *Let H be a Hilbert space and $a \in L^\infty(\mathbb{R}\times\mathbb{R};\mathscr{L}(H))$ such that $\xi \mapsto a(x,\xi)$ has an analytic extension (with values in $\mathscr{L}(H)$) in $S_\theta = \{\pm z \in \mathbb{C}; |\arg(z)| < \theta\}$ for one $\theta \in]0,\frac{\pi}{2}[$ and*

$$\sup_{z\in S_\theta}\sup_{x\in\mathbb{R}} \|a(x,z)\|_{\mathscr{L}(H)} < \infty.$$

Let $u \in \mathscr{S}(\mathbb{R};H)$ and define

$$Op(a)u(x) = \frac{1}{\sqrt{2\pi}}\int_{\mathbb{R}} e^{ix\xi}a(x,\xi)\mathscr{F}(u)(\xi)\,d\xi, \quad x \in \mathbb{R}.$$

Then $Op(a)$ extends to a bounded operator on $L^2(\mathbb{R};H)$.

Reference for the proof. The proof of this theorem is due to M. Hieber and S. Monniaux (see [20, Theorem 2.1]). This can be viewed as a parameter dependent version of the Fourier-multiplier theorem in Hilbert spaces. □

A version of Theorem 5.4 was proved by P. Portal and Ž. Štrkalj in UMD-spaces (see [38]). This allows to prove the following result, similar to Theorem 5.3 in the case of UMD-spaces (see also [41] for the autonomous case). The idea is to replace boundedness of the resolvent in the case of a Hilbert space with R-boundedness of the resolvent in the case of a Banach space with the UMD-property.

Definition 5.5. Let X be a Banach space. Let $\mathscr{A} = (A(t))_{t\in[0,T]}$ be a family of uniformly sectorial operators on X. Then $\mathscr{A}$ is said to be a family of uniformly R-sectorial operators on X if it satisfies

$$R\Big(\{(1+|\lambda|)(\lambda I + A)^{-1}; \lambda \in \Sigma_\theta, t \in [0,T]\}\Big) < \infty,$$

where the R-bound $R(\tau)$ of a set of bounded operators τ has been defined in Section 2 (Definition 2.12).

Theorem 5.6 (Portal-Štrkalj, 2006). *Let X be a UMD-space. Let $\mathscr{A} = (A(t))_{t\in[0,T]}$ be a family of uniformly R-sectorial operators on X satisfying the Acquistapace–Terreni condition* (5.2). *Then $\mathscr{A}$ enjoys the maximal L^p-regularity property for all $p \in]1,\infty[$.*

Reference for the proof. The proof of this result can be found in [38, Corollary 14 in Section 5]. □

The first theorem about maximal regularity in the non-autonomous setting was proved by S. Monniaux and J. Prüss in 1997 (see [34]) and is a generalization of the theorem of Dore–Venni (see Theorem A.14 below) when the operator A depends on t and satisfies a condition of Acquistapace–Terreni-type (see (5.2)).

Theorem 5.7. *Let X be a UMD-Banach space. Let $\mathscr{A} = (A(t))_{t\in[0,T]}$ be a family of uniformly sectorial operators with bounded imaginary powers on X satisfying the Acquistapace–Terreni condition* (5.2) *such that*

$$\sup_{t\in[0,T]} \sup_{s\in\mathbb{R}} \left\{ \frac{1}{|s|} \ln \|A(t)^{is}\|_{\mathscr{L}(X)} \right\} \in \left[0, \frac{\pi}{2}\right[.$$

Then $\mathscr{A}$ enjoys the maximal L^p-regularity property for all $p \in]1,\infty[$.

Reference for the proof. This result is due to S. Monniaux and J. Prüss and its proof, in a slightly more general setting, can be found in [34, Theorem 1]. Remark that in particular, the bound on $\{A(t)^{is}; s \in \mathbb{R}\}$ implies by Theorem A.14 that every operator $A(t)$ has the property of maximal L^p-regularity. □

The result presented now is a generalization of Theorem 3.3 in the non-autonomous setting.

Theorem 5.8. *Let $\Omega \subset \mathbb{R}^n$ and $\mathscr{A} = (A(t))_{t\in[0,T]}$ be a family of unbounded operators defined on $L^2(\Omega)$ such that $-A(t)$ generates an analytic semigroup in $L^2(\Omega)$. We assume moreover that $\mathscr{A}$ satisfies the Acquistapace–Terreni condition* (5.2) *in $L^2(\Omega)$ and that for all $t \in [0,T]$ the semigroup $(e^{-sA(t)})_{s\geq 0}$ has a kernel $p_t(s,x,y)$ with uniform Gaussian upper bounds, more precisely,*

$$e^{-sA(t)}f(x) = \int_\Omega p_t(s,x,y)f(y)\,dy, \quad x \in \Omega,\ t \in [0,T],\ s \geq 0,$$

and there exist constants $c, b > 0$ (independent of t) such that

$$|p_t(s,x,y)| \leq cg(bs,x,y), \quad x,y \in \Omega,\ t \in [0,T],\ s \geq 0,$$

where g was defined in Theorem 3.3. Then $\mathscr{A}$ enjoys the maximal L^p-regularity property in $L^q(\Omega)$ for all $p,q \in]1,\infty[$.

Reference for the proof. This result is due to M. Hieber and S. Monniaux and its proof, though in a slightly more general setting, can be found in [19, Theorem 1]. Remark that in particular, the Gaussian bound for p_t implies by Theorem 3.3 that every operator $A(t)$ has the property of maximal L^p-regularity. □

5.2 Operator families with domain constant in time and quasilinear problems

The case where the domains $D(A(t))$ of the operators $A(t)$, $t \in [0,T]$ do not depend on t, i.e., $D(A(t)) = D \subset X$, was investigated by J. Prüss and R. Schnaubelt [39], H. Amann [2], in view of applications to quasilinear evolution equations in [3], and by W. Arendt, R. Chill, S. Fornaro and C. Poupaud [5].

To state the main result in its full generality, we need to introduce the notion of relative continuity.

Definition 5.9. A function $A : [0,T] \to \mathscr{L}(D,X)$ is called relatively continuous if for each $t \in [0,T]$ and all $\varepsilon > 0$ there exist $\delta > 0$ and $\eta \geq 0$ such that for all $x \in D$,

$$\|A(t)x - A(s)x\|_X \leq \varepsilon \|x\|_D + \eta \|x\|_X$$

holds for all $s \in [0,T]$ with $|t-s| \leq \delta$.

Theorem 5.10. *Let $A : [0,T] \to \mathscr{L}(D,X)$ be strongly measurable and relatively continuous. Assume that $A(t)$ has the maximal L^p-regularity property for all $t \in [0,T]$. Then for each $x \in (X,D)_{\frac{1}{p'},p}$ and $f \in L^p(0,T;X)$ there exists a unique solution u of* (5.1) *satisfying $u \in L^p(0,T;D) \cap W^{1,p}(0,T;X)$.*

References for the proof. This theorem was first proved by J. Prüss and R. Schnaubelt in the case where $A : [0,T] \to \mathscr{L}(D,X)$ is continuous (see [39, Theorem 2.5] and also [2, Theorem 7.1]). They proved in particular that the hypotheses of the theorem imply the existence of an evolution family $(U(t,s))_{t \geq s,\, s,t \in [0,T]}$. The condition $x \in (X,D)_{\frac{1}{p'},p}$ is to compare with Remark 2.3. The theorem, in its full generality as stated above, is due to W. Arendt, R. Chill, S. Fornaro and C. Poupaud (see [5, Theorem 2.7]). They also proved the existence of an evolution family $(U(t,s))_{t \geq s,\, s,t \in [0,T]}$ and the solution u of (5.1) is given by

$$u(t) = U(t,s)u_s + \int_s^t U(t,r)f(r)\,dr, \quad t \geq s.$$

□

Non-autonomous maximal regularity results seem to be a good starting point to study quasilinear evolution equations. This has been thoroughly studied by H. Amann (see, e.g., [3]). The problem is the following: find a solution u of

$$u' + A(u)u = F(u) \text{ on } [0,T], \quad u(0) = u_0, \tag{5.3}$$

with "reasonable" conditions on $u \mapsto A(u)$ and $u \mapsto F(u)$. The idea to treat this problem is to apply a fixed point theorem on the map

$$v \mapsto u, \quad \text{where } u'(t) + A(v(t))u(t) = F(v(t)),\ t \in [0,T],\ u(0) = u_0. \tag{5.4}$$

The problem is of course to find a suitable Banach space in which one can apply the fixed point theorem, i.e., for which the solution of (5.4) has the best possible regularity properties, such as maximal regularity. The situation studied in [3] is the following. Let

$$\mathscr{E}_p = L^p(0,T;D) \cap W^{1,p}(0,T;X)$$

be the space associated to maximal L^p-regularity for (5.4). The operators $A(u)$ and the function F satisfy, for all $u \in \mathscr{E}_p$,

$$A(u) \in L^\infty(0,T;\mathscr{L}(D,X)) \quad \text{and} \quad F(u) \in L^p(0,T;X).$$

It is also assumed that the restriction of $(A(u), F(u))$ on a subinterval J depends only on the restriction of u on the interval J (i.e., A and F are Volterra operators).

Theorem 5.11. *Under the above assumptions, the problem* (5.3) *has a unique maximal solution.*

Reference for the proof. This result is due to H. Amann [3, Section 2]. □

A Appendix

We collect here some of the results used in the previous sections, mostly without proofs but with references where those can be found.

A.1 Picard's fixed point theorem for bilinear operators

Let $\mathscr{F}$ be a Banach space and let $a \in \mathscr{F}$. We want to solve the equation

$$u = a + B(u,u), \quad u \in \mathscr{F}, \tag{A.1}$$

where B is a symmetric bilinear continuous operator on $\mathscr{F} \times \mathscr{F}$. Let $\|B\|$ be the smallest constant $c > 0$ such that

$$\|B(u,v)\|_{\mathscr{F}} \le c\|u\|_{\mathscr{F}}\|v\|_{\mathscr{F}}, \quad u, v \in \mathscr{F}.$$

Theorem A.1 (Picard's fixed point theorem). *For all $a \in \mathscr{F}$ with $\|a\|_{\mathscr{F}} \le \frac{1}{4\|B\|}$, there exists a solution $u \in \mathscr{F}$ of* (A.1). *More precisely, there exists for all $k = 1, 2, \ldots$ an operator T_k, the restriction on the diagonal of a continuous k-linear operator from $\mathscr{F} \times \ldots \times \mathscr{F}$ to $\mathscr{F}$, satisfying*

(*i*) *there exists a constant $c > 0$ (even not depending on $\mathscr{F}$) such that for all $a \in \mathscr{F}$ and all $k = 1, 2, \ldots$*

$$\|T_k(a)\|_{\mathscr{F}} \le \frac{c}{\|B\|} k^{-\frac{3}{2}} (4\|B\|\|a\|_{\mathscr{F}})^k;$$

(*ii*) *u has the form*

$$u = \sum_{k=1}^{\infty} T_k(a).$$

Moreover, we have $\|u\|_{\mathscr{F}} \le \frac{1}{2\|B\|}$, and u is the unique solution of (A.1) *in the closed ball $\overline{B}_{\mathscr{F}}(0, \frac{1}{2\|B\|})$.*

Proof. We set $T_1(a) = a$ and define by induction

$$T_k(a) = \sum_{\ell=1}^{k-1} B(T_\ell(a), T_{k-\ell}(a)). \tag{A.2}$$

The fact that T_k, $k = 1, 2, \ldots$, is the restriction of a k-linear operator is clear by induction. By induction, we can also prove that for all $k = 1, 2, \ldots$, there exists a constant $\alpha_k > 0$ such that $\|T_k(a)\|_{\mathscr{F}} \leq \alpha_k \|a\|_{\mathscr{F}}^k$. By the definition of T_k, we have

$$\|T_k(a)\|_{\mathscr{F}} \leq \|B\| \sum_{\ell=1}^{k-1} \|T_\ell(a)\|_{\mathscr{F}} \|T_{k-\ell}(a)\|_{\mathscr{F}}.$$

We can then chose α_k defined by induction by

$$\alpha_1 = 1, \quad \alpha_k = \|B\| \sum_{\ell=1}^{k-1} \alpha_\ell \alpha_{k-\ell}, \quad k = 2, 3, \ldots.$$

The numbers $c_k = \|B\|^{-k+1} \alpha_k$, $k = 1, 2, \ldots$, satisfy

$$c_1 = 1, \quad c_k = \sum_{\ell=1}^{k-1} c_\ell c_{k-\ell}, \quad k = 2, 3, \ldots.$$

These numbers are Catalan's numbers with the explicit representation $c_k = \frac{(2k-2)!}{k!(k-1)!}$. It is known that there exists a constant $c > 0$ such that $c_k \leq c 4^k k^{-\frac{3}{2}}$ for all $k = 1, 2, \ldots$, which gives the bound announced in (i). It is also known that

$$\sum_{k=1}^{\infty} c_k z^k = \frac{1 - \sqrt{1 - 4z}}{2}, \quad |z| < \frac{1}{4}. \tag{A.3}$$

Let now $a \in \mathscr{F}$ with $\|a\|_{\mathscr{F}} \leq \frac{1}{4\|B\|}$. Then the series $\sum_{k=1}^{\infty} T_k(a)$ is uniformly convergent in $\mathscr{F}$ with limit u. We will prove now that u is the solution of (A.1). Indeed, we have

$$\begin{aligned} B(u, u) &= B\Big(\sum_{\ell-1}^{\infty} T_\ell(a), \sum_{m=1}^{\infty} T_m(a)\Big) = \sum_{\ell, m=1}^{\infty} B(T_\ell(a), T_m(a)) \\ &= \sum_{k=2}^{\infty} \sum_{\ell+m=k} B(T_\ell(a), T_m(a)) = \sum_{k=2}^{\infty} T_k(a) = u - T_1(a) = u - a. \end{aligned}$$

The first equality comes from the definition of u. The second equality is due to the bilinearity of B. The third equality is obtained by rearranging the double sum which is allowed because of the uniform convergence of the series. The fourth equality uses only the definition (A.2) of T_k. The last two equalities are obvious. In addition, we have

$$\begin{aligned} \|u\|_{\mathscr{F}} &\leq \sum_{k=1}^{\infty} \|T_k(a)\|_{\mathscr{F}} \leq \sum_{k=1}^{\infty} c_k \|B\|^{k-1} \|a\|_{\mathscr{F}}^k \\ &\leq \frac{1}{2\|B\|} (1 - \sqrt{1 - 4\|B\| \|a\|_{\mathscr{F}}}) \leq \frac{1}{2\|B\|}. \end{aligned}$$

The first inequality follows from the definition of u. The second inequality comes from the estimates $\|T_k(a)\|_{\mathscr{F}} \leq c_k \|B\|^{k-1} \|a\|_{\mathscr{F}}^k$ for all $k = 1, 2, \ldots$. The third inequality is in fact an equality coming from the formula (A.3). The last inequality is obvious. So far, we have proved the existence of a solution $u \in \mathscr{F}$ of (A.1) for $a \in \mathscr{F}$ with $\|a\|_{\mathscr{F}} \leq \frac{1}{4\|B\|}$. We also proved that $\|u\|_{\mathscr{F}} \leq \frac{1}{2\|B\|}$. It remains to prove that such a solution is unique in the closed ball $\overline{B}_{\mathscr{F}}(0, \frac{1}{2\|B\|})$. Assume that there exists another solution v of (A.1) in $\overline{B}_{\mathscr{F}}(0, \frac{1}{2\|B\|})$. We can distinguish two cases:

(i) If $\|u\| < \frac{1}{2\|B\|}$ or $\|v\| < \frac{1}{2\|B\|}$, then, by taking the difference of $u = a + B(u, u)$ and $v = a + B(v, v)$, we obtain $u - v = B(u - v, u + v)$ and therefore

$$\begin{aligned} \|u - v\|_{\mathscr{F}} &\leq \|B\| \|u - v\|_{\mathscr{F}} \|u + v\|_{\mathscr{F}} \\ &\leq \|B\| (\|u\|_{\mathscr{F}} + \|v\|_F) \|u - v\|_{\mathscr{F}} \\ &< \|u - v\|_{\mathscr{F}}, \end{aligned}$$

which implies directly that $u = v$.

(ii) If $\|u\|_{\mathscr{F}} = \frac{1}{2\|B\|}$ and $\|v\|_{\mathscr{F}} = \frac{1}{2\|B\|}$ then for all $n \geq 1$, we define

$$v_n = v - T_1(a) - \ldots - T_n(a).$$

By induction, we can prove that

$$\|v_n\|_{\mathscr{F}} \leq \sum_{k=n+1}^{\infty} \frac{1}{\|B\|} 4^{-k} c_k \xrightarrow[n \to \infty]{} 0$$

and therefore $v = \sum_{k=1}^{\infty} T_k(a) = u$.

This completes the proof of Theorem A.1. □

A.2 Interpolation of operators

Theorem A.2 (Riesz–Thorin theorem). *Let $(X, \mathcal{X}, \mu)$ and $(Y, \mathcal{Y}, \nu)$ be two fixed measure spaces. Let $1 \leq p_0, p_1, q_0, q_1 \leq \infty$ and $\theta \in]0, 1[$. Let*

$$T : L^{p_0}(X) + L^{p_1}(X) \to L^{q_0}(Y) + L^{q_1}(Y)$$

be a linear operator such that there exist two constants $c_0, c_1 > 0$ with

$$\|Tf\|_{L^{q_i}(Y)} \leq c_i \|f\|_{L^{p_i}(X)} \quad \text{for all } f \in L^{p_i}(X),\ i = 0, 1.$$

Then we have for all $f \in L^{p_\theta}(X)$

$$\|Tf\|_{L^{q_\theta}(Y)} \leq c_\theta \|f\|_{L^{p_\theta}(X)},$$

where $\frac{1}{p_\theta} = \frac{1-\theta}{p_0} + \frac{\theta}{p_1}$, $\frac{1}{q_\theta} = \frac{1-\theta}{q_0} + \frac{\theta}{q_1}$ and $c_\theta = c_0^{1-\theta} c_1^{\theta}$.

References for the proof. A proof of this theorem can be found in [8, Theorem 1.1.1]. Another nice reference is T. Tao's lecture on this subject (see [40, Theorem 3]). □

Definition A.3. An operator T mapping measurable functions over a measure space $(X, \mathcal{X}, \mu)$ into measurable functions over a measure space $(Y, \mathcal{Y}, \nu)$ is said to be sublinear if it maps simple functions $f : X \to \mathbb{C}$, whose support is of finite measure, to nonnegative-valued functions (modulo almost everywhere equivalence) such that the homogeneity

$$T(cf) = |c|\, Tf \quad \text{for all } c \in \mathbb{C}$$

and the pointwise sublinearity

$$|T(f+g)| \le |Tf| + |Tg|$$

are satisfied for all simple functions f, g whose support is of finite measure.

Remark A.4. If S is a linear operator, then $T = |S|$ is sublinear.

Definition A.5. A linear or sublinear operator mapping measurable functions over a measure space $(X, \mathcal{X}, \mu)$ into measurable functions over a measure space $(Y, \mathcal{Y}, \nu)$ is said to be of strong type (p, q) if there exists a constant $c > 0$ such that for all $f \in L^p(X)$

$$\|Tf\|_{L^q(Y)} \le c\|f\|_{L^p(X)}.$$

It is said to be of weak type (p, q) if there exists a constant $c > 0$ such that for all $f \in L^p(X)$

$$\sup_{t \ge 0} \left[t\mu\Big(\{x \in X; |Tf(x)| \ge t\}\Big)^{\frac{1}{q}} \right] \le c\|f\|_{L^p(X)}.$$

Theorem A.6 (Marcinkiewicz interpolation theorem). *Let $1 \le p_0, p_1, q_0, q_1 \le \infty$ and $\theta \in]0,1[$ such that $q_0 \ne q_1$ and $p_i \le q_i$ for $i = 0, 1$. Let T be a sublinear operator of weak type (p_0, q_0) and of weak type (p_1, q_1). Then T is of strong type (p_θ, q_θ) where $\frac{1}{p_\theta} = \frac{1-\theta}{p_0} + \frac{\theta}{p_1}$ and $\frac{1}{q_\theta} = \frac{1-\theta}{q_0} + \frac{\theta}{q_1}$.*

References for the proof. A proof for this theorem can be found in [8, Theorem 1.3.1]. Another nice reference is T. Tao's lecture on this subject [40, Theorem 4]. □

Theorem A.7 (Mikhlin multiplier theorem). *Let X and Y be Hilbert spaces. If the differentiable function $M : \mathbb{R} \setminus \{0\} \to \mathscr{L}(X, Y)$ satisfies*

$$\|M(t)\|_{\mathscr{L}(X,Y)} \le C \quad \textit{and} \quad \|tM'(t)\|_{\mathscr{L}(X,Y)} \le C \quad \textit{for all } t \in \mathbb{R} \setminus \{0\}$$

for some constant $C > 0$, then M is a Fourier multiplier (see Definition 2.11) in $L^p(\mathbb{R}; X)$ for all $p \in]1, \infty[$.

References for the proof. See [8, Section 6.1]. □

A.3 Calderón–Zygmund theory

Theorem A.8 (Calderón–Zygmund decomposition). *Let $f \in L^1(\mathbb{R}^n)$ and fix $\lambda > 0$. Then $f = g + \sum b_k$, where*

(*i*) $|g| \le 2^n\lambda$ *almost everywhere;*

(*ii*) $\|g\|_1 + \sum_k \|b_k\|_1 \le 3\|f\|_1$*;*

(*iii*) *there exists a family of disjoint cubes $(Q_k)_{k\in\mathbb{N}}$ of $\mathbb{R}^n$ such that*

$$\operatorname{supp} b_k \subset Q_k, \quad \int_{\mathbb{R}^n} b_k\, dx = 0 \text{ and } \sum_k |Q_k| \le \frac{1}{\lambda}\|f\|_1.$$

Proof. Let $f \in L^1(\mathbb{R}^n)$ and fix $\lambda > 0$. We may assume $\|f\|_1 = 1$. We decompose $\mathbb{R}^n$ into cubes of measure $\frac{1}{\lambda}$: $\mathbb{R}^n = \bigcup_m \tilde{Q}_{0,m}$. Then we have for all m

$$\frac{1}{|\tilde{Q}_{0,m}|}\int_{\tilde{Q}_{0,m}} |f|\, dx \le \lambda\|f\|_1 = \lambda.$$

We then decompose each cube $\tilde{Q}_{0,m}$ into cubes of measure $\frac{1}{2^n\lambda}$. We denote by $(Q_{1,m})_m$ all cubes for which

$$\frac{1}{|Q_{1,m}|}\int_{Q_{1,m}} |f|\, dx > \lambda.$$

The other cubes are denoted by $\tilde{Q}_{1,m}$. We repeat this operation with these cubes $\tilde{Q}_{1,m}$ and obtain cubes $(Q_{2,m})_m$ of measure $\frac{1}{4^n\lambda}$ for which

$$\frac{1}{|Q_{2,m}|}\int_{Q_{2,m}} |f|\, dx > \lambda,$$

the remaining cubes being denoted by $\tilde{Q}_{2,m}$. After a countable number of steps, we obtain a family of cubes $(Q_{j,m})_{j,m\in\mathbb{N}}$ renamed as $(Q_k)_{k\in\mathbb{N}}$. We now define b_k as $b_k = (f - m_{Q_k}(f))\chi_{Q_k}$ with the notation

$$m_Q(f) = \frac{1}{|Q|}\int_Q f\, dx.$$

We denote by g the quantity

$$g = f - \sum_k b_k.$$

It remains to show that g and the b_k's satisfy the conditions (i), (ii) and (iii) of the theorem. First, if $x \in \mathbb{R}^n \setminus \bigcup_k Q_k$, then for all $j \in \mathbb{N}$, there exists $m \in \mathbb{N}$ such that $x \in \tilde{Q}_{j,m}$. By construction, we have

$$\lambda \ge \frac{1}{|\tilde{Q}_{j,m}|}\int_{\tilde{Q}_{j,m}} |f|\, dy \xrightarrow[j\to\infty]{} |f(x)|$$

if x is a Lebesgue point of f. If $x \in Q_k$ for one $k \in \mathbb{N}$, then

$$|g(x)| = |m_{Q_k}(f)| \leq \frac{|\tilde{Q}_k|}{|Q_k|}\lambda = 2^n\lambda.$$

Altogether, this gives (i) since the set of points that are not Lebesgue points of f is of measure zero. Again by construction, we have

$$\operatorname{supp} b_k \subset Q_k \quad \text{and} \quad \int_{Q_k} b_k \, dx = 0.$$

Moreover, we have $\frac{1}{|Q_k|}\int_{Q_k}|f|\,dx > \lambda$. Therefore, we find $|Q_k| < \frac{1}{\lambda}\int_{Q_k}|f|\,dx$. Since the cubes Q_k are disjoint, this gives

$$\sum_k |Q_k| \leq \sum_k \frac{1}{\lambda}\int_{Q_k}|f|\,dx \leq \frac{1}{\lambda}\int_{\mathbb{R}^n}|f|\,dx = \frac{1}{\lambda}\|f\|_1.$$

This proves (iii). Finally, we have $\|g\|_1 \leq \|f\|_1$ and $\|b_k\|_1 \leq 2\int_{Q_k}|f|\,dx$ for all $k \in \mathbb{N}$, and this gives (ii). □

Remark A.9. The Calderón–Zygmund decomposition is also available in a measurable space (E, μ, d) of homogeneous type where μ is a σ-finite measure and d is a quasi-metric, i.e., there exists a constant $c > 0$ such that for any ball $B = \{y; d(x,y) < r\}$, if we denote by $2B$ the ball with the same center and twice the radius of B, it holds $\mu(2B) \leq c\mu(B)$.

A.4 Bounded imaginary powers

Definition A.10. A (linear) operator A on a Banach space X is sectorial if it is closed, densely defined, has empty kernel, dense range $R(A)$ and satisfies

$$\sup_{t>0} \|t(tI + A)^{-1}\|_{\mathscr{L}(X)} < \infty.$$

Let $x \in D(A) \cap R(A)$. Then one can define for $z \in \mathbb{C}$ with $|\mathrm{Re}(z)| < 1$

$$\begin{aligned} A^z x &= \frac{\sin \pi z}{\pi}\Big(\frac{x}{z} - \frac{1}{1+z}A^{-1}x + \int_0^1 t^{z+1}(tI+A)^{-1}A^{-1}x\,dt \\ &\quad + \int_1^\infty t^{z-1}(tI+A)^{-1}Ax\,dt\Big). \end{aligned}$$

We are now in the position to give the definition of operators with bounded imaginary powers.

Definition A.11. A sectorial operator on a Banach space X has bounded imaginary powers if the closure of the operator $(A^{is}, D(A) \cap R(A))$ defines a bounded operator on X for all $s \in \mathbb{R}$ and if $\sup_{|s|\leq 1}\|A^{is}\|_{\mathscr{L}(X)} < \infty$.

Remark A.12. If A admits bounded imaginary powers, then $(A^{is})_{s\in\mathbb{R}}$ forms a strongly continuous group on X.

Remark A.13. Conversely, it has been proved in [32] that a given strongly continuous group of type strictly less than π on a UMD-Banach space can be represented as the imaginary powers of a sectorial operator, called its analytic generator.

In the class of UMD-spaces, a positive result for operators having bounded imaginary powers has been proved by G. Dore and A. Venni.

Theorem A.14 (Dore–Venni, 1987). *Let X be a Banach space in the UMD-class. Let A be an operator with bounded imaginary powers (see Definition A.11) for which the type of the group $(A^{is})_{s\in\mathbb{R}}$ is strictly less than $\frac{\pi}{2}$. Then A has the maximal L^p-regularity property.*

Idea of the proof. The idea is to show that $\mathcal{A}+\mathcal{B}$ with domain $D(\mathcal{A})\cap D(\mathcal{B})$ is invertible, where

$$D(\mathcal{A}) = L^2(0,\infty;D(A)),\ (\mathcal{A}u)(t) = Au(t),\ t>0,$$

and

$$D(\mathcal{B}) = H_0^1(0,\infty;X),\ \mathcal{B}u = u'.$$

The operator $\mathcal{A}$ has bounded imaginary powers with angle strictly less than $\frac{\pi}{2}$ and the operator $\mathcal{B}$ has bounded imaginary powers with angle $\frac{\pi}{2}$. We define then, for $c\in]0,1[$,

$$S = \frac{1}{2i}\int_{c-i\infty}^{c+i\infty} \frac{\mathcal{A}^{-z}\mathcal{B}^{z-1}}{\sin \pi z}\,dz.$$

The purpose is to show that $\mathcal{B}S$ is bounded and that $S = (\mathcal{A}+\mathcal{B})^{-1}$, and this is done by letting $c\to 0^+$ and taking into account that the Hilbert transform is bounded in $L^2(\mathbb{R};X)$.
Another (shorter) proof uses the result presented in Remark A.13 by showing that the group $(\mathcal{A}^{-is}\mathcal{B}^{is})_{s\in\mathbb{R}}$ has a sectorial analytic generator. □

Acknowledgments. The author would like to thank the Technische Universität Berlin where the content of this survey has been presented in a series of four lectures during May 2009. Special thanks go to Petra Wittbold and Etienne Emmrich for their kind invitation and to the students and colleagues who attended the course.

References

[1] P. Acquistapace and B. Terreni, A unified approach to abstract linear nonautonomous parabolic equations, *Rend. Sem. Mat. Univ. Padova* **78** (1987), pp. 47–107.

[2] H. Amann, Maximal regularity for nonautonomous evolution equations, *Adv. Nonlinear Stud.* **4** (2004), pp. 417–430.

[3] ———, Quasilinear parabolic problems via maximal regularity, *Adv. Differential Equations* **10** (2005), pp. 1081–1110.

[4] W. Arendt, Semigroups and evolution equations: functional calculus, regularity and kernel estimates, in: *Evolutionary equations*, Handb. Differ. Equ. 1, pp. 1–85, North-Holland, Amsterdam, 2004.

[5] W. Arendt, R. Chill, S. Fornaro and C. Poupaud, L^p-maximal regularity for non-autonomous evolution equations, *J. Differential Equations* **237** (2007), pp. 1–26.

[6] P. Auscher, S. Hofmann, M. Lacey, A. McIntosh and P. Tchamitchian, The solution of the Kato square root problem for second order elliptic operators on $\mathbb{R}^n$, *Ann. of Math. (2)* **156** (2002), pp. 633–654.

[7] A. Benedek, A. P. Calderón and R. Panzone, Convolution operators on Banach space valued functions, *Proc. Nat. Acad. Sci.* **48** (1962), pp. 356–365.

[8] J. Bergh and J. Löfström, *Interpolation spaces. An introduction*, Springer-Verlag, Berlin – New York, 1976.

[9] J. Bourgain, Some remarks on Banach spaces in which martingales difference sequences are unconditional, *Ark. Math.* **22** (1983), pp. 163–168.

[10] D. L. Burkholder, A geometric condition that implies the existence of certain singular integrals of Banach-space-valued functions, in: *Conference on harmonic analysis in honor of Antoni Zygmund, Vol. I, II (Chicago, Ill., 1981)*, Wadsworth Math. Ser., pp. 270–286, Wadsworth, Belmont, CA, 1983.

[11] M. Cannone, MR1813331 (2002j:76036), *Math. Reviews* (2002), available at `http://www.ams.org/mathscinet/pdf/1813331.pdf`.

[12] T. Coulhon and X. T. Duong, Maximal regularity and kernel bounds: observations on a theorem by Hieber and Prüss, *Adv. Differential Equations* **5** (2000), pp. 343–368.

[13] T. Coulhon and D. Lamberton, Régularité L^p pour les équations d'évolution, in: *Séminaire d'Analyse Fonctionelle 1984/1985*, 26, pp. 155–165, Publ. Math. Univ. Paris VII, 1986.

[14] E. B. Davies, *Heat kernels and spectral theory*, Cambridge University Press, 1989.

[15] L. de Simon, Un'applicazione della teoria degli integrali singolari allo studio delle equazioni differenziali lineari astratte del primo ordine, *Rendiconti del Seminario Matematico della Università di Padova* **34** (1964), pp. 205–223.

[16] G. Dore and A. Venni, On the closedness of the sum of two closed operators, *Math. Z.* **196** (1987), pp. 189–201.

[17] X. T. Duong and A. McIntosh, Singular integral operators with non-smooth kernels on irregular domains, *Rev. Mat. Iberoamericana* **15** (1999), pp. 233–265.

[18] G. Furioli, P. G. Lemarié-Rieusset and E. Terraneo, Unicité dans $L^3(\mathbb{R}^3)$ et d'autres espaces fonctionnels limites pour Navier-Stokes, *Rev. Mat. Iberoamericana* **16** (2000), pp. 605–667.

[19] M. Hieber and S. Monniaux, Heat-kernels and maximal L^p-L^q-estimates: the non-autonomous case, *J. Fourier Anal. Appl.* **6** (2000), pp. 467–481.

[20] ———, Pseudo-differential operators and maximal regularity results for non-autonomous parabolic equations, *Proc. Amer. Math. Soc.* **128** (2000), pp. 1047–1053.

[21] M. Hieber and J. Prüss, Heat kernels and maximal L^p-L^q estimates for parabolic evolution equations, *Comm. Partial Differential Equations* **22** (1997), pp. 1647–1669.

[22] N. J. Kalton and G. Lancien, A solution to the problem of L^p-maximal regularity, *Math. Z.* **235** (2000), pp. 559–568.

[23] T. Kato, Strong L^p-solutions of the Navier-Stokes equation in $\mathbb{R}^m$, with applications to weak solutions, *Math. Z.* **187** (1984), pp. 471–480.

[24] P. C. Kunstmann, On maximal regularity of type L^p-L^q under minimal assumptions for elliptic non-divergence operators, *J. Funct. Anal.* **255** (2008), pp. 2732–2759.

[25] P. C. Kunstmann and L. Weis, Maximal L^p-regularity for parabolic equations, Fourier multiplier theorems and H^∞-functional calculus, in: *Functional analytic methods for evolution equations*, Lecture Notes in Math. 1855, pp. 65–311, Springer, Berlin, 2004.

[26] R. Labbas and B. Terreni, Somme d'opérateurs linéaires de type parabolique. I, *Boll. Un. Mat. Ital. B (7)* **1** (1987), pp. 545–569.

[27] ———, Sommes d'opérateurs de type elliptique et parabolique. II. Applications, *Boll. Un. Mat. Ital. B (7)* **2** (1988), pp. 141–162.

[28] D. Lamberton, Équations d'évolution linéaires associées à des semi-groupes de contractions dans les espaces L^p, *J. Funct. Anal.* **72** (1987), pp. 252–262.

[29] P. G. Lemarié-Rieusset, *Recent developments in the Navier-Stokes problem*, Chapman & Hall/CRC, Boca Raton, 2002.

[30] A. Lunardi, *Analytic semigroups and optimal regularity in parabolic problems*, Birkhäuser, Basel, 1995.

[31] Y. Meyer, Wavelets, paraproducts, and Navier-Stokes equations, in: *Current developments in mathematics, 1996 (Cambridge, MA)*, pp. 105–212, Int. Press, Boston, MA, 1997.

[32] S. Monniaux, A new approach to the Dore–Venni theorem, *Math. Nachr.* **204** (1999), pp. 163–183.

[33] ———, Uniqueness of mild solutions of the Navier-Stokes equation and maximal L^p-regularity, *C. R. Acad. Sci. Paris Sér. I Math.* **328** (1999), pp. 663–668.

[34] S. Monniaux and J. Prüss, A theorem of the Dore–Venni type for noncommuting operators, *Trans. Amer. Math. Soc.* **349** (1997), pp. 4787–4814.

[35] F. Oru, *Rôle des oscillations dans quelques problèmes d'analyse non linéaire*, Ph.D. thesis, ENS Cachan, 1998.

[36] E. M. Ouhabaz, Gaussian upper bounds for heat kernels of second-order elliptic operators with complex coefficients on arbitrary domains, *J. Operator Theory* **51** (2004), pp. 335–360.

[37] ———, *Analysis of heat equations on domains*, Princeton University Press, Princeton, 2005.

[38] P. Portal and Ž. Štrkalj, Pseudodifferential operators on Bochner spaces and an application, *Math. Z.* **253** (2006), pp. 805–819.

[39] J. Prüss and R. Schnaubelt, Solvability and maximal regularity of parabolic evolution equations with coefficients continuous in time, *J. Math. Anal. Appl.* **256** (2001), pp. 405–430.

[40] T. Tao, 245C, Notes 1: Interpolation of L^p-spaces, available at `http://terrytao.wordpress.com/2009/03/30/245c-notes-1-interpolation-of-lp-spaces/`.

[41] L. Weis, Operator-valued Fourier multiplier theorems and maximal L^p-regularity, *Math. Ann.* **319** (2001), pp. 735–758.

[42] F. Weissler, Existence and nonexistence of global solutions for a semilinear heat equation, *Israel J. Math.* **38** (1981), pp. 29–40.

Author information

Sylvie Monniaux, LATP UMR 6632 – Case Cour A – Faculté des Sciences de Saint-Jérôme – Aix-Marseille Université – Avenue Escadrille Normandie Niémen – 13397 Marseille, France.
E-mail: `sylvie.monniaux@univ-cezanne.fr`

Index

Bei Fragen zur Produktsicherheit wenden Sie sich bitte an:
If you have any questions regarding product safety,
please contact:

Walter de Gruyter GmbH
Genthiner Straße 13
10785 Berlin
productsafety@degruyterbrill.com